Cyanobacteria and Cyanotoxins

Cyanobacteria and Cyanotoxins: New Advances and Future Challenges

Special Issue Editors

Ana M. Cameán
Angeles Jos

MDPI • Basel • Beijing • Wuhan • Barcelona • Belgrade

Special Issue Editors
Ana M. Cameán
University of Sevilla
Spain

Angeles Jos
University of Sevilla
Spain

Editorial Office
MDPI
St. Alban-Anlage 66
4052 Basel, Switzerland

This is a reprint of articles from the Special Issue published online in the open access journal *Toxins* (ISSN 2072-6651) from 2018 to 2019 (available at: https://www.mdpi.com/journal/toxins/special_issues/cyanotoxins1)

For citation purposes, cite each article independently as indicated on the article page online and as indicated below:

LastName, A.A.; LastName, B.B.; LastName, C.C. Article Title. *Journal Name* **Year**, *Article Number*, Page Range.

ISBN 978-3-03921-838-7 (Pbk)
ISBN 978-3-03921-839-4 (PDF)

Contents

About the Special Issue Editors vii

Preface to "Cyanobacteria and Cyanotoxins: New Advances and Future Challenges" ix

Osmindo Rodrigues Pires Júnior, Natiela Beatriz de Oliveira, Renan J. Bosque, Maria Fernanda Nice Ferreira, Veronica Morais Aurélio da Silva, Ana Carolina Martins Magalhães, Carlos José Correia de Santana and Mariana de Souza Castro
Histopathological Evaluation of the Exposure by Cyanobacteria Cultive Containing [D-Leu^1]Microcystin-LR on *Lithobates catesbeianus* Tadpoles
Reprinted from: *Toxins* **2018**, *10*, 318, doi:10.3390/toxins11060318 1

Leticia Díez-Quijada, Remedios Guzmán-Guillén, Ana I. Prieto Ortega, María Llana-Ruíz-Cabello, Alexandre Campos, Vítor Vasconcelos, Ángeles Jos and Ana M. Cameán
New Method for Simultaneous Determination of Microcystins and Cylindrospermopsin in Vegetable Matrices by SPE-UPLC-MS/MS
Reprinted from: *Toxins* **2018**, *10*, 406, doi:10.3390/toxins10100406 13

Leticia Díez-Quijada, Ana I. Prieto, María Puerto, Ángeles Jos and Ana M. Cameán
In Vitro Mutagenic and Genotoxic Assessment of a Mixture of the Cyanotoxins Microcystin-LR and Cylindrospermopsin
Reprinted from: *Toxins* **2019**, *11*, 318, doi:10.3390/toxins11060318 29

Zheng Xu, Shu Harn Te, Cong Xu, Yiliang He and Karina Yew-Hoong Gin
Variations of Bacterial Community Composition and Functions in an Estuary Reservoir during Spring and Summer Alternation
Reprinted from: *Toxins* **2018**, *10*, 315, doi:10.3390/toxins10080315 45

Ana Regueiras, Sandra Pereira, Maria Sofia Costa and Vitor Vasconcelos
Differential Toxicity of Cyanobacteria Isolated from Marine Sponges towards Echinoderms and Crustaceans
Reprinted from: *Toxins* **2018**, *10*, 297, doi:10.3390/toxins10070297 67

Frederic Pitois, Jutta Fastner, Christelle Pagotto and Magali Dechesne
Multi-Toxin Occurrences in Ten French Water Resource Reservoirs
Reprinted from: *Toxins* **2018**, *10*, 283, doi:10.3390/toxins10070283 79

Zhenhua An, Yingying Zhang and Longshen Sun
Effects of Dietary Astaxanthin Supplementation on Energy Budget and Bioaccumulation in *Procambarus clarkii* (Girard, 1852) Crayfish under Microcystin-LR Stress
Reprinted from: *Toxins* **2018**, *10*, 277, doi:10.3390/toxins10070277 95

Murendeni Magonono, Paul Johan Oberholster, Shonhai Addmore, Makumire Stanley and Jabulani Ray Gumbo
The Presence of Toxic and Non-Toxic Cyanobacteria in the Sediments of the Limpopo River Basin: Implications for Human Health
Reprinted from: *Toxins* **2018**, *10*, 269, doi:10.3390/toxins10070269 104

Lamei Lei, Liang Peng, Yang Yang and Bo-ping Han
Development of Time-Resolved Fluoroimmunoassay for Detection of Cylindrospermopsin Using Its Novel Monoclonal Antibodies
Reprinted from: *Toxins* **2018**, *10*, 255, doi:10.3390/toxins10070255 127

Haohao Liu, Shenshen Zhang, Chuanrui Liu, Jinxia Wu, Yueqin Wang, Le Yuan, Xingde Du, Rui Wang, Phelisters Wegesa Marwa, Donggang Zhuang, Xuemin Cheng and Huizhen Zhang
Resveratrol Ameliorates Microcystin-LR-Induced Testis Germ Cell Apoptosis in Rats via SIRT1 Signaling Pathway Activation
Reprinted from: *Toxins* **2018**, *10*, 235, doi:10.3390/toxins10060235 . . . . **140**

Julia Kleinteich, Jonathan Puddick, Susanna A. Wood, Falk Hildebrand, H. Dail Laughinghouse IV, David A. Pearce, Daniel R. Dietrich and Annick Wilmotte
Toxic Cyanobacteria in Svalbard: Chemical Diversity of Microcystins Detected Using a Liquid Chromatography Mass Spectrometry Precursor Ion Screening Method
Reprinted from: *Toxins* **2018**, *10*, 147, doi:10.3390/toxins10040147 . . . . **156**

Lixia Shang, Muhua Feng, Xiangen Xu, Feifei Liu, Fan Ke and Wenchao Li
Co-Occurrence of Microcystins and Taste-and-Odor Compounds in Drinking Water Source and Their Removal in a Full-Scale Drinking Water Treatment Plant
Reprinted from: *Toxins* **2018**, *10*, 26, doi:10.3390/toxins10010026 . . . . **171**

Amber Lyon-Colbert, Shelley Su and Curtis Cude
A Systematic Literature Review for Evidence of *Aphanizomenon flos-aquae* Toxigenicity in Recreational Waters and Toxicity of Dietary Supplements: 2000–2017
Reprinted from: *Toxins* **2018**, *10*, 254, doi:10.3390/toxins10070254 . . . . **188**

Silvia Pichardo, Ana M. Cameán and Angeles Jos
In Vitro Toxicological Assessment of Cylindrospermopsin: A Review
Reprinted from: *Toxins* **2017**, *9*, 402, doi:10.3390/toxins9120402 . . . . **206**

About the Special Issue Editors

Ana M. Cameán holds a PhD in Pharmacy from the University of Sevilla (US) (1985), and has been Professor of Toxicology at US since 2005. She has developed her teaching and research career in the Toxicology Area of the Department of Food Science, Toxicology and Legal Medicine of US. She is responsible for the research group Toxicology (CTS-358) since its creation. Her research interests are in the area of Food Safety, focusing mainly on the study of cyanotoxins (microcystins, cylindrospermopsin) present in water and food, evaluating their transference through the development and validation of methods for their determination, and their toxic effects with *in vitro* and *in vivo* models, effects of cooking, histopathological alterations, and bioaccessibility. In parallel, the safety of various natural products in food packaging or as antioxidants for their approval in feed is also of interest.

Angeles Jos is Full Professor in Toxicology of in the Faculty of Pharmacy at the University of Sevilla. Born and raised in Sevilla, she has developed her professional career mainly at the University of Sevilla with postdoctoral stays at the University of Bern (Switzerland). She is Senior Scientist in the research group Toxicology (CTS-358) in the Department of Food Science, Toxicology and Legal Medicine. Her research interests are in the field of Food Safety, particularly in the hazard characterization of different toxicants present in food. Among them, cyanotoxins (mainly microcystins and cylindrospermopsin) have a pivotal role in her research, where she studies their toxic effects using both *in vitro* and *in vivo* methods, their toxic mechanisms (genotoxicity, oxidative stress, etc.), and is involved in developing analytical methods for their determination in water and food samples.

Preface to "Cyanobacteria and Cyanotoxins: New Advances and Future Challenges"

Cyanobacteria are a group of ubiquitous photosynthetic prokaryotes. Their occurrence has been increasing worldwide, due to anthropogenic activities and climate change. Several cyanobacterial species are able to synthesize a high number of bioactive molecules, among them, cyanotoxins (microcystins, cylindrospermopsin, nodularin, etc.), which are considered a health concern. For risk assessment of cyanotoxins, more scientific knowledge is required to perform adequate hazard characterization, exposure evaluation and, finally, risk characterization of these toxins. This Special Issue "Cyanobacteria and Cyanotoxins: New Advances and Future Challenges" presents new research or review articles related to different aspects of cyanobacteria and cyanotoxins, and contributes to providing new toxicological data and methods for a more realistic risk assessment. Thus, of interest are new advances and tools for the sampling and monitoring of blooms, analytical determination of cyanotoxins in different matrices such as water, food, soil, and biological samples, water treatment methods, remediation approaches, and toxicological evaluation, including *in vitro* and *in vivo* studies. Moreover, the use of different experimental models (tadpole, *L. catesbeianus*), bioassays to investigate the bioactivity profile of cyanobacteria isolated from marine sponges, toxic mechanisms at a molecular level, searching for new biomarkers, etc., are included. The variation in cyanobacteria presence is a commonplace phenomenon with an increasing trend (Asia, Africa), sometimes in conjunction to taste-and-odor compounds. Additionally, special interest is focused on research on multitoxins as most of the data refers to individual cyanotoxins although the co-occurrence of different toxins or variants has been demonstrated. Thus, one study reveals that although individual toxins were detected in 75% water samples monitored (France), and multitoxin occurrence appeared in 40% of samples (2–3 toxins), concluding the need for monitoring several classes of cyanotoxins simultaneously, instead of individual toxins. Thus, new methods have been developed not only for the detection of individual toxins, such as MC-LR, in unusual matrices such as benthic biofilms or lichens by LC–MS, or CYN (by time-resolved fluoroimmunoassay), but also to detect several cyanotoxins (MCs and CYN) in vegetables by UPLC–MS/MS. Among the results indicating potential protection against the toxicity induced by cyanotoxins, it is remarkable that one study indicates that the administration of resveratrol could ameliorate MC-LR toxicity induced in the testis of rats by stimulating some molecular signaling pathways involved in its mechanisms of toxicity. Moreover, dietary astaxanthin supplements seems to block the bioaccumulation of MC-LR in some organs of Procambarus clarkii, with practical and economic consequences for aquaculture, industry, environment, etc. Similarly, the possibility of interactions with other contaminants is worth considering, as recommended by international organizations. Finally, two review articles compile an assessment of, on the one hand, the evidence for toxicogenicity of *Aphanizomenon flos-aquae* in fresh water and blue-green algae supplements (BGAS), including recommendations; and on the other hand, the *in vitro* scientific literature dealing with CYN, indicating the new emerging toxicity and the need to perform further assays. All of the articles included in this Special Issue have been carefully selected and reviewed, and as the Guest Editors, we would like to express our gratitude to both the authors and the reviewers.

Acknowledgments: The co-editors are grateful to all the authors who contributed to this Special Issue "Cyanobacteria and Cyanotoxins: New Advances and Future Challenges". We greatly appreciate all the efforts carried out by the external peer reviewers/expertise. The co-editors also wish to thank the Ministerio de Ciencia, Innovación y Universidades (AGL2015-64558-R, MINECO/FEDER, UE), and the Faculty of Pharmacy of the Universidad of Sevilla. Finally, we are highly appreciative for the editorial support of the MPDI management team and staff.

Conflicts of interest: The authors declare no conflict of interest.

Ana M. Cameán, Angeles Jos

Special Issue Editors

toxins

MDPI

Article

Histopathological Evaluation of the Exposure by Cyanobacteria Cultive Containing [D-Leu^{1}]Microcystin-LR on *Lithobates catesbeianus* Tadpoles

Osmindo Rodrigues Pires Júnior [1,*], Natiela Beatriz de Oliveira [1], Renan J. Bosque [2], Maria Fernanda Nice Ferreira [2], Veronica Morais Aurélio da Silva [1], Ana Carolina Martins Magalhães [1], Carlos José Correia de Santana [1] and Mariana de Souza Castro [1]

[1] Toxinology Laboratory, Depto. Physiological Sciences, Institute of Biology, University of Brasilia, Brasilia 70910-900, Brazil; natiela@gmail.com (N.B.d.O.); veronicasilva@gmail.com (V.M.A.d.S.); bioana.11@gmail.com (A.C.M.M.); carlosjcsantana@gmail.com (C.J.C.d.S.); mscastro69@gmail.com (M.d.S.C.)

[2] Depto. Genetics and Morphology, Institute of Biology, University of Brasilia, Brasilia 70910-900, Brazil; rjbosque@go.olemiss.edu (R.J.B.); mfnf@unb.br (M.F.N.F.)

* Correspondence: osmindo@unb.br or osmindo@gmail.com; Tel.: +55-(61)-3107-3110

Received: 11 June 2018; Accepted: 24 July 2018; Published: 6 August 2018

Abstract: This study evaluated the effects of [D-Leu^{1}]Microcystin-LR variant by the exposure of *Lithobates catesbeianus* tadpole to unialgal culture *Microcystis aeruginosa* NPLJ-4 strain. The Tadpole was placed in aquariums and exposed to *Microcystis aeruginosa* culture or disrupted cells. For 16 days, 5 individuals were removed every 2 days, and tissue samples of liver, skeletal muscle, and intestinal tract were collected for histopathology and bioaccumulation analyses. After exposure, those surviving tadpoles were placed in clean water for 15 days to evaluate their recovery. A control without algae and toxins was maintained in the same conditions and exhibited normal histology and no tissue damage. In exposed tadpoles, samples were characterized by serious damages that similarly affected the different organs, such as loss of adhesion between cells, nucleus fragmentation, necrosis, and hemorrhage. Samples showed signs of recovery but severe damages were still observed. Neither HPLC-PDA nor mass spectrometry analysis showed any evidence of free Microcystins bioaccumulation.

Keywords: [D-Leu^{1}]Microcystin-LR; *Lithobates catesbeianus*; tadpoles; exposure; Histopathological evaluation

Key Contribution: To our knowledge, this is the first report of cytotoxic effects on amphibian tadpoles, with histopathological description of acute intoxication effects by a Microcystin variant [D-Leu^{1}]Microcystin-LR), as well as the concomitant evaluation of bioaccumulation analysis.

1. Introduction

Cyanobacterial blooms have been reported to occur in both natural and artificial water bodies and have caused severe problems for wildlife and livestock as well as humans [1,2]. Some factors contributing to a bloom formation include increased concentrations of biologically available forms of nitrogen and phosphorous in the water source, as well as high temperatures and pH, and calm weather conditions [3].

Due to the toxicity problems associated with numerous cyanobacterial species, a significant number of scientific papers mainly describe the occurrence and the toxins of toxic blooms, bioaccumulation and toxicity focusing on aquatic invertebrates, particularly mollusks and fishes. There are almost no reports of the effects of cyanotoxins on amphibians. The fact that the most amphibian tadpole species develop in waters, and the skin is the main osmoregulator organ and intimately connected with the aquatic environment, make them vulnerable to toxic action of some cyanobacteria blooms.

Rhinella (*Bufo*) *marina* tadpoles exposure for 7 days to *Cylindrospermopsis raciborskii* live culture containing cylindrospermopsin (232 μg/L) had 66% of mortality. In despite, no tadpole mortality were observed in 14 days cylindrospermopsin toxin (400 μg/L). Longer exposure to higher cylindrospermopsin concentration of *C. raciborskii* live culture resulted in an accumulation of over (400 μg/g ww) by the tadpoles [4]. In *Ambystoma mexicanum*, *Triturus vulgaris*, *Rana ridibunda* no effects were recorded during embryonic development following exposure MC-LR, -YR and saxitoxin. Although the crude extracts of cyanobacteria induced craniofacial malformations and gills hyperplasia leading to embryos death [5]. *Xenopus laevis* tadpoles fed with lyophilized cyanobacterial biomass containing Microcystin-LR (MC-LR) were unable to bioaccumulation of MC-LR, and no significantly affected in the development was observed [6]. Although Dvořáková et al. [7] demostrated in a 96 h *Xenopus laevis* embryos teratogenesis assay that MC-LR caused weak lethality but the cyanobacterial biomass containing MC-LR caused significant embryos lethality.

The fact is that there is a lack of information on amphibian biology when they are exposed to natural cyanobacteria blooms, cyanobacteria unialgal cultives or isolated cyanotoxins.

2. Results

2.1. HPLC-PDA and MALDI-TOF Analysis

In HPLC-PDA analysis of Liver, Intestinal tract and Muscle extracts of *L. catesbeianus* tadpole exposed to *M. aeruginosa* cells (Disrupted or not), no chromatographic fraction with same retention time or UV spectra (200–300 nm) of any five Microcytins (MCs) as described in Ferreira et al. [8] for *Microcystis aeruginosa* NPLJ-4 were evidenced (Figure 1). In MALDI-TOF analysis no mass components referring to any MCs described in Ferreira et al. [8] for *Microcystis aeruginosa* NPLJ-4 were found. Due to the covalent linkage between MCs and protein phosphatase, these both methods only detect free MCs available into the tissues, and not the covalently bound toxin [9], so we suggested that *Lithobates catesbeianus* tadpoles cannot accumulate microcystins as free toxins.

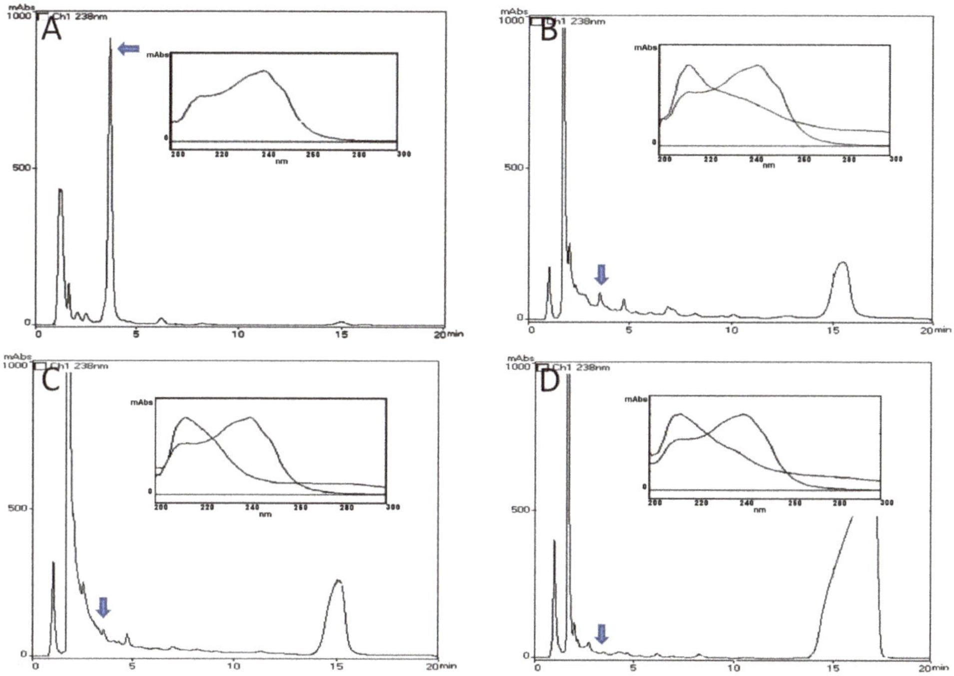

Figure 1. Chromatogram profile (238 nm) in HPLC-PDA system of (**A**) *Microcystis aeruginosa* NPLJ-4 culture, the arrow indicates [D-Leu[1]]Microcystin-LR at 4.5 min of retention time (Ferreira et al., 2010), the insert is MC spectra in range of 200–300 nm, (**B**) Liver, (**C**) Intestinal tract and (**D**) Muscle extracts of *L. catesbeianus* tadpole of 16 days exposed to *Microcystis aeruginosa* NPLJ-4 cells, the arrow indicates the retention time of 4.5 min, the insert is the comparison of MC spectra with fraction in 4.5 min of retention time.

2.2. Histology Analysis

2.2.1. Liver Histology

Control Group

The microscopic analysis indicated that *L. catesbeianus* tadpole liver is covered by mesothelium underlaid by a conjunctive tissue thin layer, the hepatic serosa, that coats the gland externally with no evidenced of the parenchyma division into well-defined lobules. Reticulum staining revealed that the parenchyma was supported by delicate reticular fibers surrounding hepatic cells plates intercalated by the sinusoids capillaries that converge to a central vein endowed with an endothelium. The hepatocytes was polyhedral in shape and their sizes vary. Nuclei were observed in the cells central region, but some of them were shifted toward the edge. The cytoplasm containing granules viewed as small vacuoles appeared little eosinophilic when analyzed by the H&E staining technique. The presence of granulocytes and single melano-macrophages, components of the reticulo-histiocytic system of the liver localized predominately in the sinusoid space, were observed in all of the individuals. Melano-macrophages are cells with diverse functions, including the synthesis of melanin, phagocytosis and free radicals neutralization and found numerous in amphibians. The interstices portal tracts were supported by abundant conjunctive tissue. Most of the tracts contain a bile duct, at least one vein branch and many arteries (Figure 2A).

Exposed to *M. aeruginosa* Cells

Macroscopically, liver showed hypertrophy with increasing exposure time to both treatments: culture of cyanobacteria and toxins after cells lyses. Changes in color and texture of the liver were also noted characterized by variation in tone dark red and firm texture going to ocher tones and looks brittle. The histological analysis showed hepatocytes highly vacuolated and at the fourth day their nucleoli were fragmented, showing characteristics of apoptosis (Figure 2B). Microcystin caused the loss of normal hepatocyte structure suggesting advanced necrosis stage, observed an eosinophilic retracted cytoplasm and the nucleus was fragmentation whereby its chromatin is distributed irregularly throughout the cytoplasm indicating karyorrhexis or karyolysis. The necrosed hepatocytes were invaded by numerous neutrophil granulocytes which were infiltrated in the parenchyma. It was also displayed a greater number of melano-macrophage filled with phagocytosed cell debris. Hepatic sinusoids presented a narrow lumen and an enlargement of the interspaces was also observed. The central veins were enlarged and fibrosed and their endothelium become progressively degenerate, and signs of hemorrhage began to appear. The portal structures formed of terminal portal veins, arterioles and biliary ducts were slightly fibrosed.

Intense perisinusoidal fibrosis was observed around the granulomatous areas. As a consequence of hepatocytes necrosis, and active hepatic regeneration, the common hepatic lobes shape changed, appearing more rounded and shortened, comparatively to the controls. The Kupffer cells appeared a yellow-golden granular pigment, suggesting lipofuscin accumulation within the cytoplasm. Lipofuscin is pigment granules composed of lipid-containing residues of lysosomal digestion, frequently observed at the biliary pole of the hepatocytes. The lesions worsened after a longer exposure and apoptotic cells increased in number.

Exposed to *M. aeruginosa* Disrupted Cells

Exposure to the cells extract of *M. aeruginosa* caused injuries often similar to those observed when exposed to intact cells. Already the three-day exhibition revealed the first cyanotoxins effects, such as sinusoids dilatation near the central vein, gaps in the vein endothelium, hepatocytes with nucleus displaced, sometimes joined the cytoplasmic membrane, and increasing of vacuoles in the hepatocytes cytoplasm. At nine days of exposure occurred the disruption in the central veins endothelium. Melano-macrophages increased in size and become polynuclear cells. The misshapen appearance of sinusoids culminated in congestion and hemorrhage. Bile ducts appeared lacerated with granulocytes in the periphery (Figure 2C) revealed the most significant liver effects caused by the action of cyanotoxin after 12 days of exposure. Between 12 and 16 days the lesions get worse with more melano-macrophages polynuclear cells and increased number of granulocytes in the connective tissue surrounding bile ducts and large vessel size. The periphery of the liver was necrotic.

Recovery

At the end of recovery time, the remaining individuals showed the volume approached the size of the control animals and reddish. There was an improvement on the periphery of the body structure when compared to that observed at the end of exposure, despite the presence of vacuolated hepatocytes and necrotic lesions. A large number of melano-macrophages cells were still present leading to the lysis of the cells. After the recovery period of ten days, staining was most evident and connective tissue and muscle reconstitute the blood vessel walls (Figure 2D).

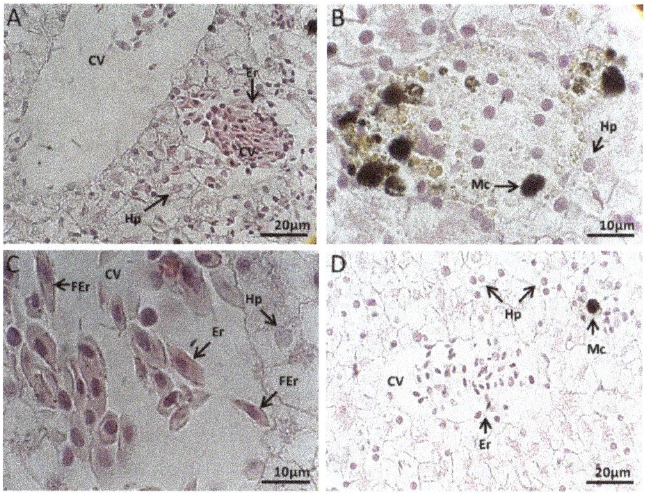

Figure 2. (**A**) *L. catesbeianus* tadpole liver control showing central lobe vein, erythrocytes and leukocytes, hepatocytes with uniform and well defined nuclei (40×). (**B**) giant macrophage performing phagocytosis (100×) in hepatic animal tissue 16 days exposed to *M. aeruginosa* cells. (**C**) Centrilobular vein view with morphologically altered erythrocyte micronuclei (100×), tadpoles 16 days exposed to *M. aeruginosa* disrupted cells. (**D**) Centrilobular vein with wrinkled appearance, hepatocytes with regular nuclei (40×), after a recovery period of 15 days without contact with the cyanobacteria extract. (CV) Central Vein; (Er) erythrocyte; (Hp) Hepatocytes; (Mc) macrophage, (FEr) Falciform erythrocyte.

2.2.2. Intestinal Tract Histology

Control Group

The simple columnar epithelium retained a brush border and apoptotic cells were seldom found. The enterocytes showed no morphological change and goblet cells were abundant. Loose connective tissue was thin and showing fibroblast-like cells. The muscular layer was quite thin. The serous layer composed of simple squamous epithelium was difficult to visualize on sections stained with HE (Figure 3A).

Exposed to *M. aeruginosa* Cells

The first analysis showed granulocytes increased number through the migration of cells from the deeper layers of tissue to the base of the epithelial tissue. The enterocytes showed lesions as the presence of cytoplasmic vacuoles. Also noticeable was the increased fibrosis in the connective tissue layer beneath the epithelium which appeared thicker. An intense blood supply was observed around the granulomatous areas. On subsequent days of exposure, the intestine showed signs of change as a greater number of apoptotic epithelial cells.

Exposed to *M. aeruginosa* Disrupted Cells

After five days of exposure to extract the absorptive cells and goblet cells that form the simple cylindrical epithelium apparently remain with integrity and preserved (Figure 3B). The enterocytes were attached to basal lamina and despite the abundant vascularization, connective tissue showed signs of fibrosis. No changes were observed in the muscle layer and serosa. The main changes brought about by the action of the extract of *M. aeruginosa* in the intestine occurred after six days of exposure. At that point, clusters of melano-macrophage were evident among the epithelial cells. These cells had an elongated aspect, vacuoles in apical part and loss of adhesion. It was observed an increase in melano-macrophages throughout the intestinal wall. After 16 days of exposure, these aspects were exacerbated by the loss of demarcation between the epithelial cells, culminating in degeneration process with nuclei displaced adhered to the cytoplasmic membrane and necrosis (Figure 3C).

Recovery

After the period of recovering, the number and size of melano-macrophages decrease. The cells adhesion and delimitation were still limited. The blood vessels remained narrow and the intestine wall appeared thin (Figure 3D).

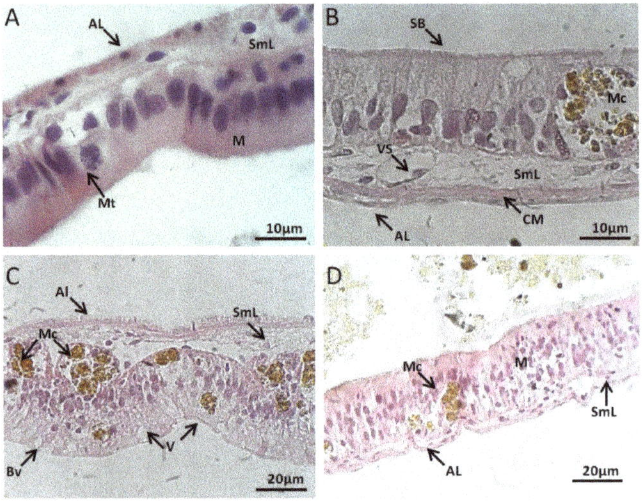

Figure 3. (**A**) *L. catesbeianus* tadpole Intestinal Tract control showing the structure of the layers and the uniformity of the cells (100×). (**B**) View of the striated border of the organ absorptive cells exposed for 16 days to *M. aeruginosa* cells (100×). (**C**) Loss of cell adhesion and macrophages performing phagocytosis after 16 of exposure to *M. aeruginosa* disrupted cells (40×). (**D**) Organ layers with better structured absorptive cells after recovery period of 15 days without contact with the cyanobacteria extract (40×). (AL) adventitious layer, (SmL) submucosal layer, (M) mucosa, (Mt) mitosis, (SB) striatum border, (BV) blood vessel, (ML) muscular layer, (Mc) macrophage, (Vc) vacuole, (V) villus.

2.2.3. Muscle Histology

Control Group

This tissue had transverse striations, multinucleated cells, peripheral nuclei and other characteristics of skeletal muscles. The muscle fibers or myocytes showed a surrounding connective tissue (Figure 4A).

Exposed to *M. aeruginosa* Cells

In muscle fibers, after two days of exposure, staining becomes weaker, and myofibrils, the smallest contractile units, were poorly visible with the light microscope. The lesions were more pronounced with exposure time, increasing the intercellular spaces between the muscle fibers and cytoplasmic derangement with no evident cross-striation (Figure 4B). At the end of the toxicity experiment, muscle fibers had signs of fragmentation and cytoplasmic lysis with cells contents release.

Exposed to *M. aeruginosa* Disrupted Cells

The weak staining also allows us to observe a loss of striations which implies the loss myofibril arrangement. At half time exposure tissue abnormalities were more advanced, when necrosis and the intercellular spaces increased. Due the loss of intracellular adhesion, muscle layers cells were separated from each other. Signs of nuclear fragmentation and necrosis were observed (Figure 4C).

Recovery

Although we observed an improvement in the stain that has become less pale, tissue recovery, after fragmentation of muscle fibers, was insignificant (Figure 4D).

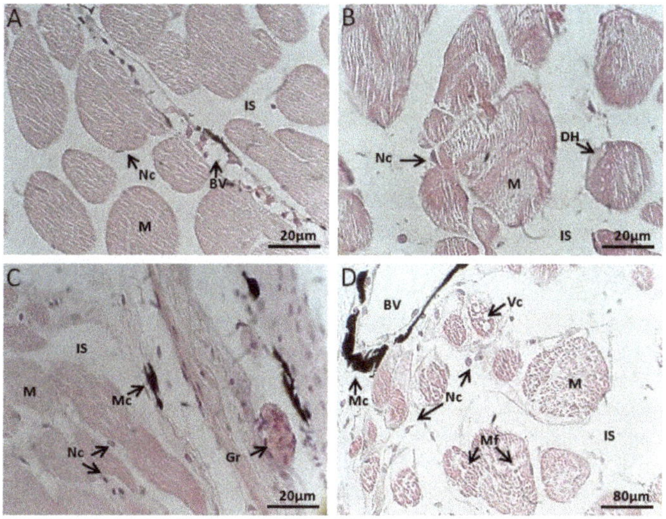

Figure 4. (**A**) *L. catesbeianus* tadpole muscle control showing structure and organization of muscle tissue (40×). (**B**) Animal tissue exposed for 16 days to *M. aeruginosa* cells (40×). (**C**) Longitudinal section of animal tissue exposed for 16 days to expose to *M. aeruginosa* disrupted cells (40×). (**D**) Transverse section of recovering animal tissue after 15 days without contact with the extract (10×). (M) myocyte, (IS) intercellular space, (Nc) nucleus, (BV) blood vessel, (HD) degradation hyaline, (Mc) macrophage, (Gr) granulocyte, (Vc) vacuole, (Mf) myofibril.

3. Discussion

There are numerous studies describing the harmful effects of Microcystins on adult fish [10,11], as well as embryonic and post-hatch development [12]. In *Cyprinus carpio*, histopathological changes were observed in the liver and gastrointestinal tract and increased in severity with post-dose time [11]. The tissues of Silver carp *Hypophthalmichthys molitrix* exposed *Microcystis aeruginosa* NPLJ4 were characterized by significant damages such as cells dissociation, necrosis and hemorrhage [8]. In spite of numerous studies with aquatic animals, there is a lack of work relating the toxic effects of cyanotoxins on amphibians. According to our observation, the exposal to a *Microcystis aeruginosa* containing D-Leu-Microcistin-LR cultures cells (disrupted or not) promotes severe damage liver, muscle and intestinal tract of *L. catesbeianus* tadpole. The organization and morphology of organs and tissues samples were basically the same, the loss of sinusoidal integrity, plasma membrane integrity, the loss of normal cell-to-cell adhesions, loss of cell shape, necrosis, apoptosis and cytoplasmic vacuolization.

MCs interact with protein phosphatases at a molecular level by forming a covalent linkage between Mdha residue and the phosphatase's cysteine residue [13–18]. Perhaps all these effects are being centralized in the destabilization of the cytoskeleton as a direct action of cyanotoxin inhibition of protein phosphatases [19,20].

In addition, recent studies have revealed that the exposure hepatocytes to MCs induce hepatocytes cytoskeleton components rearrangement or collapse, mainly microtubules, micro- and intermediate filaments [20–26], and similar effects have also been observed in others tissues [27–34]. MCs can induce cytoskeletal disruption by affects the expression of cytoskeletal and cytoskeleton associated proteins as downregulated de expression of actin and tubulin, and hyperphosphorylation several microfilament-associated proteins, like Vimentin, Ezrin and VASP [27,33].

In contrast to the results of the exposure of *Rhinella (Bufo) marina* tadpoles to live culture containing Cylindrospermosin that caused 66% mortality [4], the treatment (disrupted cells or not) caused no mortality of *L. catesbeianus* tadpoles during the exposure period and subsequent detoxification. Although, the damage level was characterized by a time-dependent exposure.

Rhinella (*Bufo*) *marina* also accumulated cylindrospermopsin only when exposed to *C. raciborskii* culture, but not when exposed to the culture lysed cells. In this work the authors argue that the difference in results is related to the way the toxin is or not absorb. The *C. raciborskii* live cells culture was eaten by tadpole and the absorption was intestinal tract, while the *C. raciborskii* lysed cells the cylindrospermopsin was not absorbed by the skin.

Even though the methodology employed in this work is limited to determination of unbound MC concentration, interestingly, *L. catesbeianus* tadpoles do not appear to accumulate Microcystins in its unbounded form in tissues since we could not identify the presence of any Microcystins in either treatment, disrupted cells or not. In a similar way, *Xenopus laevis* tadpoles fed with cyanobacterial biomass containing MC-LR did not bioaccumuled this toxin [6].

Amphibian tadpoles appear to be very susceptible to the cytotoxic effects of Microcystins, but also appear to have a high capacity for tissue regeneration after acute intoxication. To our knowledge, this is the first report of cytotoxic effects on amphibian tadpoles, with histopathological description of the effects of acute intoxication by a Microcystin variant, as well as the concomitant evaluation of the possibility of bioaccumulation

4. Materials and Methods

4.1. *Cyanobacteria Culture and the Experimental Animals*

Unialgal *M. aeruginosa* NPLJ4 culture were carried out in the Laboratory of Toxinology, University of Brasilia. 150 *Lithobates catesbeianus* tadpoles about 24–25 stage development according to Gosner [34], were donated by a toad breeding company nearby Brasilia-DF. Animals were distributed in 3 aquariums (84 L of each, 30 cm × 60 cm × 50 cm) with filtrated and dechlorined water, acclimatized for 20 days with 12/12 h light/darkness photoperiod, continuous oxygenation with submerged pumps, and temperature at 25 ± 2 °C. The tadpoles were fed with specific tadpole food, also kindly given by the toad breeding company.

4.2. *M. aeruginosa NPLJ4 Toxins Characterization*

The purification and characterization protocols used were the same of Ferreira et al., [8]. *M. aeruginosa* NPLJ4 culture extract of showed the same presence of 5 microcystins, the chromatographic fraction identified as [D-Leu^1]MC-LR showed over 90% of all microcystins detected [8]. The [D-Leu^1]MC-LR concentration for each assay was calculated in 11 mg for 840 mL culture medium containing colonies of *M. aeruginosa* NPLJ4 at the end of the exponential growing phase.

4.3. *Exposure of Tadpole to M. aeruginosa Cells or Cells Extract*

At the end of the acclimatization period, animals were submitted to: Aquaria No. 1, received 840 mL of culture medium containing colonies of *M. aeruginosa* in the final exponential growth phase. The concentration of cells in aquaria was the equivalent of 10^5 cells per mL. Aquaria No. 2, received the ultrasonicated disrupted *M. aeruginosa* cell extract from 840 mL culture medium containing colonies of *M. aeruginosa* in the final exponential growth phase. As microcystins are endotoxins, the lysis by ultrasonicated disruption had intended the release of all microcystins in the growth medium. The final concentration of [D-Leu^1]MC-LR itself were estimated in 0.13 mg/L.

During the initial period of 16 days, 5 (five) individuals, from each aquaria, were removed at 2-day intervals and euthanized by over anesthesia using a lidocaine 5% ointment applied in ventral skin, All procedures were in accordance with the Local Ethics Committee of Brasilia University (Process Number CEUA 56344/2005, project name: Microcystin Bioaccumulation in aquatic animals). The five individuals were divided into two groups, the first group consisting of three (3) individuals intended for quantification of MCs bioaccumulated by HPLC-PDA and MALDI-TOF analysis, and the second group by two (2) individuals reserved for histopathological analysis.

During the experiment the aquariums were kept continuously aerated; temperature was maintained at 25 ± 1 °C, pH ~7.0 with 12/12 h light/darkness photoperiod. After 2 days, the water was changed and the same culture volume of *M. aeruginosa* culture or cell disrupted extract was added, aiming to maintain MCs concentration in contact with animals.

After sixteen days of exposure, the remaining animals were removed to another 2 free cell aquaria and after fifteen days 5 (five) individuals were collected and euthanized following the same previously mentioned methodology to evaluate the detoxification and tissue regeneration. Animals that were not exposed to cell culture formed the control group, and were collected under the same conditions.

4.4. HPLC-PDA Analysis

To the determination of unbound MCs concentration tissue samples (~5 g for liver and intestinal tract, 10 g for muscle), were extracted three times in methanol (5 mL/g). The extracts were filtered (glass fiber membrane 1.2 µm porosity), vacuum-dried, and resuspended in deionized water (5 mL). The extract cleanup was performed with solid phase extraction cartridges (Strata C18, 5 mg; Phenomenex, Torrance, CA, USA) [8]. The resultant extract was vacuum-dried and resuspended in deionized water (2 mL). The sample was filtered through a 0.22 µm polyethylene filter (GV Millex; Millipore Corporation, Billerica, MA, USA), and analyzed by a Shimadzu LC-10A HPLC system (Shimadzu, Kyoto, Japan) equipped with a SPD M10A photodiode array detector. Conditions Synergi column (4 µm Fusion-RP80, 150 × 4.60 mm, Phenomenex, Torrance, CA, USA), mobile phase 20 mM ammonium formate in 30% acetonitrile (pH 5.0), run isocratically, flowrate 1 mL/min for 30 min, UV detection at 238 nm. Toxins identification was performed by chromatograms comparison of standard of MC-LR (Sigma-Aldrich Corporation, Saint Louis, MO, USA) and [D-Leu^1]MC-LR purificated in our laboratory, observing the following aspects: the retention time and similarity of UV spectra (200–300 nm). Concentrations of [D-Leu^1]MC-LR, were calculated from the calibration curve (R^2 = 0.9941), with detection limit of 10 ng. For bioaccumulation positive control, 1 mg of [D-Leu^1]MC-LR was added and homogenized to untreated animal tissues, our recovery was approximately 60% (Data not shown).

4.5. Determination of Unbound MCs Concentration in Tissues—MALDITOF Analysis

To examine the presence of unbound MCs, the tissue samples were also submitted to Ultraflex II™ TOF/TOF (Bruker, Bremen, Germany). Samples aliquots were dissolved in TFA 0.1% and mixed with a saturated matrix solution of α-cyano-4-hydroxycinnamic acid (1:3, *v*/*v*) and directly applied onto a target (AnchorChip™, Bruker Daltonics, Billerica, MA, USA). The mass spectrometry was operating in positive reflector mode for MALDI-TOF, or LIFT mode for MALDI-TOF/TOF. Calibration was performed externally with ions of angiotensin I, angiotensin II, substance P, bombesin, insulin b-chain and adrenocorticotropic hormones (clip 1–17 and clip 18–39). Each spectrum was produced by accumulating data from 200 consecutive laser shots.

4.6. Histology Analysis

Histological material for the evaluation of the cyanotoxin effects on tissues and organs morphology was fixed in 10% neutral buffered formalin (pH 7.0) for 24 h, dehydrated, in xylene and paraffin-embedded routinely. Sections of 3–5 μm thick were deparaffinized in xylene, rehydrated and stained with Hematoxylin & Eosin (H&E). Detection of histopathological changes were achieved by evaluation of general architecture of organs, cellular morphology and blood vessel histology in tissue sections viewed using optical microscope. The system for capturing images consisted of camera CCD-Iris and capture plate PixelView Station, Image Pro Express 4.0, Media Cybernetics, Scopephoto Version x86, 3.1.1.615 (ScopeTek, Goleta, CA, USA).

Author Contributions: O.R.P.J. conceived and designed the experiments; V.M.A.d.S. and N.B.d.O. performed the experiments; R.J.B. and M.F.N.F. analyzed the histopathology data; M.d.S.C. performed the MALDI-TOF analysis; A.C.M.M. and C.J.C.d.S. wrote the paper.

Funding: This research received no external funding.

Acknowledgments: We thank the CNPq and CAPES for scholarships, FINATEC for financial support, and the RANDER Indústria Comércio e Exportação for kindly providing the *Lithobates catesbeanus* tadpoles.

Conflicts of Interest: The authors declare no conflict of interest.

References

1. Botes, D.P.; Tuinman, A.A.; Wessels, P.L.; Viljoen, C.C.; Kruger, H. The structure of cyanoginosin-LA, a cyclic heptapeptide toxin from the *Microcystis aeruginosa*. *J. Chem. Soc. Perkin Trans.* **1984**, 2311–2318. [CrossRef]
2. Runnegar, M.T.C.; Falconer, I.R. The in vivo and in vitro biological effects of the peptide hepatotoxin from the bluegreen alga *Microcystis aeruginosa*. *S. Afr. J. Sci.* **1982**, *78*, 363–366.
3. Van Halderen, A.; Harding, W.R.; Wessels, J.C.; Schneider, D.J.; Heine, E.W.R.; Van der Merwe, J.; Fourie, J.M. Cyanobacterial (blue-green algae) poisoning of livestock in the Western' Cape Province of South Africa. *J. S. Afr. Vet. Assoc.* **1995**, *66*, 260–264. [PubMed]
4. White, S.H.; Duivenvoorden, L.J.; Fabbro, L.D.; Eaglesham, G.K. Mortality and toxin bioaccumulation in Bufo marinus following exposure to *Cylindrospermopsis raciborskii* cell extracts and live cultures. *Environ. Pollut.* **2007**, *147*, 158–167. [CrossRef] [PubMed]
5. Oberemm, A.; Becker, J.; Codd, G.A.; Steinberg, C. Effects of cyanobacterial toxins and aqueous crude extracts of cyanobacteria on the development of fish and amphibians. *Environ. Toxicol.* **1999**, *14*, 77–88. [CrossRef]
6. Fisher, W.J.; Dietrich, D.R. Toxicity of the cyanobacterial cyclic heptapepitide toxins microcystin-LR and -RR in early stages of the African clawed frog (*Xenopus laevis*). *Aquat. Toxicol.* **2000**, *49*, 189–198. [CrossRef]
7. Dvořáková, D.; Dvořáková, K.; Bláha, L.; Maršálek, B.; Knotková, Z. Effects of Cyanobacterial biomass and purified microcystins on malformations in *Xenopus laevis*: Teratogenesis assay (FETAX). *Environ. Toxicol.* **2002**, *17*, 547–555. [CrossRef] [PubMed]
8. Ferreira, M.F.N.; Oliveira, V.M.; Oliveira, R.; Cunha, P.V.; Grisolia, C.K.; Pires, O.R., Jr. Histopathological Effects of [D-Leu1] Microcystin-LR variants on liver, skeletal muscle and intestinal tract of *Hypophthalmichthys molitrix* (Valenciennes, 1844). *Toxicon* **2010**, *55*, 1255–1262. [CrossRef] [PubMed]
9. Neffling, M.-R.; Lance, E.; Meriluoto, J. Detection of free and covalently bound microcystins in animal tissues by liquid chromatography–tandem mass spectrometry. *Environ. Pollut.* **2010**, *158*, 948–952. [CrossRef] [PubMed]
10. Kotak, B.G.; Zurawell, R.W.; Prepas, E.E.; Holmes, C.F.B. Microcystin-LR concentrations in aquatic food web compartments from lakes of varying trophic status. *Can. J. Fish. Aquat. Sci.* **1996**, *53*, 1974–1985. [CrossRef]
11. Fischer, W.J.; Dietrich, D.R. Pathological and biochemical characterization of microcystin-induced hepatopancreas and kidney damage in carp (*Cyprinus carpio*). *Toxicol. Appl. Pharmacol.* **2000**, *164*, 73–81. [CrossRef] [PubMed]
12. Jacquet, C.; Thermes, V.; De Luze, A.; Puiseux-Dao, S.; Bernard, C.; Joly, J.S.; Bourrat, F.; Edery, M. Effects of microcystin-LR on development of medaka fish embryos (*Oryzias latipes*). *Toxicon* **2004**, *43*, 141–147. [CrossRef] [PubMed]
13. Runnegar, M.; Berndt, N.; Kong, S.M.; Lee, E.Y.; Zhang, L.F. In vivo and in vitro binding of microcystin to protein phosphatase 1 and 2A. *Biochem. Biophys. Res. Commun.* **1995**, *216*, 162–169. [CrossRef] [PubMed]

14. Bagu, J.R.; Sykes, B.D.; Craig, M.M.; Holmes, C.F. A molecular basis for different interactions of marine toxins with protein phosphatase-1 Molecular models for bound motuporin, microcystins, okadaic acid, and calyculin A. *J. Biol. Chem.* **1997**, *272*, 5087–5097. [CrossRef] [PubMed]
15. Goldberg, J.; Huang, H.B.; Kwon, Y.G.; Greengard, P.; Nairn, A.C.; Kuriyan, J. Three-dimensional structure of the catalytic subunit of protein serine/threonine phosphatase-1. *Nature* **1995**, *376*, 745. [CrossRef] [PubMed]
16. MacKintosh, R.W.; Dalby, K.N.; Campbell, D.G.; Cohen, P.T.; Cohen, P.; MacKintosh, C. The cyanobacterial toxin microcystin binds covalently to cysteine-273 on protein phosphatase 1. *FEBS Lett.* **1995**, *371*, 236–240. [PubMed]
17. Campos, M.; Fadden, P.; Alms, G.; Qian, Z.; Haystead, T.A. Identification of protein phosphatase-1-binding proteins by microcystin-biotin affinity chromatography. *J. Biol. Chem.* **1996**, *271*, 28478–28484. [CrossRef] [PubMed]
18. Eriksson, J.E.; Grönberg, L.; Nygård, S.; Slotte, J.P.; Meriluoto, J.A. Hepatocellular uptake of 3H-dihydromicrocystin-LR, a cyclic peptide toxin. *Biochim. Biophys. Acta (BBA) Biomembr.* **1990**, *1025*, 60–66. [CrossRef]
19. Rudolph-Böhner, S.; Mierke, D.F.; Moroder, L. Molecular structure of the cyanobacterial tumor-promoting microcystins. *FEBS Lett.* **1994**, *349*, 319–323. [CrossRef]
20. Batista, T.; de Sousa, G.; Suput, J.S.; Rahmani, R.; Suput, D. Microcystin-LR causes the collapse of actin filaments in primary human hepatocytes. *Aquat. Toxicol.* **2003**, *65*, 85–91. [CrossRef]
21. Blankson, H.; Grotterod, E.M.; Seglen, P.O. Prevention of toxin-induced cytoskeletal disruption and apoptotic liver cell death by the grapefruit flavonoid, naringin. *Cell. Death Differ.* **2000**, *7*, 739–746. [CrossRef] [PubMed]
22. Chen, D.N.; Zeng, J.; Wang, F.; Zheng, W.; Tu, W.W.; Zhao, J.S.; Xu, J. Hyperphosphorylation of intermediate filament proteins is involved in microcystinLR-induced toxicity in HL7702 cells. *Toxicol. Lett.* **2012**, *214*, 192–199. [CrossRef] [PubMed]
23. Ding, W.X.; Shen, H.M.; Ong, C.N. Critical role of reactive oxygen species formation in microcystin-induced cytoskeleton disruption in primary cultured hepatocytes. *J. Toxicol. Environ. Health Part A* **2001**, *64*, 507–519. [CrossRef] [PubMed]
24. Eriksson, J.E.; Paatero, G.I.; Meriluoto, J.A.; Codd, G.A.; Kass, G.E.; Nicotera, P.; Orrenius, S. Rapid microfilament reorganization induced in isolated rat hepatocytes by microcystin-LR, a cyclic peptide toxin. *Exp. Cell Res.* **1989**, *185*, 86–100. [CrossRef]
25. Espina, B.; Louzao, M.C.; Cagide, E.; Alfonso, A.; Vieytes, M.R.; Yasumoto, T.; Botana, L.M. The methyl ester of okadaic acid is more potent than okadaic acid in disrupting the actin cytoskeleton and metabolism of primary cultured hepatocytes. *Br. J. Pharmacol.* **2010**, *159*, 337–344. [CrossRef] [PubMed]
26. Zeng, J.; Tu, W.-W.; Lazar, L.; Chen, D.-N.; Zhao, J.-S.; Xu, J. Hyperphosphorylation of microfilament-associated proteins is involved in microcystin-LR-induced toxicity in HL7702 cells. *Environ. Toxicol.* **2015**, *30*, 981–988. [CrossRef] [PubMed]
27. Frangez, R.; Zuzek, M.C.; Mrkun, J.; Suput, D.; Sedmak, B.; Kosec, M. Microcystin-LR affects cytoskeleton and morphology of rabbit primary whole embryo cultured cells in vitro. *Toxicon* **2003**, *41*, 999–1005. [CrossRef]
28. Khan, S.A.; Wickstrom, M.L.; Haschek, W.M.; Schaeffer, D.J.; Ghosh, S.; Beasley, V.R. Microcystin-LR and kinetics of cytoskeletal reorganization in hepatocytes, kidney cells, and fibroblasts. *Nat. Toxins* **1996**, *4*, 206–214. [CrossRef]
29. Wickstrom, M.L.; Khan, S.A.; Haschek, W.M.; Wyman, J.F.; Eriksson, J.E.; Schaeffer, D.J.; Beasley, V.R. Alterations in microtubules, intermediate filaments, and microfilaments induced by microcystin-LR in cultured cells. *Toxicol. Pathol.* **1995**, *23*, 326–337. [CrossRef] [PubMed]
30. Kozdeba, M.; Borowczyk, J.; Zimolag, E.; Wasylewski, M.; Dziga, D.; Madeja, Z.; Drukala, J. Microcystin-LR affects properties of human epidermal skin cells crucial for regenerative processes. *Toxicon* **2014**, *80*, 38–46. [CrossRef] [PubMed]
31. Rymuszka, A. Microcystin-LR induces cytotoxicity and affects carp immune cells by impairment of their phagocytosis and the organization of the cytoskeleton. *J. Appl. Toxicol.* **2013**, *33*, 1294–1302. [CrossRef] [PubMed]
32. Chen, L.; Zhang, X.; Zhou, W.; Qiao, Q.; Liang, H.; Li, G.; Wang, J.; Cai, F. The interactive effects of cytoskeleton disruption and mitochondria dysfunction lead to reproductive toxicity induced by microcystin-LR. *PLoS ONE* **2013**, *8*, e53949. [CrossRef] [PubMed]

33. Zhou, M.; Tu, W.W.; Xu, J. Mechanisms of microcystin-LR-induced cytoskeletal disruption in animal cells. *Toxicon* **2015**, *101*, 92–100. [CrossRef] [PubMed]
34. Gosner, K.L. A simplified table for staging anuran embryos and larvae with notes on identification. *Herpetologica* **1960**, *16*, 183–190.

toxins

MDPI

Article

New Method for Simultaneous Determination of Microcystins and Cylindrospermopsin in Vegetable Matrices by SPE-UPLC-MS/MS

Leticia Díez-Quijada [1], Remedios Guzmán-Guillén [1,*], Ana I. Prieto Ortega [1], María Llana-Ruíz-Cabello [1], Alexandre Campos [2], Vítor Vasconcelos [2,3], Ángeles Jos [1] and Ana M. Cameán [1]

1 Area of Toxicology, Faculty of Pharmacy, University of Sevilla, C/Profesor García González 2, 41012 Sevilla, Spain; ldiezquijada@us.es (L.D.-Q.); anaprieto@us.es (A.I.P.O.); mllana@us.es (M.L.-R.-C.); angelesjos@us.es (A.J.); camean@us.es (A.M.C.)

2 CIIMAR/CIMAR—Interdisciplinary Centre of Marine and Environmental Research, University of Porto, Terminal de Cruzeiros do Porto de leixões, Av General Norton de Matos, 4450-208 Matosinhos, Portugal; amoclclix@gmail.com (A.C.); vmvascon@fc.up.pt (V.V.)

3 Faculty of Sciences, University of Porto, Rua do Campo Alegre, 4169-007 Porto, Portugal

* Correspondence: rguzman1@us.es; Tel.: +34-954-556-762

Received: 7 September 2018; Accepted: 6 October 2018; Published: 8 October 2018

Abstract: Cyanotoxins are a large group of noxious metabolites with different chemical structure and mechanisms of action, with a worldwide distribution, producing effects in animals, humans, and crop plants. When cyanotoxin-contaminated waters are used for the irrigation of edible vegetables, humans can be in contact with these toxins through the food chain. In this work, a method for the simultaneous detection of Microcystin-LR (MC-LR), Microcystin-RR (MC-RR), Microcystin-YR (MC-YR), and Cylindrospermopsin (CYN) in lettuce has been optimized and validated, using a dual solid phase extraction (SPE) system for toxin extraction and ultra-performance liquid chromatography-tandem mass spectrometry (UPLC-MS/MS) for analysis. Results showed linear ranges (5–50 ng g^{-1} f.w.), low values for limit of detection (LOD) (0.06–0.42 ng g^{-1} f.w.), and limit of quantification (LOQ) (0.16–0.91 ng g^{-1} f.w.), acceptable recoveries (41–93%), and $\%RSD_{IP}$ values for the four toxins. The method proved to be robust for the three variables tested. Finally, it was successfully applied to detect these cyanotoxins in edible vegetables exposed to cyanobacterial extracts under laboratory conditions, and it could be useful for monitoring these toxins in edible vegetables for better exposure estimation in terms of risk assessment.

Keywords: microcystins; cylindrospermopsin; method validation; UPLC-MS/MS; lettuce

Key Contribution: A novel optimization and validation of a multitoxin method for MCs and CYN in vegetables is presented. The method is linear, sensitive, accurate, and robust for 4 cyanotoxins, and is applied to real vegetable samples.

1. Introduction

Eutrophication and climate change may promote the proliferation and expansion of harmful cyanobacterial blooms in freshwater, estuarine, and marine ecosystems [1]. An increasing number of cyanobacteria can produce toxic metabolites named cyanotoxins [2], which comprise a large variety of compounds with various structural and physicochemical properties [3]. Among these cyanotoxins, microcystins (MCs) and cylindrospermopsin (CYN) are amongst the most studied because of their widespread distribution. MCs, mainly produced by *Microcystis*, have the common cyclic heptapeptidic structure of cyclo(-D-Ala-L-X-D-MeAsp-L-Z-Adda-D-Glu-Mdha), in which X and Z are variable L-amino acids that give the name to the molecule [4]. Because of these modifications, and the methylation/desmethylation of several functional groups, more than 100 MC variants have been reported to date [5]. Hence, the most common congeners are MC-LR, MC-RR, and MC-YR, resulting from the presence of the L-form of leucine (L), arginine (R), or tyrosine (Y) in position 2, and R in position 4 [6]. On the other hand, CYN is an alkaloid consisting of a tricyclic guanidine moiety combined with hidroxymethiluracil, which can be produced by *Cylindrospermopsis raciborskii* and *Chrysosporum ovalisporum* [7]. Concerning the mechanism of action, MCs are hepatotoxins and tumor promoters due to their strong potent inhibition of protein phosphatases and the effect on cell signaling pathways [4]. CYN is well known by its inhibition of protein and glutathione synthesis, induction of oxidative stress, and cytochrome P450 seems to mediate its toxicity; its pro-genotoxic activity has also been studied [6,8]. Although these toxic effects on animals are well documented, fewer studies have focused on their effects on vegetables, either in leaves, roots, and stems, such as oxidative stress, alterations in growth, germination, and development, and in mineral and vitamin contents [4,9–16].

It is important to take in mind that these cyanotoxins are not produced alone and isolated in aquatic environments, since a coincidence in time and space of cyanobacterial blooms can produce several cyanotoxins in the same freshwater and marine ecosystem and the same cyanobacteria species can produce different toxins also. In surface waters used as an irrigation source, total MCs concentrations from 50 μg L^{-1} up to 6500 μg L^{-1} have been reported [1] and CYN environmental concentrations in surface waters ranged from 1 to 800 μg L^{-1} [17–21]. Thus, humans may be orally exposed to cyanotoxins by drinking contaminated water, through the consumption of cyanotoxin-containing freshwater fish, crops, and food supplements, or by ingesting water during recreational activities [6].

Indeed, edible vegetables may accumulate MCs in the range 1.03–2352.2 μg kg^{-1} d.w. [4,10,12–15,22,23] and CYN in the range 2.71–49,000 μg kg^{-1} f.w. [9–11,24] by direct contact with contaminated irrigation water, which represents an additional risk to public health. The data available in the literature show provisional Tolerable Daily Intakes (TDI) of 0.04 μg kg^{-1} of body weight (b.w.) for MCs [25] and 0.03 μg kg^{-1} b.w. for CYN [26]. The upper limit in drinking water of 1 μg L^{-1} was set for MC-LR [25] and, although there is still no legislation regarding CYN, the same value has been proposed for this toxin [26]. However, although no limits have been established for these toxins in edible vegetables, it would be of interest to have adequate tools for their accurate determination in these matrices.

Mass spectrometry (MS) techniques are a powerful tool for the analysis of biotoxins [27], and it has been employed in the case of MCs detection in natural blooms, cyanobacterial strains, fish, and other biological samples [28–30]. Specifically, ultra-performance liquid chromatography-tandem mass spectrometry (UPLC-MS/MS) allows excellent specificity and sensitivity for cyanotoxins' detection and quantification in waters and in more complicated matrices, becoming the preferred technique for cyanotoxin analysis [16,31–37]. Previous studies were carried out in our laboratory for the determination of CYN in water [38], fish [39], and vegetables [40]. However, the separate isolation and identification of toxins belonging to each discrete class is laborious, expensive, and time consuming. Solid phase extraction (SPE) allows the cleaning of the sample for recovery and extraction of different analytes; however, the use of the same type of SPE cartridge is normally not adequate for the recovery of different cyanotoxins due to their differential physicochemical characteristics. Therefore, a multitoxin analytical approach is necessary for the simultaneous screening, detection, identification, and quantification of these toxins at low concentration levels [3], and in complex matrices, such as food items, following the recommendations of the European Food Safety Authority (EFSA) [41]. Different analytical methods for the simultaneous determination of cyanotoxins in waters are available, with [3,13,33,42,43] or without SPE [2,34,44,45]. By contrast, studies carried out to determine cyanotoxins simultaneously in vegetables are scarcer, and have not been validated. In addition, some of them have been performed with Enzyme-Linked ImmunoSorbent Assay (ELISA) [10], or have not looked at CYN [4,46,47]. For these reasons, and considering that MCs and CYN may coexist in natural environments, the aim of this work was to develop and validate a simple, sensitive, and robust analytical method for the simultaneous detection of the most common congeners of MCs (MC-LR, MC-RR, and MC-YR) and CYN in edible vegetables, such as lettuce, in a single analysis. The proposed method employed a dual SPE cartridge system in combination with UPLC-MS/MS, and has been optimized and tested according to international guidelines [48–51]. The present method has been designed to prove its applicability for the simultaneous analysis of these cyanotoxins in vegetable samples intended for human consumption.

2. Results and Discussion

2.1. Setup of the UPLC-MS/MS

Before assessing the effectiveness of the MC and CYN extraction method, the UPLC-MS/MS system was set up for this purpose. The transitions employed for MC-LR were 996.5/135.0, 996.5/213.1, and 996.5/996.5; for MC-RR: 520.2/135.0 and 1039.5/135.0; for MC-YR: 1046.5/135.0, 1046.5/213.0, and 1046.5/1046.5; and, finally, for CYN: 416.2/194.0 and 416.2/176.0; choosing the first ones for quantitation and the others as confirmatory, for each toxin. With respect to MC-LR, MC-RR, and MC-YR, the signals at m/z 135 and 213 were the main product ions for these MCs. The transition with m/z 520.2 corresponds to the double-charged protonated molecular ion, $[M + 2H]^{2+}$. The product ion at m/z 135 was identified as the $[phenyl\text{-}CH_2\text{-}CH(OCH_3)]^+$ ion formed by the rupture of the Adda residue between C-8 and C-9 and the m/z 213 was $[Glu\text{-}Mdha + H^+]$ lined up by the rupture of the α-linked glutamic (D-Glu) acid and *N*-methyldehydroalanine (Mdha) residue [4,52].

Regarding CYN, the signals at m/z 194 and 176 correspond to the loss of SO_3 and H_2O from the fragment ion at m/z 274, which represents the loss of the [6-(2-hydroxi-4-oxo-3-hydropyrimidyl)] hydroxymethinyl moiety from the CYN structure [53].

2.2. Calibration Study

To perform a calibration study, it is necessary to subject known amounts of the quantity to the measurement process, monitoring the measurement response over the expected working range [51]. The reply as a function of the concentration were calculated from MC-LR, MC-RR, MC-YR, and CYN standards prepared in fresh lettuce leaves extracts and were measured by four 12-point calibration curves with a linear range within 0.2–75 μg toxins L^{-1} (equal to 0.2–75 ng toxins g^{-1} f.w. lettuce). Twelve points were required for an accurate linearity considering the low working concentrations in the present study. The regression equations obtained were (a) y: $127.47x + 1.0247$ ($r^2 = 0.9996$) for MC-LR; (b) y: $157.73x + 1.4145$ ($r^2 = 0.9991$) for MC-RR; (c) y: $103.83x + 0.1822$ ($r^2 = 0.9988$), for MC-YR; and (d) y: $86.079x + 0.7383$ ($r^2 = 0.9999$) for CYN (Figure 1).

2.2.1. Linearity and Goodness of the Fit

Linearity of an analytical method with regards to the analysis of a number of samples varying in analyte concentrations followed by regression statistics were performed [51]. Twelve different concentrations of the mixture solution containing MCs and CYN were spiked to blank extracts of lettuce leaves ($n = 3$) and the samples were submitted to the present validated method. According to Huber et al. [54] the calibration plot, which represents the signal response/analyte concentration relationship versus analyte concentrations, was assessed by replicate analysis ($n = 3$) of the extracts at 0.2–75 μg toxins L^{-1}. The target line in Figure 2 represents the median signal/concentration calculated for all the concentrations evaluated. The results obtained for each concentration assayed are within the median value ±5% for MC-LR, MC-RR, MC-YR, and CYN, demonstrating the linear range of the proposed method.

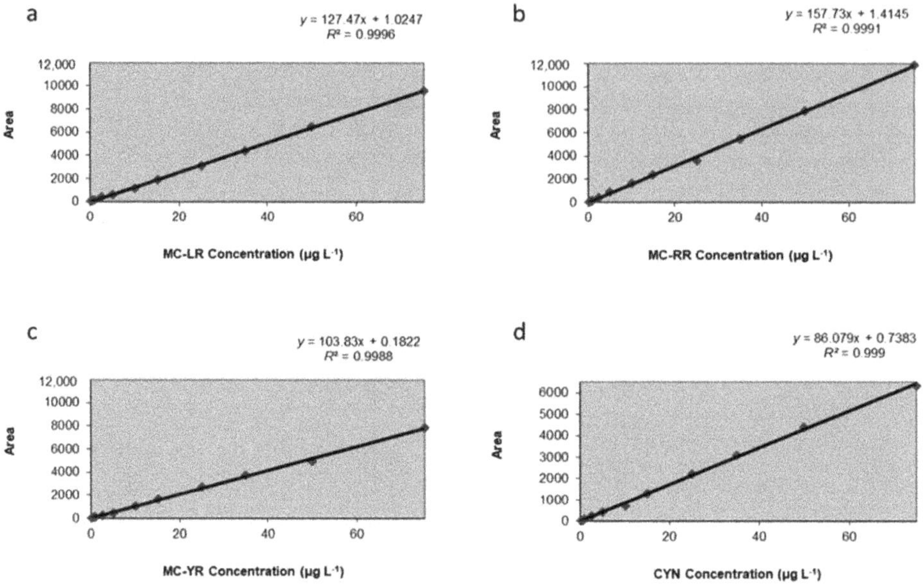

Figure 1. Calibration curves obtained for (**a**) Microcystin-LR (MC-LR); (**b**) MC-RR; (**c**) MC-YR; and (**d**) Cylindrospermopsin (CYN) in lettuce.

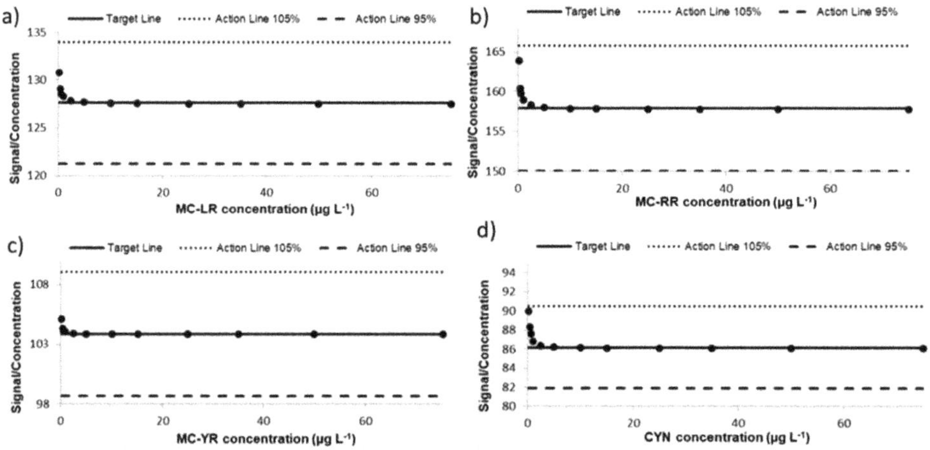

Figure 2. Response linearity in lettuce (Huber plot) for (**a**) MC-LR; (**b**) MC-RR; (**c**) MC-YR; and (**d**) CYN.

With the same signal data, the goodness of the fit was calculated. A lack of the fit *F* ratio was obtained using adequate analysis of variance (ANOVA) of the regression lines. The results showed that *F* ratio data were lower than the tabulated value for the corresponding degrees of freedom in this study (<19.4): 0.96 for MC-LR, 1.88 for MC-RR, 0.75 for MC-YR, and 0.76 for CYN, showing linear calibration functions for all toxins.

2.2.2. Sensitivity

For method validation, it is usually enough to give a sign of the level at which detection becomes problematic and quantification is adequate in terms of the precision, repeatability, and trueness.

For this objective, the limits of detection (LOD) and quantification (LOQ) were determined with 10 independent samples, according to the equation, $Y_{LOD/LOQ} = Y_{blank} + nS_{blanck}$, where Y_{blank} and S_{blank} are the average value of the blank signals and its corresponding standard deviation, and n is a constant value (3 for LOD and 10 for LOQ). Afterwards, LOD and LOQ values were converted in concentration using the calibration functions obtained previously. The LOD and LOQ values are presented in Table 1. Both the LOD and LOQ were lower than the guideline value of 1 μg L^{-1} proposed for MCs by WHO [25] and for CYN by Humpage and Falconer [26]. These results were lower or of the same order than those reported by Li et al. [4] in different vegetable matrices. The LOQ presented in the present work were in the lower limit of the ranges reported by those authors for MC-LR, -RR, and -YR in all the vegetable matrices. However, our results could not be compared with the same matrix (lettuce), in particular because these authors did not report the values for each matrix individually. Moreover, the present work reports for the first time, adequate LOD and LOQ levels for CYN in vegetables. More recently, Manubolu et al. [55] optimized an extraction method for MC-LR and MC-RR in different matrices, including vegetables. Nevertheless, the study conducted by these authors has two important limitations: The applicability of the method for the extraction of other MC congeners, and the quantification of the toxins at low concentrations (reported LOD < 26 ng g^{-1} f.w. and LOQ < 72 ng g^{-1} f.w.). Both analytical parameters have been improved in the present study. Furthermore, in the present multitoxin method that permits the simultaneous analysis of MCs variants and CYN, it presents some advantages over other methods that determine MCs or CYN individually, such as saving time, samples, materials, and resources, and diminishing solvent volumes, which could represent less environmental pollution.

Table 1. Estimations of within-day repeatability (S_W), between-day repeatability (S_B), intermediate precision (intra-laboratory reproducibility, S_{IP}), and its relative standard deviation (%RSD_{IP}) for MC-LR, MC-RR, MC-YR, and CYN, at three concentration levels, in three different days, assayed in lettuce samples. Limits of detection (LOD) and quantitation (LOQ) for the lettuce matrix. RSD_{AOAC} (%): 16–22% for 5 μg L^{-1} and 8–16% for 20 and 50 μg L^{-1}. Acceptable Recovery Range (%) by AOAC: 40–115% for 5 μg L^{-1} and 60–110% for 20 and 50 μg L^{-1}.

	Validation Parameters						
	Toxin Concentration Level (μg L^{-1})	**S_W**	**S_B**	**S_{IP}**	**RSD_{IP} (%)**	**LOD (ng g^{-1} f.w. [1])**	**LOQ (ng g^{-1} f.w. [1])**
MC-LR	5	0.20	1.10	0.66	21.68	0.06	0.16
	20	2.06	2.20	2.11	13.21		
	50	4.62	8.85	6.35	15.12		
MC-RR	5	0.14	0.21	0.17	8.31	0.23	0.50
	20	1.38	2.36	1.77	9.54		
	50	3.30	8.47	5.58	12.58		
MC-YR	5	0.23	0.74	0.46	19.86	0.42	0.91
	20	1.32	1.79	1.49	11.14		
	50	2.67	4.93	3.59	9.64		
CYN	5	0.28	0.65	0.44	19.30	0.07	0.19
	20	0.82	1.20	0.96	6.92		
	50	1.65	5.85	3.64	11.62		

[1] f.w.: Fresh weight. Microcystin-LR (MC-LR), MC-RR, MC-YR, and Cylindrospermopsin (CYN).

2.2.3. Matrix Effects

The matrix effect impacts the quantitation greatly and it depends on various factors, such as the properties of the analyte and the composition and amount of matrix [4]. Mild or no matrix effects simplifies calibration enormously if the calibration standards can be prepared as simple solutions of the analyte without matrix modifiers. For this, it is necessary to evaluate the effects of a possible general matrix mismatch in the validation process previously [56]. The lettuce matrix effect was studied for the four cyanotoxins following the recommendations of Li et al. [4]. The results showed matrix effect values of 4.23% (MC-LR), 17.17% (MC-RR), −12.32% (MC-YR), and −11.40% (CYN). Effects are considered mild (| 10 | < ME < | 20 |) for MC-RR, MC-YR, and CYN, or it can even be ignored (| 10 | < ME < | 10 |) in the case of MC-LR. Similar results were obtained by Li et al. [4] in the lettuce matrix for these three MCs (range from −13% to −5%); however, the present study shows for the first time that the presence of a new analyte in the mixture (CYN) does not increase the matrix effect.

2.3. Accuracy Study

For method validation, the accuracy of results is studied by considering both systematic and random errors. Thus, accuracy is analyzed as an entity with two components, such as trueness and precision [48,57].

2.3.1. Precision

Precision is a measure of the closeness of agreement between mutually independent measurement results obtained under specified conditions and it is generally dependent on the analyte concentration [51]. According to the International Conference on Harmonisation Guidelines (ICH) [58], the measure of precision includes three concepts: Repeatability, intermediate precision, and reproducibility. Repeatability indicates the variability when measurements are carried out by a single analyst with the same material, method, and equipment over a short period of time. On the other hand, intermediate precision assesses the variation in results when measurements are made with the same material, method, and in the same laboratory over an extended period, and therefore represents more variability than repeatability. Otherwise, reproducibility represents the variability in results when measurements are made in different laboratories [51].

The values of repeatability (within-day and between-day, S_w and S_B), intermediate precision (intralaboratory reproducibility, S_{IP}), and S_{IP} relative standard deviations (%RSD_{IP}) were calculated analyzing three replicates of lettuce leaf extracts spiked with the standard mixture solution containing MCs and CYN at different concentrations (5, 20, and 50 μg L^{-1}) on the same day, following the ICH guidelines, and over a period of three consecutive days. Considering three different days as the main source of variation, the estimations of these precision parameters were obtained by performing an analysis of variance (ANOVA) for each validation standard according to González and Herrador [48] and González et al. [49]. The relative standard deviations (%RSD_{IP}) obtained for each toxin were compared to the expected values issued by the AOAC Peer Verified Methods Program [49,50,54] for the different concentrations assayed (16–22% for 5 μg L^{-1} or 8–16% for 20 and 50 μg L^{-1}). For all studied toxins, the %RSD_{IP} values presented were lower or of the same order than RSD_{AOAC} tabulated values at the three concentration levels assayed, so that the proposed method can be considered as precise. The results are shown in Table 1.

2.3.2. Recovery

Trueness is the closeness of agreement between a test result and the accepted reference value of the property being measured and it can be investigated by spiking and recovery [56]. Total recovery for any validation method is defined as the ratio between the observed estimation of the validation standard concentrations and the true value, T, expressed as a percentage or fraction [38]. The recoveries obtained in the present study were in the range of 62–84% (MC-LR), 41–93% (MC-RR), 47–74% (MC-YR), and 45–69% (CYN) (Figure 3). The adequacy of these results was checked with the recovery values accepted for each concentration range, which are 40–115% for 5 μg L^{-1} and 60–110% for 20 and 50 μg L^{-1} [50]. As can be observed in Figure 3, MC-LR showed the best recoveries at lower concentrations (5 μg L^{-1}), whereas at higher concentrations (20 and 50 μg L^{-1}), it seems that better recoveries are obtained for MC-RR followed by MC-LR. In general, slightly better results have been obtained for MCs compared to CYN. Previously, high recoveries (104% for 20 μg L^{-1}) were shown for CYN in a validated method for determination of this toxin in lettuce [40]; however, in the present work, CYN recovery percentages resulted in compromised favor of the extraction and detection of more hydrophobic cyanotoxins, such as MCs.

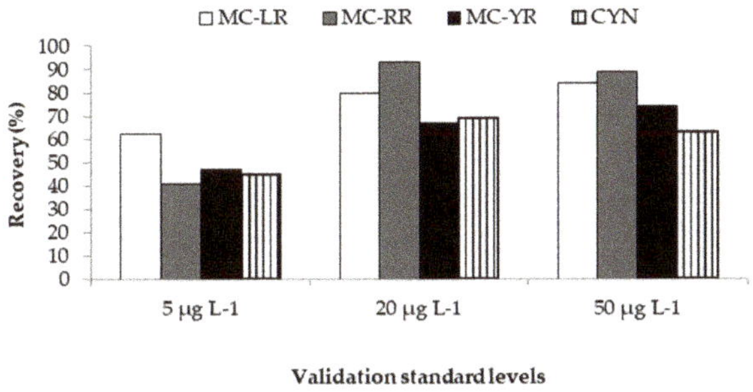

Figure 3. Recovery percentages for the three validation standards and cyanotoxins assayed.

Similar recoveries were obtained by Li et al. [4] for MCs (-LR and -RR), although these authors showed a better recovery for MC-YR. However, our study also includes the detection of another cyanotoxin (CYN), with acceptable recovery levels obtained, and which has not been included in the work from those authors. Similarly, other authors have also shown very good recoveries for MC-LR and -RR in lettuce [55]; this could be explained by two reasons: The low number of toxins analysed and the high concentrations they employed for spiking the samples (250 and 1000 ng g^{-1} f.w.) in comparison with ours (5–50 ng g^{-1} f.w.).

2.4. Robustness

The robustness, also called ruggedness, of an analytical method represents the resistance of the results to change after minor deviations are made in the experimental conditions described in the procedure. Thus, it is tested by deliberately introducing small changes to the procedure and examining the effect on the results [56]. The t values obtained for each cyanotoxin were compared with the 95% confidence level two-tailed tabulated value (t_{tab} = 2.306) corresponding with eight degrees of freedom obtained in the present study (Table 2). All the t values obtained were less than 2.306, so the present validated method can be considered as robust against the three different factors assayed at 20 μg L^{-1} for, MC-LR, MC-RR, MC-YR, and CYN. Although this is an important parameter to consider when validating analytical methods [51], as far as we know, no studies have validated robust methods for various cyanotoxins in any matrix, including vegetables.

Table 2. Coding rules for combination of the parameters in the robustness study and t values obtained for each parameter after the significance t test was applied.

Combined Variables			Toxins	*t* Values
F1	High (+)	15 min	MC-LR MC-RR	1.793 0.232
	Low (−)	10 min	MC-YR CYN	1.996 1.241
F2	High (+)	15 min	MC-LR MC-RR	0.059 0.381
	Low (−)	10 min	MC-YR CYN	0.042 0.358
F3	High (+)	1 min and 15 s	MC-LR MC-RR	0.346 0.234
	Low (−)	1 min	MC-YR CYN	1.055 2.231

F1: Sonication time of the samples; F2: Stirring time of the samples; and F3: Time for the sample to pass through the cartridge.

2.5. Application to Real Samples: Edible Vegetables Exposed to MC and CYN-Producing Extracts

The optimized and validated method was applied for the detection and quantification of the main variants of MCs (MC-LR, -RR, and -YR) and CYN in lettuce and spinach, as described in Section 4.5. LOD and LOQ levels from the proposed method permit the detection of MC-LR and CYN in these samples (Figure 4). No MC congeners were detected in the leaves of both vegetables analyzed, only low MC-LR levels were detected (0.22–1.31 ng g^{-1} f.w.) in roots, whereas CYN concentrations in the leaves ranged between 10–120 ng g^{-1} f.w. and in roots, between 24–110 ng g^{-1} f.w.

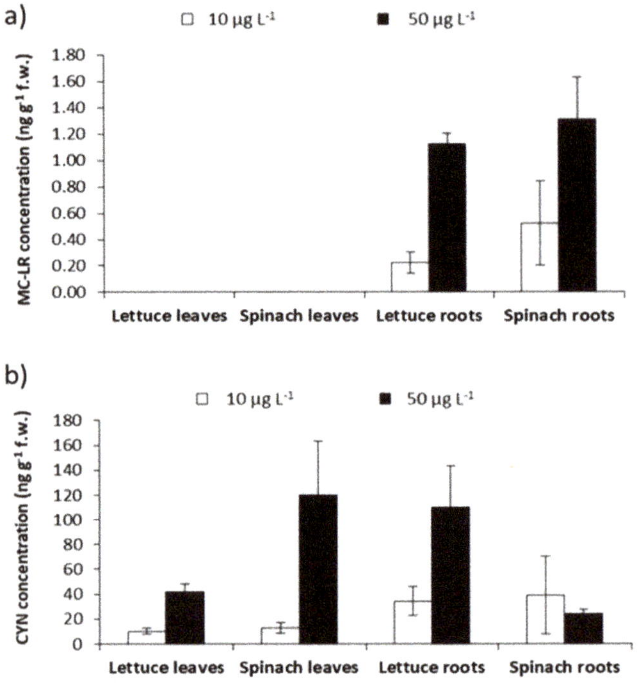

Figure 4. (**a**) MC-LR and (**b**) CYN concentrations detected in the leaves and roots of lettuce and spinach exposed to cyanobacterial extracts under laboratory conditions.

3. Conclusions

For the first time, a method for the simultaneous detection of MCs (MC-LR, -RR, and -YR) and CYN has been optimized and validated in lettuce (*L. sativa*) by SPE-UPLC-MS/MS, showing acceptable linearity, sensitivity, precision, recovery, and robustness for all toxins. This method has been successfully applied to real vegetable samples intended for human consumption. Due to the simultaneous presence of different cyanotoxins in the environment, it is indispensable to have adequate validated methods for their accurate detection in a more realistic exposure scenario in terms of health risk assessment.

4. Materials and Methods

4.1. Chemicals and Reagents

Three congeners of MCs (MC-LR, MC-RR, and MC-YR) (99% purity) and Cylindrospermopsin standard (95% purity) were purchased from Enzo Life Sciences (Lausen, Switzerland). Deionized water (18.2 MΩ cm resistivity) was obtained from a Milli-Q water purification system (Millipore, Bedford,

MA, USA). HPLC-grade methanol, dichloromethane (DCM), formic acid (FA), acetonitrile, and sodium hydroxide (NaOH) were supplied by Merck (Darmstadt, Germany). For the SPE, C18 cartridges were Bakerbond® (500 mg, 6 mL), purchased from Dicsa (Andalucía, España), and graphitized carbon cartridges were BOND ELUT® (500 mg, 6 mL), supplied by Agilent Technologies (Amstelveen, The Netherlands). For UHPLC–MS/MS analyses, reagents were of LC–MS grade: Water and acetonitrile were supplied by VWR International (Fontenay-sous-Bois, France) and formic acid by Fluka (Steinheim, Germany). A standard multitoxin solution containing 100 μg L^{-1} of each cyanotoxin (MC-LR, MC-RR, MC-YR, and CYN) was prepared in 20% MeOH to be further diluted to three different concentrations (5, 20, and 50 μg L^{-1}), used as working solutions. Lettuce control samples (without toxins) were obtained from a local supermarket, ready for human consumption.

4.2. Toxin Extraction from Lettuce Leaves and SPE

Considering previous experiments in which the lyophilization process did not affect the recovery of toxins [40], the addition of toxins to the lettuce leaves (fortification process before extraction) was performed with fresh weight lettuce and then they were lyophilized. First, before testing the extraction efficiency for the four cyanotoxins, the UPLC–MS/MS method was set up for this purpose, acquiring mass spectra and adjusting mobile phase strength for commercially available standard solutions of MC-LR, MC-RR, MC-YR, and CYN. Then, matrix-matched calibration curves were prepared for each of the four cyanotoxins by directly spiking extracts of control fresh lettuce leaves with the desired concentrations of the multitoxin solution to obtain a linear range of 0.2–75 μg L^{-1}, equivalent to 0.2–75 ng g^{-1} f.w. lettuce. To evaluate the efficiency of the proposed extraction and clean up methods, control fresh lettuce leaves (1.06 ± 0.05 g f.w.) were spiked with 1 mL of a multitoxin solution containing a mixture of the four cyanotoxins, at three concentration levels: 5, 20, and 50 μg L^{-1}, leading to 5, 20, and 50 ng g^{-1} f.w. lettuce. Afterwards, leaves were lyophilized and toxins were extracted. For this purpose, 70% and 80% MeOH were assayed, considering the different percentages of MeOH (20–100%) used by other authors when extracting these toxins simultaneously from water samples [3,33,42], as no multitoxin methods are available in vegetables. As 80% MeOH yielded the best recovery results (data not shown), the studies continued with this MeOH concentration. Then, the lyophilized lettuce leaves (0.05 ± 0.002 g d.w.) were extracted with 6 mL of 80% MeOH, homogenized in an ultraturrax (1 min), sonicated (15 min), and stirred in an orbital shaker (15 min). The mixture was centrifuged (3700 rpm, 15 min) and the supernatant collected for the clean-up; a test was performed with the supernatant at different pH (7, 9, and 11), and pH 11 was selected because it yielded the best results (data not shown), allowing the neutralization and adsorption of the toxins, which are basic, to the sorbent material in the cartridge. The purification process was developed considering the methods described by Li et al. [4] and Zervou et al. [3]. The use of reverse phase C18 cartridges is suitable for the extraction of moderately polar organic compounds, such as MCs, from aqueous matrices. However, due to the hydrophilic nature of CYN, it cannot be extracted by SPE with C_{18} cartridges, but PGC cartridges have been successfully used. So, an assembly of a C_{18} Bakerbond® cartridge (500 mg, 6 mL, Dicsa (Andalucía, Spain) and a BOND ELUT® Carbon cartridge (Agilent Technologies, Amstelveen, The Netherlands)) was employed. Moreover, because CYN could not be retained in the C_{18} cartridge, this was set at the top and PGC at the bottom, and the order was reversed (PGC on top and C_{18} on bottom) for elution to avoid MCs retention in the PGC column. After adjusting the supernatant pH to 11, the following reagents were passed through the assembled cartridges: 6 mL DCM, 6 mL 100% MeOH, 6 mL H_2O (pH 11), and sample (pH 11); then, cartridges were dried for 5 min and the order was inverted for elution of toxins with 10 mL DCM/MeOH (40/60) + 0.5% FA. The addition of DCM and FA to MeOH in the elution solvent is crucial to simultaneously extract CYN and MCs, due to their different polarities. Then, the extract was evaporated to dryness in a rotary evaporator and resuspended in 1 mL 20% MeOH for its analysis by UPLC-MS/MS. Three different percentages of MeOH were tested as redissolving solvent (20%, 50%, and 80% MeOH); 50% MeOH was rejected in the first place because it yielded the lowest recoveries for the four cyanotoxins; peak splitting was observed for CYN with 50% and 80% MeOH and its recovery was very low; so, finally, 20% MeOH was selected due to the best elution and recovery of the four cyanotoxins (data not shown).

4.3. Chromatographic Conditions

Chromatographic separation was performed using a UPLC Acquity (Waters) coupled to a Xevo TQ-S micro (Waters, Milford, MA, USA) consisting of a triple quadrupole mass spectrometer equipped with an electrospray ion source operated in positive mode. UPLC analyses were performed on a 100 × 2.1 mm XSelect HSS T3 2.5 μm column, at a flow rate of 0.45 mL min^{-1}. A binary gradient consisting of (A) water and (B) acetonitrile, both containing 0.1% formic acid (v/v) was employed, and the injection volume was 5 μL. The elution profile was: 2% B (0.8 min), linear gradient to 70% B (6.2 min), 100% B (1 min), and, finally, 2% B (2 min). Multiple Reaction Monitoring (MRM) was applied, where the parent ions and fragments ions were monitored at Q1 and Q3, respectively. For UPLC-ESI-MS/MS analyses, the mass spectrometer was set to the following optimised tune parameters: Capillary voltage: 1.0 kV; source temperature: 500 °C; source desolvation gas flow: 1000 L/h; and source cone gas flow: 50 L h^{-1}.

4.4. Analytical Criteria for Method Validation

For validation of the extraction and quantification method, several analytical parameters were calculated, such as linearity, sensitivity, precision, and recovery, considering the guidelines from Eurachem [51], and from González and Herrador [49], and the AOAC [50], For this purpose, three validation standards were employed for each cyanotoxin, performing the measures in triplicate each day for three consecutive days, covering the optimal working range. One mL solutions with three different cyantoxins concentrations (5, 20, and 50 μg L^{-1}) were added to lettuce leaves to obtain 5, 20, and 50 ng g^{-1} f.w., respectively. Precision and recovery were obtained by applying a one-factor analysis of variance (ANOVA), as explained in the Results and Discussion section, and then the results were compared with the respective tabulated reference values for each toxin concentration level. Besides, a robustness assay was also conducted to evaluate the ability of the method to stay unaffected despite small variations inherent in the analytical procedure in some parameters tested with the Student's t test, according to Youden's procedure (1967) [59]. This study was performed by spiking the leaf samples with an intermediate concentration of 20 μg L^{-1} multitoxin solution (MC-LR, MC-RR, MC-YR, and CYN) (equivalent to 20 ng g^{-1} f.w.), and the parameters were: (F1) Sonication time, (F2) stirring time, and (F3) time for the sample to pass through the cartridge. By combination of these parameters, eight different possibilities were assessed (Table 3). The weight of every factor is determined as the divergence of the medium results obtained at the level +1 and obtained at the level −1.

Table 3. Possible combinations (C1–C8) of parameters for the robustness study.

Combination Possibilities	F1	F2	F3
C1 (+++)	15 min	15 min	1 min
C2 (++−)	15 min	15 min	1 min 15 s
C3 (+−+)	15 min	10 min	1 min
C4 (+−−)	15 min	10 min	1 min 15 s
C5 (−++)	10 min	15 min	1 min
C6 (−+−)	10 min	15 min	1 min 15 s
C7 (−−+)	10 min	10 min	1 min
C8 (−−−)	10 min	10 min	1 min 15 s

F1: Sonication time of the samples; F2: Stirring time of the samples; and F3: Time for the sample to pass through the cartridge.

4.5. Exposure of Edible Vegetables Under Laboratory Conditions and Analysis of Toxins by the Validated Method

Plants of lettuce (*Lactuca sativa*) and spinach (*Spinacia oleracea*) were obtained from a local market (Porto, Portugal) as sprouts. Before their cultivation in a hydroponic system, all remaining soil was removed from the roots by washing with deionized water. Then, plants were introduced into opaque glass jars, ensuring that the roots were completely immersed in Jensen culture medium [60] at pH 6.5, as explained in Freitas et al. [61]. After an acclimation period of one week with white fluorescent light (14–10 h, light–dark period) and 21 $\pm$ 1 °C, plants were exposed in the medium to a solution containing MCs and CYN extracted from *M. aeruginosa* and *C. ovalisporum* cultures, respectively, at concentrations of 10 or 50 $\mu g\ L^{-1}$ (n = 5 per condition assayed). The culture medium with the solution containing MCs and CYN was changed three times a week for 21 d. Five plants per species were not exposed to the toxins (control groups). After the 21-d exposure period, plants were washed with distilled water, frozen, and lyophilized (Telstar Lyoquest) for analysis of MCs and CYN in leaves and roots, following the validated method presented in this work.

The *M. aeruginosa* culture (LEGE 91094) and the *C. ovalisporum* culture (LEGE X-001), isolated from Lake Kinneret, Israel [62], were grown in Z8 medium in the Interdisciplinary Centre of Marine and Environmental Research, CIIMAR (Porto, Portugal) [63]. MCs and CYN extraction was performed by the methods of Pinheiro et al. [64] and Welker et al. [65], respectively. Analysis by HPLC-PDA showed 0.2 mg MC-LR g^{-1} at 9.75 min and 2.9 mg CYN g^{-1} at 6.305 min.

Author Contributions: Conceptualization, A.M.C. and A.J.; methodology, L.D.-Q., R.G.-G., A.I.P.O. and M.L.-R.-C.; validation, L.D.-Q., R.G.-G. and A.I.P.O.; formal analysis, L.D.-Q., R.G.-G. and A.I.P.O.; investigation, L.D.-Q., R.G.-G. and A.I.P.O.; writing-original draft preparation, L.D.-Q., R.G.-G. and A.I.P.O.; writing-review & editing, A.M.C., A.J., A.C. and V.V.; visualization, L.D.-Q., R.G.-G. and A.I.P.O.; resources, A.M.C., A.J., A.C. and V.V.; supervision, A.M.C., A.J., A.C. and V.V.; project administration, A.M.C., A.J., A.C. and V.V.; funding acquisition, A.M.C., A.J., A.C. and V.V.

Funding: This research was funded by the SPANISH MINISTERIO DE ECONOMÍA Y COMPETITIVIDAD (AGL2015-64558-R, MINECO/FEDER, UE); by the FPI grant number BES-2016-078773 awarded to Leticia Díez-Quijada Jiménez; by the FCT project UID/Multi/04423/2013, and the post-doctoral grant (SFRH/BPD/103683/2014) from FCT awarded to Alexandre Campos.

Acknowledgments: Spanish Ministerio de Economía y Competitividad for the project AGL2015-64558-R, MINECO/FEDER, UE, and for the grant FPI (BES-2016-078773) awarded to Leticia Díez-Quijada Jiménez. CIIMAR members acknowledge FCT project UID/Multi/04423/2013 and the post-doctoral grant (SFRH/BPD/103683/2014) from FCT awarded to Alexandre Campos.

Conflicts of Interest: The authors declare no conflict of interest.

References

1. Corbel, S.; Mougin, C.; Bouaïcha, N. Cyanobacterial toxins: Modes of actions, fate in aquatic and soil ecosystems, phytotoxicity and bioaccumulation in agricultural crops. *Chemosphere* **2014**, *96*, 1–15. Available online: https://www.sciencedirect.com/science/article/pii/S0045653513010400 (accessed on 6 September 2018). [CrossRef] [PubMed]
2. Rodriguez, I.; Fraga, M.; Alfonso, A.; Guillebault, D.; Medlin, L.; Baudart, J.; Jacob, P.; Helmi, K.; Meyer, T.; Breitenbach, U.; et al. Monitoring of freshwater toxins in European environmental waters by using novel multi-detection methods. *Environ. Toxicol. Chem.* **2017**, *36*, 645–654. Available online: https://setac.onlinelibrary.wiley.com/doi/epdf/10.1002/etc.3577 (accessed on 6 September 2018). [CrossRef] [PubMed]
3. Zervou, S.K.; Christophoridis, C.; Kaloudis, T.; Triantis, T.M.; Hiskia, A. New SPE-LC-MS/MS method for simultaneous determination of multi-class cyanobacterial and algal toxins. *J. Hazard. Mater.* **2017**, *323*, 56–66. [CrossRef] [PubMed]
4. Li, Y.-W.; Zhan, X.-J.; Xiang, L.; Deng, Z.-S.; Huang, B.-H.; Wen, H.-F.; Sun, T.-F.; Cai, Q.-Y.; Lin, H.; Mo, C.-H. Analysis of trace microcystins in vegetables using matrix solid-phase dispersion followed by high performance liquid chromatography triple-quadrupole mass spectrometry detection. *J. Agric. Food Chem.* **2014**, *62*, 11831–11839. Available online: https://www.ncbi.nlm.nih.gov/pubmed/25393522 (accessed on 6 September 2018). [CrossRef] [PubMed]

5. Puddick, J.; Prinsep, M.R.; Wood, S.A.; Cary, S.C.; Hamilton, D.P.; Holland, P.T. Further characterization of glycine-containing microcystins from the McMurdo dry valleys of Antarctica. *Toxins* **2015**, *7*, 493–515. Available online: http://www.mdpi.com/2072-6651/7/2/493 (accessed on 6 September 2018). [CrossRef] [PubMed]
6. Buratti, F.M.; Manganelli, M.; Vichi, S.; Stefanelli, M.; Scardala, S.; Testai, E.; Funari, E. Cyanotoxins: Producing organisms, occurrence, toxicity, mechanism of action and human health toxicological risk evaluation. *Arch. Toxicol.* **2017**, *91*, 1049–1130. Available online: https://link.springer.com/article/10.1007/s00204-016-1913-6 (accessed on 6 September 2018). [CrossRef] [PubMed]
7. Ohtani, I.; Moore, R.E.; Runnegar, M.T.C. Cylindrospermopsin: A potent hepatotoxin from the blue-green alga *Cylindrospermopsis raciborskii*. *J. Am. Chem. Soc.* **1992**, *114*, 7941–7942. Available online: https://pubs.acs.org/doi/abs/10.1021/ja00046a067 (accessed on 6 September 2018). [CrossRef]
8. Gutiérrez-Praena, D.; Jos, A.; Pichardo, S.; Moreno, I.M.; Cameán, A.M. Presence and bioaccumulation of microcystins and cylindrospermopsin in food and the effectiveness of some cooking techniques at decreasing their concentrations: A review. *Food Chem. Toxicol.* **2013**, *53*, 139–152. Available online: https://www.sciencedirect.com/science/article/pii/S0278691512008083 (accessed on 6 September 2018). [CrossRef]
9. Silva, P.; Vasconcelos, V. Allelopathic effect of *Cylindrospermopsis raciborskii* extracts on the germination and growth of several plant species. *Chem. Ecol.* **2010**, *26*, 263–271. Available online: https://doi.org/10.1080/02757540.2010.495060 (accessed on 6 September 2018). [CrossRef]
10. Prieto, A.; Campos, A.; Cameán, A.; Vasconcelos, V. Effects on growth and oxidative stress status of rice plants (*Oryza sativa*) exposed to two extracts of toxin-producing cyanobacteria (*Aphanizomenon ovalisporum* and *Microcystis aeruginosa*). *Ecotoxicol. Environ. Saf.* **2011**, *74*, 1973–1980. Available online: https://www.sciencedirect.com/science/article/pii/S0147651311001734 (accessed on 6 September 2018). [CrossRef] [PubMed]
11. Kittler, K.; Schreiner, M.; Krumbein, A.; Manzei, S.; Koch, M.; Rohn, S.; Maul, R. Uptake of the cyanobacterial toxin cylindrospermopsin in *Brassica vegetables*. *Food Chem.* **2012**, *133*, 875–879. Available online: https://www.sciencedirect.com/science/article/pii/S0308814612001598 (accessed on 6 September 2018). [CrossRef]
12. Hereman, T.C.; Bittencourt-Oliveira, M.C. Bioaccumulation of Microcystins in lettuce. *J. Phycol.* **2012**, *48*, 1535–1537. Available online: https://onlinelibrary.wiley.com/doi/abs/10.1111/jpy.12006 (accessed on 6 September 2018). [CrossRef] [PubMed]
13. Romero-Oliva, C.S.; Contardo-Jara, V.; Block, T.; Pflugmacher, S. Accumulation of microcystin congeners in different aquatic plants and crops—A case study from lake Amatitlán, Guatemala. *Ecotoxicol. Environ. Saf.* **2014**, *102*, 121–128. Available online: https://www.sciencedirect.com/science/article/pii/S0147651314000359 (accessed on 6 September 2018). [CrossRef] [PubMed]
14. Bittencourt-Oliveira, M.C.; Cordeiro-Araújo, M.K.; Chia, M.A.; Arruda-Neto, J.D.; de Oliveira, E.T.; dos Santos, F. Lettuce irrigated with contaminated water: Photosynthetic effects, antioxidative response and bioaccumulation of microcystin congeners. *Ecotoxicol. Environ. Saf.* **2016**, *128*, 83–90. Available online: https://www.sciencedirect.com/science/article/pii/S0147651316300446 (accessed on 6 September 2018). [CrossRef] [PubMed]
15. Cordeiro-Araújo, M.K.; Chia, M.A.; Arruda-Neto, J.D.T.; Tornisielo, V.L.; Vilca, F.Z.; Bittencourt-Oliveira, M.C. Microcystin-LR bioaccumulation and depuration kinetics in lettuce and arugula: Human health risk assessment. *Sci. Total Environ.* **2016**, *566*, 1379–1386. Available online: https://www.sciencedirect.com/science/article/pii/S0048969716311421 (accessed on 6 September 2018). [CrossRef]
16. Guzmán-Guillén, R.; Maisanaba, S.; Prieto Ortega, A.I.; Valderrama-Fernández, R.; Jos, A.; Cameán, A.M. Changes on cylindrospermopsin concentration and characterization of decomposition products in fish muscle (*Oreochromis niloticus*) by boiling and steaming. *Food Control* **2017**, *77*, 210–220. [CrossRef]
17. Saker, M.L.; Eaglesham, G.K. The accumulation of cylindrospermopsin from the cyanobacterium *Cylindrospermopsis raciborskii* in tissues of the Redclaw crayfish *Cherax quadricarinatus*. *Toxicon* **1999**, *37*, 1065–1077. [CrossRef]
18. Shaw, G.R.; Seawright, A.A.; Moore, M.A.; Lam, P.K.S. Cylindrospermopsin, a cyanobacterial alkaloid: Evaluation of its toxicological activity. *Ther. Drug Monit.* **2000**, *22*, 89–92. [CrossRef] [PubMed]

19. Rucker, J.; Stuken, A.; Nixdorf, B.; Fastner, J.; Chorus, I.; Wiedner, C. Concentration of particulate and dissolved cylindrospermopsin in 21 *Aphanizomenon*-dominated temperate lakes. *Toxicon* **2007**, *50*, 800–809. Available online: http://www.ncbi.nlm.nih.gov/pubmed/17804031 (accessed on 6 September 2018). [CrossRef] [PubMed]
20. Messineo, V.; Bogialli, S.; Melchiorre, S.; Sechi, N.; Lugliè, A.; Casiddu, P.; Mariani, M.A.; Padedda, B.M.; Di Corcia, A.; Mazza, R.; et al. Cyanobacterial toxins in Italian freshwaters. *Limnologica* **2009**, *39*, 95–106. Available online: https://www.sciencedirect.com/science/article/pii/S0075951108000662 (accessed on 6 September 2018). [CrossRef]
21. Cartmell, C.; Evans, D.M.; Elwood, J.M.L.; Fituri, H.S.; Murphy, P.J.; Caspari, T.; Poniedzialek, B.; Rzymski, P. Synthetic analogues of cyanobacterial alkaloid cylindrospermopsin and their toxicological activity. *Toxicol. In Vitro* **2017**, *44*, 172–181. Available online: http://www.ncbi.nlm.nih.gov/pubmed/28705760 (accessed on 6 September 2018). [CrossRef] [PubMed]
22. Crush, J.R.; Briggs, L.R.; Sprosen, J.M.; Nichols, S.N. Effect of irrigation with lake water containing microcystins on microcystin content and growth of ryegrass, clover, rape, and lettuce. *Environ. Toxicol.* **2008**, *23*, 246–252. Available online: http://www.ncbi.nlm.nih.gov/pubmed/18214908 (accessed on 6 September 2018). [CrossRef] [PubMed]
23. Drobac, D.; Tokodi, N.; Kiprovski, B.; Malenčić, D.; Važić, T.; Nybom, S.; Meriluoto, J.; Svirčev, Z. Microcystin accumulation and potential effects on antioxidant capacity of leaves and fruits of *Capsicum annuum*. *J. Toxicol. Environ. Health (Part A)* **2017**, *80*, 145–154. Available online: https://www.tandfonline.com/doi/full/10.1080/15287394.2016.1259527 (accessed on 6 September 2018). [CrossRef] [PubMed]
24. Cordeiro-Araújo, M.K.; Chia, M.A.; Bittencourt-Oliveira, M.C. Potential human health risk assessment of cylindrospermopsin accumulation and depuration in lettuce and arugula. *Harm. Algae* **2017**, *68*, 217–223. Available online: http://www.ncbi.nlm.nih.gov/pubmed/27267723 (accessed on 6 September 2018). [CrossRef]
25. World Health Organization. Cyanobacterial Toxins: Microcystin-LR in Drinking-Water Background Document for Development of WHO Guidelines for Drinking-Water Quality. 2003. Available online: http://www.who.int/water_sanitation_health/dwq/chemicals/cyanobactoxins.pdf (accessed on 6 September 2018).
26. Humpage, A.R.; Falconer, I.R. Oral toxicity of the cyanobacterial toxin cylindrospermopsin in male Swiss albino mice: Determination of no observed adverse effect level for deriving a drinking water guideline value. *Environ. Toxicol.* **2003**, *18*, 94–103. Available online: http://www.ncbi.nlm.nih.gov/pubmed/12635097 (accessed on 6 September 2018). [CrossRef] [PubMed]
27. Flores, C.; Caixach, J. An integrated strategy for rapid and accurate determination of free and cell-bound microcystins and related peptides in natural blooms by liquid chromatography-electrospray-high resolution mass spectrometry and matrix-assisted laser desorption/ionization ionizationtime-of-flight/time-of-flight mass spectrometry using both positive and negative ionization modes. *J. Chromatogr. A* **2015**, *1407*, 76–89. Available online: https://www.sciencedirect.com/science/article/pii/S0021967315008699 (accessed on 6 September 2018). [CrossRef] [PubMed]
28. Cameán, A.; Moreno, I.M.; Ruiz, M.J.; Picó, Y. Determination of microcystins in natural blooms and cyanobacterial strains cultures by matrix solid-phase dispersion and liquid chromatography-mass spectrometry. *Anal. Bioanal. Chem.* **2004**, *380*, 537–544. Available online: http://www.ncbi.nlm.nih.gov/pubmed/15365676 (accessed on 6 September 2018). [CrossRef] [PubMed]
29. Moreno, I.M.; Molina, R.; Jos, Á.; Picó, Y.; Cameán, A.M. Determination of microcystins in fish by solvent extraction and liquid chromatography. *J. Chromatogr. A* **1080**, *1080*, 199–203. Available online: https://www.sciencedirect.com/science/article/pii/S0021967305010228 (accessed on 6 September 2018). [CrossRef]
30. Ruiz, M.J.; Cameán, A.M.; Moreno, I.M.; Picó, Y. Determination of microcystins in biological samples by matrix solid-phase dispersion (MSDP) and liquid chromatography-mass spectrometry (LC-MS). *J. Chromatogr. A* **2005**, *1073*, 257–262. Available online: https://www.ncbi.nlm.nih.gov/pubmed/15909527 (accessed on 6 September 2018). [CrossRef] [PubMed]
31. Geis-Asteggiante, L.; Lehotay, S.J.; Fortis, L.L.; Paoli, G.; Wijey, C.; Heinzen, H. Development and validation of a rapid method for microcystins in fish and comparing LC-MS/MS results with ELISA. *Anal. Bioanal. Chem.* **2011**, *401*, 2617–2630. Available online: https://www.ncbi.nlm.nih.gov/pubmed/21881880 (accessed on 6 September 2018). [CrossRef] [PubMed]

32. Corbel, S.; Mougin, C.; Nélieu, S.; Delarue, G.; Bouaïcha, N. Evaluation of the transfer and the accumulation of microcystins in tomato (*Solanum lycopersicum* cultivar MicroTom) tissues using a cyanobacterial extract containing microcystins and the radiolabeled microcystin-LR (^{14}C-MC-LR). *Sci. Total Environ.* **2016**, *541*, 1052–1058. Available online: https://www.sciencedirect.com/science/article/pii/S0048969715308214 (accessed on 6 September 2018). [CrossRef] [PubMed]
33. Greer, B.; McNamee, S.E.; Boots, B.; Cimarelli, L.; Guillebault, D.; Helmi, K.; Marcheggiani, S.; Panaiotov, S.; Breitenbach, U.; Akçaalan, R.; et al. A validated UPLC-MS/MS method for the surveillance of ten aquatic biotoxins in European brackish and freshwater systems. *Harm. Algae* **2016**, *55*, 31–40. Available online: https://www.sciencedirect.com/science/article/pii/S156898831530086X (accessed on 6 September 2018). [CrossRef] [PubMed]
34. Pekar, H.; Westerberga, E.; Brunoa, O.; Laanec, A.; Perssond, K.M.; Sundstromf, L.F.; Thim, A.M. Fast, rugged and sensitive ultra-high pressure liquid chromatography tandem mass spectrometry method for analysis of cyanotoxins in raw water and drinking water- First findings of anatoxins, cylindrospermopsins and microcystin variants in Swedish source waters and infiltration ponds. *J. Chromatogr. A* **2016**, *1429*, 265–276. Available online: https://www.ncbi.nlm.nih.gov/pubmed/26755412 (accessed on 6 September 2018). [CrossRef] [PubMed]
35. Prieto, A.I.; Guzmán-Guillén, R.; Valderrama-Fernández, R.; Jos, A.; Cameán, A.M. Influence of Cooking (Microwaving and Broiling) on Cylindrospermopsin Concentration in Muscle of Nile Tilapia (*Oreochromis niloticus*) and Characterization of Decomposition Products. *Toxins* **2017**, *9*, 177–190. Available online: http://www.ncbi.nlm.nih.gov/pubmed/28587145 (accessed on 6 September 2018). [CrossRef] [PubMed]
36. Theunis, M.; Naessens, T.; Verhoeven, V.; Hermans, N.; Apers, S. Development and validation of a robust high-performance liquid chromatographic method for the analysis of monacolins in red yeast rice. *Food Chem.* **2017**, *234*, 33–37. Available online: https://www.sciencedirect.com/science/article/pii/S0308814617307148 (accessed on 6 September 2018). [CrossRef] [PubMed]
37. Turner, A.D.; Waack, J.; Lewis, A.; Edwards, C.; Lawton, L. Development and single-laboratory validation of a UHPLC-MS/MS method for quantitation of microcystins and nodularin in natural water, cyanobacteria, shellfish and algal supplement tablet powders. *J. Chromatogr. B* **2018**, *1074*, 111–123. Available online: https://www.sciencedirect.com/science/article/pii/S157002321731869X (accessed on 6 September 2018). [CrossRef] [PubMed]
38. Guzmán-Guillén, R.; Prieto, A.I.; Gónzalez, A.G.; Soria-Díaz, M.E.; Cameán, A.M. Cylindrospermopsin determination in water by LC-MS/MS: Optimization and validation of the method and application to real samples. *Environ. Toxicol. Chem.* **2012**, *31*, 2233–2238. Available online: https://www.ncbi.nlm.nih.gov/pubmed/22825923 (accessed on 6 September 2018). [CrossRef]
39. Guzmán-Guillén, R.; Moreno, I.M.; Prieto, A.I.; Soria-Díaz, M.E.; Vasconcelos, V.M.; Cameán, A.M. CYN determination in tissues from fresh water fish by LC–MS/MS: Validation and application in tissues from subchronically exposed tilapia (*Oreochromis niloticus*). *Talanta* **2015**, *131*, 452–459. Available online: https://www.ncbi.nlm.nih.gov/pubmed/25281126 (accessed on 6 September 2018). [CrossRef]
40. Prieto, A.I.; Guzmán-Guillén, R.; Díez-Quijada, L.; Campos, A.; Vasconcelos, V.; Jos, Á.; Cameán, A.M. Validation of a method for cylindrospermopsin determination in vegetables: Application to real samples such as lettuce (*Lactuca sativa* L.). *Toxins* **2018**, *10*, 1–15. Available online: http://www.readcube.com/articles/10.3390/toxins10020063 (accessed on 6 September 2018). [CrossRef] [PubMed]
41. Testai, E.; Buratti, F.M.; Funari, E.; Manganelli, M.; Vichi, S.; Arnich, N.; Biré, R.; Fessard, V.; Sialehaamoa, A. Review and analysis of occurrence, exposure and toxicity of cyanobacteria toxins in food. *EFSA Support. Publ.* **2016**, *13*, 1–309. Available online: https://efsa.onlinelibrary.wiley.com/doi/abs/10.2903/sp.efsa.2016.EN-998 (accessed on 6 September 2018). [CrossRef]
42. Yen, H.K.; Lin, T.F.; Liao, P.C. Simultaneous detection of nine cyanotoxins in drinking water using dual solid-phase extraction and liquid chromatography-mass spectrometry. *Toxicon* **2011**, *58*, 209–218. Available online: https://www.sciencedirect.com/science/article/pii/S0041010111001905 (accessed on 6 September 2018). [CrossRef] [PubMed]

43. Zamyadi, A.; MacLeod, S.L.; Fan, Y.; McQuaid, N.; Dorner, S.; Sauvé, S.; Prévost, M. Toxic cyanobacterial breakthrough and accumulation in a drinking water plant: A monitoring and treatment challenge. *Water Res.* **2012**, *46*, 1511–1523. Available online: https://www.sciencedirect.com/science/article/pii/S0043135411006841 (accessed on 6 September 2018). [CrossRef] [PubMed]
44. Oehrle, S.A.; Southwell, B.; Westrick, J. Detection of various freshwater cyanobacterial toxins using ultra-performance liquid chromatography tandem mass spectrometry. *Toxicon* **2010**, *55*, 965–972. Available online: https://www.ncbi.nlm.nih.gov/pubmed/19878689 (accessed on 6 September 2018). [CrossRef] [PubMed]
45. Szlag, D.C.; Sinclair, J.L.; Southwell, B.; Westrick, J.A. Cyanobacteria and cyanotoxins occurrence and removal from five high-risk conventional treatment drinking water plants. *Toxins* **2015**, *7*, 2198–2220. Available online: http://www.mdpi.com/2072-6651/7/6/2198 (accessed on 6 September 2018). [CrossRef] [PubMed]
46. Trifirò, G.; Barbaro, E.; Gambaro, A.; Vita, V.; Clausi, M.T.; Franchino, C.; Palumbo, M.P.; Floridi, F.; De Pace, R. Quantitative determination by screening ELISA and HPLC-MS/MS of microcystins LR, LY, LA, YR, RR, LF, LW, and nodularin in the water of Occhito lake and crops. *Anal. Bioanal. Chem.* **2016**, *408*, 7699–7708. Available online: https://link.springer.com/article/10.1007/s00216-016-9867-3 (accessed on 6 September 2018). [CrossRef] [PubMed]
47. Qian, Z.-Y.; Li, Z.-G.; Ma, J.; Gong, T.-T.; Xian, Q.-M. Analysis of trace microcystins in vegetables using matrix solid-phase dispersion followed by high performance liquid chromatography triple-quadrupole mass spectrometry detection. *Talanta* **2017**, *173*, 101–106. Available online: https://www.sciencedirect.com/science/article/pii/S0039914017306070 (accessed on 6 September 2018). [CrossRef] [PubMed]
48. González, A.G.; Herrador, M.A. A practical guide to analytical method validation, including measurement uncertainty and accuracy profiles. *Trends Anal. Chem.* **2007**, *26*, 227–238. Available online: https://www.sciencedirect.com/science/article/pii/S0165993607000118 (accessed on 6 September 2018). [CrossRef]
49. González, A.G.; Herrador, M.A.; Asuero, A.G. Intra-laboratory assessment of method accuracy (trueness and precision) by using validation standards. *Talanta* **2010**, *82*, 1995–1998. Available online: http://www.ncbi.nlm.nih.gov/pubmed/20875607 (accessed on 6 September 2018). [CrossRef]
50. AOAC International; AOAC Official Methods of Analysis. *Guidelines for Standard Method Performance Requirements*; Appendix F; AOAC International: Rockville, MD, USA, 2016; Available online: http://www.eoma.aoac.org/app_f.pdf (accessed on 6 September 2018).
51. Eurachem Guide: The Fitness for Purpose of Analytical Methods—A Laboratory Guide to Method Validation and Related Topics. Available online: www.eurachem.org (accessed on 2 October 2018).
52. Yuan, M.; Namikoshi, M.; Otsuki, A.; Rinehart, K.L.; Sivonen, K.; Watanabe, M.F. Low-energy collisionally activated decomposition and structural characterization of cyclic heptapeptide Microcystins by electrospray ionization mass spectrometry. *J. Mass Spectrom.* **1999**, *34*, 33–43. Available online: https://www.ncbi.nlm.nih.gov/pubmed/10028690 (accessed on 6 September 2018). [CrossRef]
53. Dell'Aversano, C.; Eaglesham, G.K.; Quilliam, M.A. Analysis of cyanobacterial toxins by hydrophilic interaction liquid chromatography–mass spectrometry. *J. Chromatogr. A* **2004**, *1028*, 155–164. Available online: http://www.ncbi.nlm.nih.gov/pubmed/14969289 (accessed on 6 September 2018). [CrossRef]
54. Huber, L. *Validation and Qualification in Analytical Laboratories*; Interpharm: East Englewood, CO, USA, 1998; pp. 1–288.
55. Manubolu, M.; Lee, J.; Riedi, K.M.; Kua, Z.X.; Collart, L.P.; Collart, S.P.; Ludsin, S.A. Optimization of extraction Methods for quantification of microcystin-LR and microcystin-RR in fish, vegetables, and soil matrices using UPLC-MS/MS. *Harm. Algae* **2018**, *76*, 47–57. Available online: https://www.sciencedirect.com/science/article/pii/S1568988318300593 (accessed on 6 September 2018). [CrossRef] [PubMed]
56. Thompson, M.; Ellison, S.L.R.; Wood, R. Harmonized guidelines for single-laboratory validation of methods of analysis (IUPAC Technical Report). *Pure Appl. Chem.* **2002**, *74*, 835–855. Available online: https://www.degruyter.com/view/j/pac.2002.74.issue-5/pac200274050835/pac200274050835.xml (accessed on 6 September 2018). [CrossRef]
57. Taverniers, I.; Van Bockstaele, E.; De Loose, M. Trends in quality in the analytical laboratory, II: Analytical method validation and quality assurance. *Trends Anal. Chem.* **2004**, *23*, 535–552. Available online: https://www.sciencedirect.com/science/article/pii/S0165993604030031 (accessed on 6 September 2018). [CrossRef]

58. ICH Harmonised Tripartite Guideline, Validation of Analytical Procedures: Text and Methodology, ICH Working Group, November 2005. Available online: http://www.ich.org/fileadmin/Public_Web_Site/ICH_Products/Guidelines/Quality/Q2_R1/Step4/Q2_R1__Guideline.pdf (accessed on 6 September 2018).
59. Youden, W.J. *Statistical Techniques for Collaborative Tests*; Association of Official Analytical Chemists: Washington, DC, USA, 1967; pp. 1–64.
60. Jensen, M.H.; Malter, A.J. Chapter 7: Water Supply, Water Quality and Mineral Nutrition. In *Protected Agriculture: A Global Review*, 1st ed.; The World Bank: Washington, DC, USA, 1995; pp. 65–69.
61. Freitas, M.; Azevedo, J.; Pinto, E.; Neves, J.; Campos, A.; Vasconcelos, V. Effects of microcystin-LR, cylindrospermopsin and a microcystin-LR/cylindrospermopsin mixture on growth, oxidative stress and mineral content in lettuce plants (*Lactuca sativa* L.). *Ecotoxicol. Environ. Saf.* **2015**, *116*, 59–67. [CrossRef] [PubMed]
62. Banker, R.; Carmeli, S.; Hadas, O.; Teltsch, B.; Porat, R.; Sukenik, A. Identification of cylindrospermopsin in *Aphanizomenon ovalisporum* (cyanophyceae) isolated from Lake Kinneret, Israel. *J. Phycol.* **1997**, *33*, 613–616. Available online: https://onlinelibrary.wiley.com/doi/abs/10.1111/j.0022-3646.1997.00613.x (accessed on 6 September 2018). [CrossRef]
63. Campos, A.; Araújo, P.; Pinheiro, C.; Azevedo, J.; Osório, H.; Vasconcelos, V. Effects on growth, antioxidant enzyme activity and levels of extracellular proteins in the green alga *Chlorella vulgaris* exposed to crude cyanobacterial extracts and pure microcystin and cylindrospermopsin. *Ecotoxicol. Environ. Saf.* **2013**, *94*, 45–53. Available online: https://www.ncbi.nlm.nih.gov/pubmed/23726538 (accessed on 6 September 2018). [CrossRef] [PubMed]
64. Pinheiro, C.; Azevedo, J.; Campos, A.; Loureiro, S. Absence of negative allelopathic effects of cylindrospermopsin and microcystin-LR on selected marine and freshwater phytoplankton species. *Hydrobiologia* **2013**, *705*, 27–42. Available online: https://link.springer.com/article/10.1007/s10750-012-1372-x (accessed on 6 September 2018). [CrossRef]
65. Welker, M.; Bickel, H.; Fastner, J. HPLC-DAD detection of cylindrospermopsin-Opportunities and limits. *Water Res.* **2002**, *36*, 4659–4663. Available online: https://www.sciencedirect.com/science/article/pii/S004313540200194X (accessed on 6 September 2018). [CrossRef]

toxins

MDPI

Article

In Vitro Mutagenic and Genotoxic Assessment of a Mixture of the Cyanotoxins Microcystin-LR and Cylindrospermopsin

Leticia Díez-Quijada, Ana I. Prieto, María Puerto, Ángeles Jos * and Ana M. Cameán

Area of Toxicology, Faculty of Pharmacy, University of Sevilla, C/Profesor García González 2, 41012 Sevilla, Spain; ldiezquijada@us.es (L.D.-Q.); anaprieto@us.es (A.I.P.); mariapuerto@us.es (M.P.); camean@us.es (A.M.C.)
* Correspondence: angelesjos@us.es; Tel.: +34-954-556-762

Received: 10 April 2019; Accepted: 31 May 2019; Published: 4 June 2019

Abstract: The co-occurrence of various cyanobacterial toxins can potentially induce toxic effects different than those observed for single cyanotoxins, as interaction phenomena cannot be discarded. Moreover, mixtures are a more probable exposure scenario. However, toxicological information on the topic is still scarce. Taking into account the important role of mutagenicity and genotoxicity in the risk evaluation framework, the objective of this study was to assess the mutagenic and genotoxic potential of mixtures of two of the most relevant cyanotoxins, Microcystin-LR (MC-LR) and Cylindrospermopsin (CYN), using the battery of in vitro tests recommended by the European Food Safety Authority (EFSA) for food contaminants. Mixtures of 1:10 CYN/MC-LR (CYN concentration in the range 0.04–2.5 μg/mL) were used to perform the bacterial reverse-mutation assay (Ames test) in *Salmonella typhimurium*, the mammalian cell micronucleus (MN) test and the mouse lymphoma thymidine-kinase assay (MLA) on L5178YTk$^{\pm}$ cells, while Caco-2 cells were used for the standard and enzyme-modified comet assays. The exposure periods ranged between 4 and 72 h depending on the assay. The genotoxicity of the mixture was observed only in the MN test with S9 metabolic fraction, similar to the results previously reported for CYN individually. These results indicate that cyanobacterial mixtures require a specific (geno)toxicity evaluation as their effects cannot be extrapolated from those of the individual cyanotoxins.

Keywords: genotoxicity; mutagenicity; Cylindrospermopsin; Microcystin-LR; mixture

Key Contribution: A genotoxic and mutagenic assessment of cyanotoxin binary mixtures of CYN and MC-LR was performed by a battery of in vitro tests. Results showed a similar response to CYN individually. Thus, evaluation of mixtures is required as interactions can occur.

1. Introduction

Nowadays, a proliferation of cyanobacterial species can be seen globally because of water eutrophication and climate change, leading to an increasing occurrence of cyanotoxins [1–3]. Cyanotoxins are toxic secondary metabolites produced by various species of cyanobacteria, which involved an ample variety of compounds with different structural and physicochemical properties [4]. Humans may be exposed to cyanotoxins via different routes, but oral exposure by means of contaminated water and foods (fish, crops, vegetables and food supplements) is by far the most important [5,6]. Microcystins (MCs) and cylindrospermopsins (CYN) are among the most frequently investigated cyanotoxins due to their toxicity and extensive distribution.

MCs are cyclic heptapeptides and 246 variants were identified so far [7], with Microcystin-LR (MC-LR) as the reference congener. The liver is the main target organ in MC-LR toxicity because of its uptake into hepatocytes by the organic anion transport system [8]. MC-LR inhibits the protein serine/threonine phosphatases by covalent binding, especially PP1 and PP2. Thus, the proteins are hyperphosphorylated leading to the modification of cytoskeleton and disruption of actin filaments [9]. In addition, MCs induce oxidative stress [1,10], disrupt different enzymatic activities [11,12] and induce apoptosis [13]. MC-LR was classified as possible human carcinogen (Group 2B) by the International Agency of Research on Cancer (IARC) [14]. It can produce genotoxic effects in vitro and in vivo [15], although the mechanisms involved are not yet completely understood [16].

Cylindrospermopsins are guanidine alkaloid hepatotoxins with five known analogues [17]. Cylindrospermopsin (CYN) has zwitterionic characteristics, thus being highly water soluble and chemically stable at high temperatures and a wide range of pH [18,19]. For these reasons, humans can be more likely exposed to CYN than to other cyanotoxins as up to 90% of total CYN is presented in surrounding waters. Although the liver and kidney are target organs of CYN, other organs such us lungs, heart, thymus, stomach, spleen, intestinal tract, skin, nervous, immune, vascular and lymphatic systems could also be damaged [1,20–22].

The absorption mechanism of CYN is not totally elucidated, but it was shown that paracellular transport is involved in the intestinal uptake [1,23]. The main mechanisms of CYN toxicity is the irreversible inhibition of protein synthesis [24,25] and glutathione (GSH) depletion [26] related to the oxidative stress induced by CYN [27–29]. Moreover, the bioactivation of CYN by cytochrome P-450 plays an important role in its mechanism of toxicity [30]. CYN was shown to induce DNA fragmentation and DNA strands breaks [31–38]. However, it was not yet classified by its carcinogenic potential by the IARC.

Both cyanotoxins have been extensively studied individually, but there are very few studies that evaluate their combined effects, as indicated by the European Food Safety Authority (EFSA) [5]. The simultaneous occurrence of MCs and CYN was reported repeatedly [39,40]. They have different chemical structures and mechanisms of action, thus interaction phenomena such as synergism, antagonism or toxicity potentiation must be considered. Moreover, a risk assessment can be greatly influenced when diverging from individual toxin exposure to a multi-toxin exposure scenario. Gutiérrez-Praena et al. [41] found an antagonistic effect of CYN and MC-LR when investigating the cytotoxicity of binary mixtures in comparison to the individual toxins in HepG2 cells. Hercog et al. [42] observed a genotoxic potential of CYN/MC-LR mixtures comparable to that of CYN alone when using the micronucleus (MN) and comet assays in the same experimental model.

The exploration of the genotoxic potential of CYN/MC-LR applicable to food and feed safety assessment is of great current interest. EFSA has indicated the need for further data on the toxicity of cyanotoxins mixtures [5] following recommended genotoxicity testing strategies [43].

Thus, the purpose of this research was to assess the mutagenic and genotoxic potential of the CYN/MC-LR mixtures trough a complete battery of different in vitro tests. This battery included: (1) The bacterial reverse-mutation assay in five strains of *Salmonella typhimurium* (Ames test, OECD 471 [44]) which detects gene mutations in the absence and presence of the microsomal fraction S9; (2) the Micronucleus test (MN, OECD 487 [45]) on L5178Y $Tk^{+/-}$ cells that detects clastogenic and aneugenic chromosome aberrations in the absence and presence of the microsomal fraction S9; (3) the standard and enzyme modified comet assays with restriction enzymes (Endonuclease III (Endo III) and Formamide pyrimidine glycosylase (FPG)) that detect DNA strand breaks and oxidative DNA damage in Caco-2 cells; (4) the mouse lymphoma thymidine-kinase assay (MLA, OECD 490 [46]) on L5178Y $Tk^{+/-}$ cells to detect gene mutations in the timidine kinase (Tk) locus in the absence and presence of the microsomal fraction S9. The microsomal fraction S9 was used to assess if CYN/MC-LR genotoxicity is due to metabolic bioactivation of these toxins or due to the parent compounds.

2. Results

2.1. Ames Test

No signals of toxicity and/or test solutions instability were observed during the test performance. CYN/MC-LR mixtures did not induce changes in any of the *S. typhimurium* strains without S9 fraction (Table 1). On the contrary, a significant increase in the number of revertants per plate was observed with TA97A, TA102 and TA135 strains. However, a MI higher than 2 was not obtained in any of the assayed experimental conditions. Solvent controls (MetOH 2% and DMSO) did not induce statistical significant changes versus the negative controls.

2.2. Micronucleus Test

In the absence of S9 fraction, CYN/MC-LR mixtures did not increase the number of binucleated cells with MN in any of the concentration assayed (Table 2). However, a significant reduction of the cytokinesis-block proliferation index (CBPI) was observed at the highest concentration (1.35 μg/mL CYN + 13.5 μg/mL MC-LR). Positive controls for clastogens (MMC) and aneugens (colchicine) showed a significant increase in the frequency of binucleated cells with micronuclei (BNMN) ($p < 0.01$).

In the presence of S9 fraction, CYN/MC-LR induced an increase of BNMN (%) when compared to the negative control, but only at 1 μg/mL CYN + 10 μg/mL MC-LR this change was statistically significant ($p < 0.01$).

2.3. Mouse Lymphoma Thymidine-Kinase Assay (MLA)

Results of the MLA are shown in Tables 3–5. None of the evaluated CYN/MC-LR mixture concentrations induced a mutagenic response in the absence or presence of S9 fraction, neither after a short treatment (4 h) nor a long treatment (24 h). Concurrent vehicle control did not show changes in comparison to negative control (data not shown).

2.4. Standard and Enzyme-Modified Comet Assays

Caco-2 cells exposure to CYN/MC-LR mixtures did not result in DNA strand breaks in the standard comet assay after 24 and 48 h (Figure 1a). In addition, an oxidative damage induced genotoxicity was not observed as the experiments performed with Endo III and FPG enzymes did not show a significant increase of % DNA in tail (Figure 1b,c). Results for the solvent control were similar to the negative control (data not shown) and only positive controls showed a significant ($p < 0.001$) genotoxicity.

Table 1. Effect of CYN-MC-LR mixtures on the Ames test in three independent experiments by triplicate. Data are given as mean ± SD revertants/plate. * $p < 0.05$. ** $p < 0.01$ in comparison to negative control.

Concentration (µg/mL)		TA97A				TA98				TA100				TA102				TA1535			
		−S9	MI	+S9	MI	−S9	MI	+S9	MI	−S9	MI	+S9	MI	−S9	MI	+S9	MI	−S9	MI	+S9	MI
Pure CYN-MC-LR mixture	Negative controls	231 ± 42	-	244 ± 5	-	21 ± 2	-	24 ± 9	-	117 ± 25	-	135 ± 14	-	215 ± 12	-	292 ± 11	-	293 ± 23	-	273 ± 33	-
	0.125–1.25	297 ± 37	1.4	319 ± 51	1.3	19 ± 2	0.9	18 ± 8	0.8	136 ± 40	1.2	153 ± 21	1.1	230 ± 36	1.1	440 ± 29 **	1.5	327 ± 25	1.1	376 ± 54 *	1.3
	0.25–2.5	165 ± 28	0.8	334 ± 49 **	1.4	20 ± 1	1.0	17 ± 7	0.7	144 ± 12	1.2	166 ± 24	1.2	217 ± 29	1.0	380 ± 33 **	1.3	311 ± 10	1.1	411 ± 54 **	1.4
	0.5–5	213 ± 15	1.0	290 ± 58	1.2	26 ± 9	1.3	20 ± 10	0.9	154 ± 13	1.3	143 ± 19	1.1	251 ± 17	1.2	296 ± 18	10.	309 ± 42	1.1	336 ± 18 *	1.1
	1–10	168 ± 10	0.8	234 ± 43	1.0	21 ± 2	1.0	19 ± 9	1.0	146 ± 18	1.2	130 ± 10	1.0	134 ± 12	0.6	383 ± 44 **	1.3	250 ± 43	0.9	464 ± 44 **	1.6
	2–20	205 ± 31	1.0	295 ± 25	1.2	19 ± 5	0.9	25 ± 8	1.0	104 ± 31	0.9	143 ± 19	1.1	151 ± 1	0.7	397 ± 32 **	1.4	276 ± 15	0.9	476 ± 52 **	1.6
	Positive controls	613 ± 66 **	2.9	527 ± 19 **	2.2	883 ± 55 **	42.0	960 ± 53 **	40.9	816 ± 11 **	7.0	583 ± 39 **	4.3	950 ± 118 **	4.4	671 ± 22 **	2.3	833 ± 25 **	2.8	659 ± 39 **	2.2
MeOH 2%		176 ± 25	0.8	316 ± 32	1.3	17 ± 5	0.8	25 ± 13	1.1	92 ± 13	0.8	87 ± 29	0.6	192 ± 8	0.8	280 ± 12	0.6	313 ± 9	1.1	233 ± 35	0.9
DMSO		209 ± 66	1.3	184 ± 38	0.8	25 ± 2	1.2	30 ± 6	1.3	115 ± 5	1.0	113 ± 17	0.8	250 ± 65	1.2	231 ± 35	0.8	342 ± 63	1.2	298 ± 16	1.1

Negative control: Milli Q water. Control solvent: MeOH 2% and DMSO. Positive controls without S9 for TA97A: 9-aminoacridine (50 µg/plate), TA98: 2-nitrofluorene (0.1 µg/plate), TA100 and TA1535: NaN3 (1.5 µg/plate) and TA102: mytomicin C (2.5 µg/plate). Positive control for all strains with S9: 2-aminofluorene (20 µg/plate).

Table 2. Percentage of binucleated cells with micronuclei (BNMN) and cytokinesis-block proliferation index (CBPI) in cultured mouse lymphoma cells L5178YTk$^{+/-}$ exposed to CYN+MC-LR mixture ($n = 3$). The genotoxicity assay was performed in the absence and presence of the metabolic fraction S9. The values are expressed as mean ± SD. ** $p < 0.01$, *** $p < 0.001$ in comparison to negative control group values.

Experimental Group	Absence of S9				Presence of S9			
	Exposure Time (h)	Concentrations (µg/mL)	BNMN (%) ± SD	CBPI ± SD	Exposure Time (h)	Concentrations (µg/mL)	BNMN (%) ± SD	CBPI ± SD
Negative control	24	-	2.3 ± 0.5	1.9 ± 0.1	4	-	2.5 ± 1.0	1.8 ± 0.1
Positive control	24	Mitomycin C 0.0625	10.5 ± 4.1 ***	1.5 ± 0.1 ***	4	Cyclophosfamide 8	8.3 ± 1.9 **	1.8 ± 0.1
		Colchicine 0.0125	9.6 ± 1.7 ***	1.8 ± 0.0				
CYN+MC-LR	24	0.084–0.84	1.8 ± 1.5	1.9 ± 0.0	4	0.125–1.25	4.8 ± 2.6	1.8 ± 0.1
	24	0.168–1.68	2.3 ± 1.0	1.9 ± 0.0	4	0.250–2.5	4.0 ± 1.4	1.8 ± 0.1
	24	0.336–3.36	2.5 ± 0.6	1.8 ± 0.0	4	0.5–5	5.8 ± 1.5	1.8 ± 0.1
	24	0.672–6.72	1.3 ± 0.5	1.7 ± 0.1	4	1–10	8.8 ± 4.2 **	1.8 ± 0.1
	24	1.35–13.5	0.8 ± 1.0	1.3 ± 0.3 ***	4	2–20	4.8 ± 0.5	1.8 ± 0.1

Clastogen and aneugen positive controls: mitomicyn C (0.0625 µg/mL) and colchicine (0.0125 µg/mL), respectively.

Table 3. Toxicity and mutagenicity of CYN/MC-LR in L5178YTk$^{+/-}$ cells after 4 h without S9 fraction by the mouse lymphoma thymidine-kinase assay (MLA) (n = 2). [a]: Total mutant frequency divided into small/large (S/L) colony mutant frequencies. The induced mutant frequency (IMF) was determined according to the formula IMF = MF-SMF, where MF is the test culture mutant frequency and SMF is the spontaneous mutant frequency. *** $p < 0.001$.

Concentration (μg/mL)	Relative Total Growth		Percent Plating Efficiency		Mutant Frequency (× 10^{-6})		MF (S/L) [a]		IMF (MF-SMF) (× 10^{-6})	
	Experiment 1	Experiment 2	Experiment 1	Experiment 2	Experiment 1	Experiment 2	Experiment 1	Experiment 2	Experiment 1	Experiment 2
0	100	100	91	124	107	152	51/56	33/41	-	-
0.04 CYN-0.4 MC	77	90	98	98	126	143	95/48	86/57	56	70
0.08 CYN-0.8 MC	98	100	93	70	202	157	111/91	102/55	95	83
0.16 CYN-1.6 MC	82	86	102	82	71	162	44/27	100/62	−14.4	89
0.33 CYN-3.3 MC	64	72	98	91	165	150	84/81	80/70	58	76
0.67 CYN-6.7 MC	57	58	95	88	174	144	106/68	60/84	67	71
MMS (10 μg/mL)	46	70	69	82	728 ***	738 ***	407/321	424/314	621	664

Positive controls: methylmethanesulfonate, MMS 10 μg/mL without S9 fraction and cyclophosphamide, CP 3 μg/mL with S9 fraction.

Table 4. Toxicity and mutagenicity of CYN/MC-LR in L5178YTk$^{+/-}$ cells after 4 h with S9 fraction by the mouse lymphoma thymidine-kinase assay (MLA) (n = 2). [a]: Total mutant frequency divided into small/large (S/L) colony mutant frequencies. The induced mutant frequency (IMF) was determined according to the formula IMF = MF-SMF, where MF is the test culture mutant frequency and SMF is the spontaneous mutant frequency. *** $p < 0.001$.

Concentration (μg/mL)	Relative Total Growth		Percent Plating Efficiency		Mutant Frequency (× 10^{-6})		MF (S/L) [a]		IMF (MF-SMF) (× 10^{-6})	
	Experiment 1	Experiment 2	Experiment 1	Experiment 2	Experiment 1	Experiment 2	Experiment 1	Experiment 2	Experiment 1	Experiment 2
0	100	100	93	102	155	146	96/59	82/64	-	-
0.04 CYN-0.4 MC	96	84	82	84	94	100	43/51	42/58	−61	−46
0.08 CYN-0.8 MC	82	72	91	91	95	95	50/45	50/45	−60	−51
0.16 CYN-1.6 MC	58	51	95	100	98	95	48/50	49/47	−57	−51
0.33 CYN-3.3 MC	58	56	102	100	98	105	56/43	60/45	−57	−41
0.67 CYN-6.7 MC	26	31	118	113	120	132	62/58	77/55	−35	−14
1.35 CYN-13.5 MC	16	16	130	116	70	91	29/41	38/53	−85	−55
CP (3 μg/mL)	99	81	65	73	480 ***	433 ***	228/252	213/220	325	286

Positive controls: methylmethanesulfonate, MMS 10 μg/mL without S9 fraction and cyclophosphamide, CP 3 μg/mL with S9 fraction.

Table 5. Toxicity and mutagenicity of CYN/MC-LR in L5178YTk$^{+/-}$ cells after 24 h without S9 fraction by the mouse lymphoma thymidine-kinase assay (MLA) (n = 2). [a]: Total mutant frequency divided into small/large (S/L) colony mutant frequencies. The induced mutant frequency (IMF) was determined according to the formula IMF = MF-SMF, where MF is the test culture mutant frequency and SMF is the spontaneous mutant frequency. *** $p < 0.001$.

Concentration (µg/mL)	Relative Total Growth		Percent Plating Efficiency		Mutant Frequency (× 10^{-6})		MF (S/L) [a]		IMF (MF-SMF) (× 10^{-6})	
	Experiment 1	Experiment 2	Experiment 1	Experiment 2	Experiment 1	Experiment 2	Experiment 1	Experiment 2	Experiment 1	Experiment 2
0	100	100	113	124	170	170	106/72	87/92	-	-
0.04 CYN-0.4 MC	103	115	90	87	107	78.9	62/45	48/30	−71	−100
0.08 CYN-0.8 MC	91	102	102	93	121	124	50/71	72/52	−57	−55
0.16 CYN-1.6 MC	79	96	76	108	143	100	81/66	56/44	−35	−79
0.33 CYN-3.3 MC	71	74	116	104	115	168	64/51	109/59	−63	−12
0.67 CYN-6.7 MC	39	39	127	104	113	195	74/39	77/118	−66	16
MMS (10 µg/mL)	52	66	35	34	778 ***	897 ***	370/408	459/438	599	718

Positive controls: methylmethanesulfonate, MMS 10 µg/mL without S9 fraction and cyclophosphamide, CP 3 µg/mL with S9 fraction.

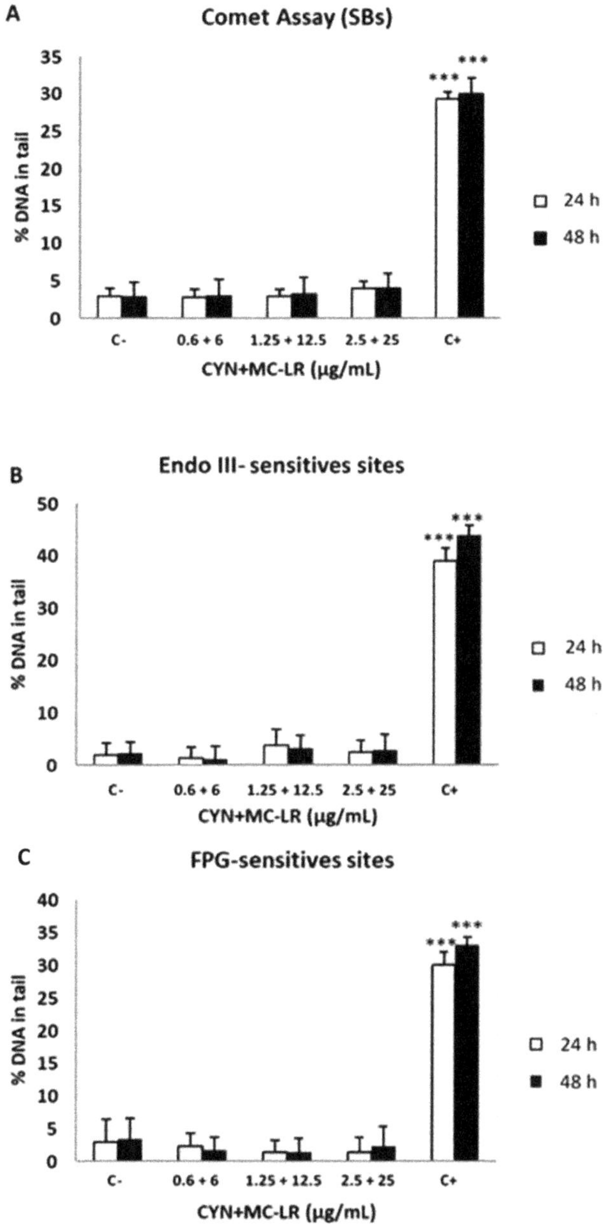

Figure 1. DNA damage in Caco-2 cells after exposure to CYN+MC-LR mixtures for 24 and 48 h. Results expressed as the formation of strand breaks (**a**) and oxidative DNA damage as Endo III-sensitive sites (**b**) and FPG-sensitive sites (**c**) ($n = 3$). The level of DNA strand-breaks (SBs), oxidized pyrimidines and oxidized purines are expressed as % DNA in tail. All values are expressed as mean ± SD. Negative control (C-): culture medium. Positive controls (C+): 100 µM H_2O_2 for the standard comet assay and Endo III-sensitive sites, and 2 µM of Ro 19-8022 photosensitizer with light irradiation for FPG-sensitive sites. *** $p < 0.001$.

3. Discussion

The data on the genotoxicity of a chemical is of key importance as it drives the type of human risk assessment to be performed. While a genotoxic chemical and health-based guidance value is usually set, for an unavoidable chemical, that is, a genotoxic carcinogen, the Margin of Exposure approach is usually applied [47]. For the generation and evaluation of data on genotoxic potential, the EFSA [43] recommends a step-wise approach for the generation and evaluation of data on genotoxic potential that begins with a basic battery of in vitro tests, including a bacterial reverse mutation assay and an in vitro MN assay. Moreover, further in vitro assays should be conducted in case of inconclusive, conflicting or equivocal results. The need for using several assays is justified as it is considered that there is no single mutagenicity test which can detect all kinds of potential human mutagens with 100% accuracy or prediction. This was shown to be true as mutagenesis itself is multifactorial [48].

Moreover, the genotoxicity evaluation of chemical mixtures is of great current interest and the EFSA has recently published a statement on the topic [49]. Thus, the Scientific Committee advocates for chemically fully defined mixtures, a component based approach, i.e., assessing all components individually using all suitable information including read across and quantitative structure–activity relationship (QSAR) considerations about their genotoxic potential, following the Scientific Committee guidance already mentioned [43]. In the present case, there are available data on CYN genotoxicity following EFSA recommendations [38], while MC-LR, was classified by the IARC in group 2B [14]. Moreover, the two single toxicity studies dealing with CYN/MC-LR mixtures have shown an antagonistic effect regarding cytotoxicity [41] and genotoxicity [42] in HepG2 cells. In addition, the genotoxicity of CYN/MC-LR mixtures has not been previously evaluated following a complete battery of in vitro tests, and a potential antagonic result for the mixture could affect the risk evaluation.

The first assay included in the basic battery was the Ames test. The mixture did not show a mutagenic response at the conditions tested, similar to previous results obtained for CYN [35]. In both cases, TA102 was one of the most responsive strains although the mutagenic indexes (MI) was always lower than 2. As CYN concentrations were similar in both studies, the results obtained suggest that MC-LR does not contribute to the genotoxicity of the mixture. This agrees with Sieroslawska [50] who found no effects in the Ames microplate format mutagenicity assay for pure MC-LR, pure CYN and neither for a mixture CYN/MC-LR/Anatoxin-a (1 μg/mL each).

A MN test is included in the basic battery to cover potential structural and numerical chromosome aberrations in addition to the Ames test. Chromosomal abnormalities, such as increased chromosomal breakage or chromosomal loss, are associated with enhanced risk of carcinogenesis and progression of neoplastic transformation [51]. In the case of the CYN/MC-LR mixture, an increase of MN was only observed with S9 fraction, similar to CYN in an individual exposure [38]. Moreover, single CYN showed this enhancement from lower concentrations (0.25 μg/mL) whereas the mixture showed this effect at 1 μg/mL CYN (+10 μg/mL MC-LR). This finding suggests that MC-LR ameliorates in this case the CYN response. However, in the scientific literature, there are contradictory data on the genotoxic potential of MC-LR by the MN assay. Thus, Abramsson-Zetterberg et al. [52] did not observe changes in vitro (in human lymphocytes, up to 2.0 mg extract of freeze-dried cyanobacteria per ml cell culture) and in vivo (in mice up to 55 μg/kg bw pure MC-LR by i.p. administration). On the contrary, Dias et al. [15] found that MC-LR treatment (5 and 20 μM) caused a significant induction in the MN frequency in kidney- (Vero-E6) and liver-derived (HepG2) cell lines and, interestingly, a similar positive effect was observed in mouse reticulocytes (37.5 μg MCLR/kg, i.p. route). Huang et al. [53] found that MC-LR induced a 1.6-fold increase in MN frequency in a human–hamster hybrid AL cell line after 30 days of exposure to 0.1 μg/mL (but no changes after 1 and 3 days of exposure). Regarding cyanobacterial mixtures, there is a single study that explored the MN induction of a CYN/MC-LR mixture and found that 0.5 μg/mL CYN + 1 μg/mL MC-LR induced a significant increase of MN in HepG2 cells [42].

Additional in vitro methods were applied (MLA and Comet assay), following the recommendations of [43], because the results obtained with the Ames test and the MN assay did not allow confirmation of the genotoxicity (or absence of genotoxicity) of the mixture.

The MLA results did not provide new evidence as no changes were observed at any of the conditions tested. Puerto et al. [38] also did not find a mutagenic response when single CYN exposure was evaluated. Zhan et al. [54] performed the TK gene mutation assay in the TK6 human lymphoblastoid cell line for MC-LR and found TK mutation in a concentration-dependent manner. The MLA is the most extensively used of the different in vitro mammalian gene-mutation assays [55]. Both MN assay and MLA are performed in the same experimental model, the L5178YTk$^{+/-}$ cells, recommended in the Organization for Economic Co-operation and Development OECD guidelines. It seems that MN assay is more sensitive, or that the potential mutagenicity of the evaluated cyanotoxins is related mostly with chromosomal aberrations and to a lesser extent, with gene (point) mutations. However, the MLA detects intragenic events, mainly point mutations, and also loss of heterozygosity. This can result from the entire Tk gene loss, leading to karyotypically visible deletions and rearrangements of the Tk$^{+/-}$ bearing chromosome [56]. These features make the MLA especially useful to evaluate the ability of chemicals to induce a broad variety of mutational events [57].

Similarly, the Comet assay also did not evidence DNA damage induced by the CYN/MC-LR mixture in any of the procedures performed, that is, the standard assay and the modified version to detect oxidative DNA-damage. CYN alone showed the same response in similar conditions: Experimental model, concentrations and times of exposure [38]. Other authors, however, have observed genotoxic effects for CYN in the Comet assay both in vitro [31,33,34] and in vivo [58,59]. MC-LR single exposure was also reported to induce DNA strand breaks by the comet assay in vitro [15,60–62] and in vivo [15]. There is a single study [42] that showed DNA strand breaks induction by cyanobacterial mixtures CYN/MC-LR in HepG2 cells after 24 h exposure, but to lesser degree than CYN. Once more, it seems that MC-LR ameliorates the genotoxicity induced by CYN.

Overall, it is difficult to derive any statement about the (geno)toxicity of CYN/MC-LR mixtures because the available studies in the scientific literature for the individual toxins mostly use different model systems and exposure concentrations. This is the first time that a thorough investigation using 4 different mutagenicity and genotoxicity assays has been performed for cyanobacterial mixtures and the results indicate that the mixture does not show a higher genotoxicity compared to CYN. However, taking into account that MC-LR was classified in the group 2B by the IARC due to its tumour promotion mechanism [14], caution is required when trying to elucidate its role in the mixture toxicity.

As Zouaoui et al. [63] highlighted, the type of interactions among toxins could be related with the different chemical structures and properties, and the competition or not, for the same cell receptor. It is, therefore, required to explore the cyanotoxins mechanisms of action when they are alone or in mixtures. In this case, the investigated cyanotoxins showed different toxicity mechanisms but also share others, such as the oxidative stress induction. Thus, Gutiérrez-Praena et al. [41] suggested that the depletion of GSH could be related with the antagonistic response as it could decrease the uptake ratio of CYN. Other authors such as Hercog et al. [42] pointed out to their different kinetics as MC-LR and CYN are detoxified and toxified, respectively, after [30,64] and also to the compromise of DNA repair mechanisms induced by MC-LR [65]. In any case, further studies would be required to fully understand the mechanisms involved in the toxicity of mixtures. Moreover, despite using the battery proposed by EFSA [43], considering the results obtained (positive effects only in one of the four tests performed) and the limitations of in vitro genotoxicity tests to predict the in vivo situation suggested by Nesslany [66], the further step would be to assess in vivo the genotoxicity of cyanobacterial mixtures.

4. Conclusions

The in vitro mutagenicity and genotoxicity showed by CYN/MC-LR mixtures do not differ substantially from that observed for CYN tested individually. This effect was evident only when S9 fraction was used, indicating the relevance of CYN on the mixture toxicity at the conditions tested.

The increased knowledge of cyanotoxins mixture genotoxic potential would contribute to perform more realistic risk evaluations.

5. Materials and Methods

5.1. Chemicals and Reagents

Cylindrospermopsin (95% purity) and Microcystin-LR (99% purity) standards were provided by Alexis Corporation (Lausen, Switzerland). Chemicals for different assays were supplied by Gibco (Biomol, Sevilla, Spain), Sigma -Aldrich (Madrid, Spain), C-Viral S.L. (Sevilla, Spain) and Moltox (Trinova, Biochem, Germany).

5.2. Cells and Culture Conditions

Five *Salmonella typhimurium* histidine-auxotrophic strains TA97A, TA98, TA100, TA102 and TA1535 were used for the Ames test. L5178Y $Tk^{+/-}$ mouse lymphoma cells used for the MN test and MLA were originally provided by Dr. Oliver Gillardeux (Safoni-Synthélabo, Paris, France). Caco-2 cell line, used for standard and enzyme-modified comet assays, come from a human colon adenocarcinoma (ATCC© HTB-37). L5178Y $Tk^{+/-}$ cells and Caco-2 cells were maintained in an incubator with 5% CO_2 and 95% relative humidity at 37 °C.

5.3. Test Solutions

Stock solution of CYN (1000 μg/mL) and MC-LR (4000 μg/mL) were prepared in milliQ sterile water and water: MeOH, respectively and stored at less than 4 °C. The exposure concentration solutions were prepared by dilution in sterile MilliQ water (Ames test), RPMI 1640 medium (MN and MLA assays) or MEM medium (standard and enzyme-modified comet assays). Test concentrations were selected individually for every test as they need to fulfil toxicity criteria in each of the experimental models used. The selected concentrations of MC-LR were 10 times higher than that of CYN since MC-LR is normally more abundant in nature [1,2,67].

5.4. Bacterial Reverse Mutation Test (Ames Test)

The Ames test was performed following the OECD Guideline 471 [44] and Maron et al. [68] with minor modifications as follows. Five *Salmonella typhimurium* histidine-auxotrophic strains (TA97, TA98, TA100, TA102 and TA1535) obtained from TRINOVA BIOCHEM GmbH (Germany) were cultured following the provider instructions. The mutagenic activity of CYN/MC-LR mixtures was assessed in the absence and presence of the external metabolic activation system from rat livers (S9 fraction). Each experiment was conducted with five growing concentrations of CYN/MC-LR mixtures (0.125–2 μg/mL CYN and 1.25–20 μg/mL MC-LR) selected according to the results obtained by Puerto et al. [38] when CYN mutagenicity was assessed by the Ames test. Also, a negative control (distilled sterile water), solvent controls (MeOH and DMSO) and a positive control for each strain in accordance with the presence or absence of S9 fraction were included. Nine-aminoacridine (50 μg/plate) was the positive control for TA97A without S9 fraction; 2-Nitrofluorene (2-NF) (0.1 μg/plate) for TA98; sodium azide (NaN_3) (1 μg/plate) for TA100 and TA1535; and mitomycin C (MMC) (2.5 μg/plate) for TA102. The positive control in the presence of S9 fraction was 2-aminofluorene (2-AF) (20 μg/plate) for all strains. At least 3 independent experiments were performed using triplicate plates for each test concentration. Results are expressed as revertant colonies and mutagenic indexes (MI).

5.5. Micronucleus Test (MN)

The MN test was carried out following the OECD guideline 487 [45]. L5178Y $Tk^{+/-}$ cells were seeded at a density of 2.0×10^5 cell/mL and exposed to five different concentrations of CYN/MC-LR mixture (0.084–1.35 μg/mL CYN and 0.84–13.5 μg/mL MC-LR in the absence of S9 fraction for 24 h, and 0.125–2 μg/mL CYN and 1.25–20 μg/mL MC-LR for 4 h in the presence of S9 fraction). These concentrations were selected taking into account previous results obtained in cytotoxicity assays and carried out according to the OECD Guideline 487 [45]. The RPMI medium was used as negative control; MeOH as vehicle control; and 0.0625 μg/mL MMC and 0.0125 colchicine (without S9 fraction) and 8 μg/mL cyclophosphamide (CP) (with S9 fraction) as positive controls. Cells were exposed to CYN/MC-LR mixtures (4 or 24 h, with and without S9 mix, respectively), then exposed to cythochalasin B (Cyt-B) (6 μg/mL) for 20 h to block cytokinesis and obtain binucleated cells. Afterward, cells were exposed to a hypotonic treatment with KCl and fixed. Subsequently, cells were dripped on slides and stained with Giemsa 10%. Quantification of binucleated cells with micronuclei (BNMN) and cytokinesis-block proliferation index (CBPI) were carried out following the OECD 487 guideline [45] by analysing at least 2000 binucleated cells/concentration.

5.6. Mouse Lymphoma Thymidine-Kinase Assay (MLA)

The MLA assay was performed in agreement to OECD Guideline 490 [46] and Maisanaba et al. [69]. Each experiment includes a negative control (fresh media), a solvent control (MeOH), a positive control (methylmetanosulfonate, MMS 10 μg/mL in absence of S9 fraction and cyclophosphamide, CP 3 μg/mL in presence of S9 fraction), five concentrations of CYN/MC-LR mixture in the absence of S9 fraction for 4 and 24 h assays (0.04–0.67 μg/mL CYN and 0.4–6.7 μg/mL MC-LR) and six concentrations in the presence of S9 fraction for 4 h assay (0.04–1.35 μg/mL CYN and 0.4–13.5 μg/mL MC-LR). These concentrations were selected in accordance with previous tests performed to define the cytotoxicity of CYN/MC-LR mixtures by the relative total growth (RTG) after 4 and 24 h of treatment without S9 fraction. According to the ICH Expert Working Group [70], the highest concentration chosen for the mutagenicity test must be higher than 10–20% of RTG. RTG values were employed to determine the acceptability of the toxicity at each concentration. Cells were seeded at 10^4 cells/mL in 96-well plates (two replicates per experimental group) to assess the viability and mutagenicity. The mutation analysis cells were exposed to 4 μg/mL trifluorothymidine (TFT), and both the viability plates and the mutagenicity plates were incubated at 37 °C and 5% CO_2 for 12 days. Afterwards, viable colonies and TFT mutation colonies were counted. Thiazolyl blue tetrazolium (MTT) (2.5 mg/mL) was added to wells to facilitate the counting of mutant colonies, and the plates were incubated for 4 h. According to Honma et al. [71], the size of the colonies were described as small (less than 1/3 of well diameter) or large (higher than 1/3 of well diameter) colonies. Moreover, the induced mutant frequency (IMF) was also analyzed.

5.7. Standard and Enzyme-Modified Comet Assay

The standard comet assay was carried out to evaluate genotoxicity, and a modified version of this assay with endonuclease III (Endo III) and formamidopyrimidine (FPG), which recognise oxidized pyrimidines and purines, was performed to determine oxidative DNA damage, respectively.

The standard and enzyme-modified comet assays were carried out to assess the genotoxicity of CYN/MC-LR mixtures, as previously described by Collins et al. [72] and Llana-Ruiz-Cabello et al. [73]. Caco-2 cell line was selected as cyanotoxins are food contaminants and it is a commonly used enterocytic model in toxicological studies [74–77]. Cells were seeded at 3.5×10^5 cells/mL into 24-well tissue culture plates and treated with increasing concentrations of CYN/MC-LR mixtures (0.6–2.5 μg/mL CYN and 6–25 μg/mL MC-LR) for 24 h and 48 h, according to the value obtained in the most sensitive cytotoxicity endpoint assayed [76]. Cells were treated with a negative control (medium) and a positive control (H_2O_2 100 μM) for standard comet assay and Endo III sensitives sites and Ro 19-8022 (2 μM) for FPG-sensitive sites. After exposure time, cells were washed, trypsinized and re-suspended in phosphate buffer saline (PBS) at 2.5×10^5 cell/mL. Cells suspensions were mixed with 1% (*w*/*v*) low-melting-agarose in PBS and placed on agarose precoated glass slides. Afterwards, lysis, incubation with Endo III and FPG (in the case of modified comet assay), denaturing, electrophoresis, neutralization, washing, fixation, dying, staining with SYBR Gold and quantification of nuclei were performed.

Olympus BX61 (fluorescence microscope) with the comet assay IV software (Perceptive Instruments, UK) available at the Microscopy Service of the University of Seville (CITIUS) was used to score the cells. The results were expressed as mean % DNA in tail respect to the negative control group. The % DNA in tail represents the amount of DNA breakage. Both types of comet assays (standard and modified) were performed in at least three independent experiments and using a triplicate/experiment.

5.8. Statistical Analysis

The statistical analysis was performed with Graph-Pad InStat software (Graph-Pad Software Inc., La Jolla, CA, USA). The non-parametric Wilcoxon matched-pairs signed-rank test was employed to compare the exposed samples with the negative control. Differences were considered significant at * $p < 0.05$, ** $p < 0.01$ and *** $p < 0.001$, respectively.

Author Contributions: Conceptualization, A.M.C. and Á.J.; methodology, L.D.-Q., M.P. and A.I.P.; software, L.D.-Q., M.P. and A.I.P.; formal analysis, L.D.-Q., M.P., A.I.P., Á.J. and A.M.C.; investigation, L.D.-Q., M.P. and A.I.P.; resources, A.M.C. and Á.J.; writing—original draft preparation, L.D.-Q., Á.J. and A.M.C.; writing—review and editing, A.M.C. and Á.J.; supervision, A.M.C. and Á.J.; project administration, A.M.C. and Á.J.; funding acquisition, A.M.C. and Á.J.

Funding: This research was funded by the SPANISH MINISTERIO DE ECONOMÍA Y COMPETITIVIDAD (AGL2015-64558-R, MINECO/FEDER, UE); by the FPI grant number BES-2016-078773 awarded to Leticia Díez-Quijada Jiménez.

Acknowledgments: Spanish Ministerio de Economía y Competitividad for the project AGL2015-64558-R, MINECO/FEDER, UE, and for the grant FPI (BES-2016-078773) awarded to Leticia Díez-Quijada Jiménez.

Conflicts of Interest: The authors declare no conflict of interest.

References

1. Buratti, F.M.; Manganelli, M.; Vichi, S.; Stefanelli, M.; Scardala, S.; Testai, E.; Funari, E. Cyanotoxins: Producing organisms, occurrence, toxicity, mechanism of action and human health toxicological risk evaluation. *Arch. Toxicol.* **2017**, *91*, 1049–1130. [CrossRef] [PubMed]
2. Diez-Quijada, L.; Puerto, M.; Gutierrez-Praena, D.; Llana-Ruiz-Cabello, M.; Jos, A.; Camean, A.M. Microcystin-RR: Occurrence, content in water and food and toxicological studies. A review. *Environ. Res.* **2019**, *168*, 467–489. [CrossRef]
3. Diez-Quijada, L.; Prieto, A.I.; Guzman-Guillen, R.; Jos, A.; Camean, A.M. Occurrence and toxicity of microcystin congeners other than MC-LR and MC-RR: A review. *Food Chem. Toxicol.* **2019**, *125*, 106–132. [CrossRef] [PubMed]
4. Codd, G.A.; Meriluoto, J.; Metcalf, J.S. Introduction: Cyanobacteria, cyanotoxins, their human impact, and risk management. *Handb. Cyanobact. Monit. Cyanotoxin Anal.* **2016**, 1–8. [CrossRef]
5. Testai, E.; Buratti, F.M.; Funari, E.; Manganelli, M.; Vichi, S.; Arnich, N.; Biré, R.; Fessard, V.; Sialehaamoa, A. Review and analysis of occurrence, exposure and toxicity of cyanobacteria toxins in food. *EFSA Support. Publ.* **2016**, *13*, 1–309. [CrossRef]
6. Roy-Lachapelle, A.; Solliec, M.; Bouchard, M.F.; Sauvé, S. Detection of Cyanotoxins in Algae Dietary Supplements. *Toxins* **2017**, *9*, 76. [CrossRef]
7. Spoof, L.; Catherine, A. Appendix 3: Tables of microcystins and nodularins. *Handb. Cyanobact. Monit. Cyanotoxin Anal.* **2016**, 526–537. [CrossRef]
8. Fischer, W.J.; Altheimer, S.; Cattori, V.; Meier, P.J.; Dietrich, D.R.; Hagenbuch, B. Organic anion transporting polypeptides expressed in liver and brain mediate uptake of microcystin. *Toxicol. Appl. Pharmacol.* **2005**, *203*, 257–263. [CrossRef]
9. MacKintosh, C.; Beattie, K.A.; Klumpp, S.; Cohen, P.; Codd, G.A. Cyanobacterial microcystin-LR is a potent and specific inhibitor of protein phosphatases 1 and 2A from both mammals and higher plants. *FEBS Lett.* **1990**, *264*, 187–192. [CrossRef]
10. Prieto, A.I.; Jos, A.; Pichardo, S.; Moreno, I.; Cameán, A.M. Protective role of vitamin E on the microcystin-induced oxidative stress in tilapia fish (*Oreochromis niloticus*). *Environ. Toxicol. Chem.* **2008**, *27*, 1152–1159. [CrossRef]

11. Moreno, I.; Mate, A.; Repetto, G.; Vázquez, C.; Cameán, A.M. Influence of microcystin-LR on the activity of membrane enzymes in rat intestinal mucosa. *J. Physiol. Biochem.* **2003**, *59*, 293–299. [CrossRef] [PubMed]
12. Atencio, L.; Moreno, I.; Prieto, A.I.; Moyano, R.; Molina, A.M.; Camean, A.M. Acute effects of microcystins MC-LR and MC-RR on acid and alkaline phosphatase activities and pathological changes in intraperitoneally exposed tilapia fish (*Oreochromis sp.*). *Toxicol. Pathol.* **2008**, *36*, 449–458. [CrossRef] [PubMed]
13. Valério, E.; Vasconcelos, V.; Campos, A. New insights on the mode of action of microcystins in animal cells-a review. *Mini Rev. Med. Chem.* **2016**, *16*, 1032–1041. [CrossRef]
14. Ingested Nitrate and Nitrite, and Cyanobacterial Peptide Toxins. Available online: https://www.ncbi.nlm.nih.gov/books/NBK326544/pdf/Bookshelf_NBK326544.pdf (accessed on 10 April 2019).
15. Dias, E.; Louro, H.; Pinto, M.; Santos, T.; Antunes, S.; Pereira, P.; Silva, M.J. Genotoxicity of microcystin-LR in in vitro and in vivo experimental models. *BioMed. Res. Int.* **2014**. [CrossRef] [PubMed]
16. Žegura, B. An overview of the mechanisms of microcystin-LR genotoxicity and potential carcinogenicity. *Mini Rev. Med. Chem.* **2016**, *16*, 1042–1062. [CrossRef]
17. Kokociński, M.; Cameán, A.M.; Carmeli, S.; Guzmán-Guillén, R.; Jos, Á.; Mankiewicz-Boczek, J.; Metcalf, J.S.; Moreno, I.M.; Prieto, A.I.; Sukenik, A. Cylindrospermopsin and congeners. *Handb. Cyanobact. Monit. Cyanotoxin Anal.* **2017**, 127–137. [CrossRef]
18. Chiswell, R.K.; Shaw, G.R.; Eaglesham, G.; Smith, M.J.; Norris, R.L.; Seawright, A.A.; Moore, M.R. Stability of cylindrospermopsin, the toxin from the cyanobacterium, Cylindrospermopsis raciborskii: Effect of pH, temperature, and sunlight on decomposition. *Environ. Toxicol. Int. J.* **1999**, *14*, 155–161. [CrossRef]
19. Falconer, I.R.; Humpage, A.R. Cyanobacterial (blue-green algal) toxins in water supplies: Cylindrospermopsins. *Environ. Toxicol.* **2006**, *21*, 299–304. [CrossRef]
20. Guzmán-Guillén, R.; Puerto, M.; Gutiérrez-Praena, D.; Prieto, A.; Pichardo, S.; Jos, Á.; Campos, A.; Vasconcelos, V.; Cameán, A. Potential use of chemoprotectants against the toxic effects of cyanotoxins: A review. *Toxins* **2017**, *9*, 175. [CrossRef]
21. Hinojosa, M.; Gutiérrez-Praena, D.; Prieto, A.; Guzmán-Guillén, R.; Jos, A.; Cameán, A. Neurotoxicity induced by microcystins and cylindrospermopsin: A review. *Sci. Total Environ.* **2019**, *668*, 547–565. [CrossRef]
22. Poniedziałek, B.; Rzymski, P.; Kokociński, M. Cylindrospermopsin: Water-linked potential threat to human health in Europe. *Environ. Toxicol. Pharmacol.* **2012**, *34*, 651–660. [CrossRef] [PubMed]
23. Pichardo, S.; Devesa, V.; Puerto, M.; Vélez, D.; Cameán, A.M. Intestinal transport of Cylindrospermopsin using the Caco-2 cell line. *Toxicol. In Vitro* **2017**, *38*, 142–149. [CrossRef] [PubMed]
24. Terao, K.; Ohmori, S.; Igarashi, K.; Ohtani, I.; Watanabe, M.; Harada, K.; Ito, E.; Watanabe, M. Electron microscopic studies on experimental poisoning in mice induced by cylindrospermopsin isolated from blue-green alga *Umezakia natans*. *Toxicon* **1994**, *32*, 833–843. [CrossRef]
25. Froscio, S.M.; Humpage, A.R.; Burcham, P.C.; Falconer, I.R. Cylindrospermopsin-induced protein synthesis inhibition and its dissociation from acute toxicity in mouse hepatocytes. *Environ. Toxicol. Int. J.* **2003**, *18*, 243–251. [CrossRef] [PubMed]
26. Runnegar, M.T.; Kong, S.-M.; Zhong, Y.-Z.; Lu, S.C. Inhibition of reduced glutathione synthesis by cyanobacterial alkaloid cylindrospermopsin in cultured rat hepatocytes. *Biochem. Pharmacol.* **1995**, *49*, 219–225. [CrossRef]
27. Gutiérrez-Praena, D.; Pichardo, S.; Jos, Á.; Cameán, A.M. Toxicity and glutathione implication in the effects observed by exposure of the liver fish cell line PLHC-1 to pure cylindrospermopsin. *Ecotoxicol. Environ. Saf.* **2011**, *74*, 1567–1572. [CrossRef] [PubMed]
28. Puerto, M.; Jos, A.; Pichardo, S.; Gutiérrez-Praena, D.; Cameán, A.M. Acute effects of pure cylindrospermopsin on the activity and transcription of antioxidant enzymes in tilapia (*Oreochromis niloticus*) exposed by gavage. *Ecotoxicology* **2011**, *20*, 1852–1860. [CrossRef]
29. Poniedziałek, B.; Rzymski, P.; Karczewski, J. The role of the enzymatic antioxidant system in cylindrospermopsin-induced toxicity in human lymphocytes. *Toxicol. In Vitro* **2015**, *29*, 926–932. [CrossRef]
30. Norris, R.; Seawright, A.; Shaw, G.; Senogles, P.; Eaglesham, G.; Smith, M.; Chiswell, R.; Moore, M. Hepatic xenobiotic metabolism of cylindrospermopsin in vivo in the mouse. *Toxicon* **2002**, *40*, 471–476. [CrossRef]
31. Humpage, A.R.; Fontaine, F.; Froscio, S.; Burcham, P.; Falconer, I.R. Cylindrospermopsin genotoxicity and cytotoxicity: Role of cytochrome P-450 and oxidative stress. *J. Toxicol. Environ. Health A* **2005**, *68*, 739–753. [CrossRef]

32. Humpage, A.R.; Fenech, M.; Thomas, P.; Falconer, I.R. Micronucleus induction and chromosome loss in transformed human white cells indicate clastogenic and aneugenic action of the cyanobacterial toxin, cylindrospermopsin. *Mutat. Res.* **2000**, *472*, 155–161. [CrossRef]
33. Štraser, A.; Filipič, M.; Žegura, B. Genotoxic effects of the cyanobacterial hepatotoxin cylindrospermopsin in the HepG2 cell line. *Arch. Toxicol.* **2011**, *85*, 1617–1626. [CrossRef] [PubMed]
34. Žegura, B.; Gajski, G.; Štraser, A.; Garaj-Vrhovac, V. Cylindrospermopsin induced DNA damage and alteration in the expression of genes involved in the response to DNA damage, apoptosis and oxidative stress. *Toxicon* **2011**, *58*, 471–479. [CrossRef] [PubMed]
35. Štraser, A.; Metka, F.; Matjaž, N.; Bojana, Ž. Double strand breaks and cell-cycle arrest induced by the cyanobacterial toxin cylindrospermopsin in HepG2 cells. *Mar. Drugs* **2013**, *11*, 3077–3090.
36. Sieroslawska, A.; Rymuszka, A. Cylindrospermopsin induces oxidative stress and genotoxic effects in the fish CLC cell line. *J. Appl. Toxicol.* **2015**, *35*, 426–433. [CrossRef] [PubMed]
37. Pichardo, S.; Cameán, A.M.; Jos, Á.M. *In vitro* toxicological assessment of cylindrospermopsin: A review. *Toxins* **2017**, *9*, 402. [CrossRef] [PubMed]
38. Puerto, M.; Prieto, A.I.; Maisanaba, S.; Gutierrez-Praena, D.; Mellado-Garcia, P.; Jos, A.; Camean, A.M. Mutagenic and genotoxic potential of pure Cylindrospermopsin by a battery of in vitro tests. *Food Chem. Toxicol.* **2018**, *121*, 413–422. [CrossRef] [PubMed]
39. Bittencourt-Oliveira, M.; Carmo, D.; Piccin-Santos, V.; Moura, A.N.; Aragão-Tavares, N.K.; Cordeiro-Araújo, M.K. Cyanobacteria, microcystins and cylindrospermopsin in public drinking supply reservoirs of Brazil. *An. Acad. Bras. Cienc.* **2014**, *86*, 297–310. [CrossRef]
40. Jančula, D.; Straková, L.; Sadílek, J.; Maršálek, B.; Babica, P. Survey of cyanobacterial toxins in Czech water reservoirs—the first observation of neurotoxic saxitoxins. *Environ. Sci. Pollut. Res. Int.* **2014**, *21*, 8006–8015. [CrossRef] [PubMed]
41. Gutiérrez-Praena, D.; Guzmán-Guillén, R.; Pichardo, S.; Moreno, F.J.; Vasconcelos, V.; Jos, Á.; Cameán, A.M. Cytotoxic and morphological effects of microcystin-LR, cylindrospermopsin, and their combinations on the human hepatic cell line HepG2. *Environ. Toxicol.* **2018**, *34*, 240–251. [CrossRef] [PubMed]
42. Hercog, K.; Maisanaba, S.; Filipič, M.; Jos, Á.; Cameán, A.M.; Žegura, B. Genotoxic potential of the binary mixture of cyanotoxins microcystin-LR and cylindrospermopsin. *Chemosphere* **2017**, *189*, 319–329. [CrossRef] [PubMed]
43. EFSA, S.C. Scientific opinion on genotoxicity testing strategies applicable to food and feed safety assessment. *EFSA J.* **2011**, *9*. [CrossRef]
44. OECD Guidelines for the Testing of Chemicals, Bacterial Reverse Mutation Test. Available online: https://www.oecd.org/chemicalsafety/risk-assessment/1948418.pdf (accessed on 1 April 2019).
45. OECD Guidelines for the Testing of Chemicals, In Vitro Mammalian Cell Micronucleus Test. Available online: https://ntp.niehs.nih.gov/iccvam/suppdocs/feddocs/oecd/oecd-tg487-2014-508.pdf (accessed on 1 April 2019).
46. OECD Guidelines for the Testing of Chemicals, In Vitro Mammalian Cell Gene Mutation Tests Using the Thymidine Kinase Gene. Available online: https://www.oecd-ilibrary.org/docserver/9789264264908-en.pdf?expires=1559275406&id=id&accname=guest&checksum=323552A68EC3041C6EA6F9E8A7ACF632 (accessed on 1 April 2019).
47. EFSA. Opinion of the Scientific Committee on a request from EFSA related to a harmonised approach for risk assessment of substances which are both genotoxic and carcinogenic. *EFSA J.* **2005**, *3*, 282. [CrossRef]
48. Kamath, G.H.; Rao, K. Genotoxicity guidelines recommended by International Conference of Harmonization (ICH). *Methods Mol. Biol.* **2013**, *1044*, 431–458. [PubMed]
49. EFSA, S.C.; More, S.; Bampidis, V.; Benford, D.; Boesten, J.; Bragard, C.; Halldorsson, T.; Hernandez-Jerez, A.; Hougaard-Bennekou, S.; Koutsoumanis, K. Genotoxicity assessment of chemical mixtures. *EFSA J.* **2019**, *17*, 5519. [CrossRef]
50. Sieroslawska, A. Assessment of the mutagenic potential of cyanobacterial extracts and pure cyanotoxins. *Toxicon* **2013**, *74*, 76–82. [CrossRef] [PubMed]
51. Guerin, M.R. Energy Sources of Polycyclic Aromatic Hydrocarbons. Available online: https://www.osti.gov/servlets/purl/7303055 (accessed on 1 April 2019).
52. Abramsson-Zetterberg, L.; Sundh, U.B.; Mattsson, R. Cyanobacterial extracts and microcystin-LR are inactive in the micronucleus assay in vivo and in vitro. *Mutat. Res.* **2010**, *699*, 5–10. [CrossRef]

53. Huang, P.; Xu, A. Genotoxic effects of microcystin-LR in mammalian cells. In Proceedings of the 2009 3rd International Conference on Bioinformatics and Biomedical Engineering, Beijing, China, 11–16 June 2009.
54. Zhan, L.; Sakamoto, H.; Sakuraba, M.; Wu, D.-S.; Zhang, L.-S.; Suzuki, T.; Hayashi, M.; Honma, M. Genotoxicity of microcystin-LR in human lymphoblastoid TK6 cells. *Mutat. Res.* **2004**, *557*, 1–6. [CrossRef]
55. Moore, M.M.; Honma, M.; Clements, J.; Bolcsfoldi, G.; Cifone, M.; Delongchamp, R.; Fellows, M.; Gollapudi, B.; Jenkinson, P.; Kirby, P. Mouse lymphoma thymidine kinase gene mutation assay: International Workshop on Genotoxicity Tests Workgroup report—Plymouth, UK 2002. *Mutat. Res.* **2003**, *540*, 127–140. [CrossRef]
56. Chen, T.; Harrington-Brock, K.; Moore, M.M. Mutant frequencies and loss of heterozygosity induced by N-ethyl-N-nitrosourea in the thymidine kinase gene of L5178Y/Tk$^{+/-}$-3.7.2C mouse lymphoma cells. *Mutagenesis* **2002**, *17*, 105–109. [CrossRef]
57. Demir, E.; Kaya, B.; Soriano, C.; Creus, A.; Marcos, R. Genotoxic analysis of four lipid-peroxidation products in the mouse lymphoma assay. *Mutat. Res.* **2011**, *726*, 98–103. [CrossRef] [PubMed]
58. Shen, X.; Lam, P.; Shaw, G.; Wickramasinghe, W. Genotoxicity investigation of a cyanobacterial toxin, cylindrospermopsin. *Toxicon* **2002**, *40*, 1499–1501. [CrossRef]
59. Bazin, E.; Huet, S.; Jarry, G.; Hégarat, L.L.; Munday, J.S.; Humpage, A.R.; Fessard, V. Cytotoxic and genotoxic effects of cylindrospermopsin in mice treated by gavage or intraperitoneal injection. *Environ. Toxicol.* **2012**, *27*, 277–284. [CrossRef] [PubMed]
60. Žegura, B.; Sedmak, B.; Filipič, M. Microcystin-LR induces oxidative DNA damage in human hepatoma cell line HepG2. *Toxicon* **2003**, *41*, 41–48. [CrossRef]
61. Lankoff, A.; Krzowski, Ł.; Głąb, J.; Banasik, A.; Lisowska, H.; Kuszewski, T.; Góźdź, S.; Wójcik, A. DNA damage and repair in human peripheral blood lymphocytes following treatment with microcystin-LR. *Mutat. Res.* **2004**, *559*, 131–142. [CrossRef] [PubMed]
62. Žegura, B.; Gajski, G.; Štraser, A.; Garaj-Vrhovac, V.; Filipič, M. Microcystin-LR induced DNA damage in human peripheral blood lymphocytes. *Mutat. Res.* **2011**, *726*, 116–122. [CrossRef] [PubMed]
63. Zouaoui, N.; Mallebrera, B.; Berrada, H.; Abid-Essefi, S.; Bacha, H.; Ruiz, M.-J. Cytotoxic effects induced by patulin, sterigmatocystin and beauvericin on CHO–K1 cells. *Food Chem. Toxicol.* **2016**, *89*, 92–103. [CrossRef] [PubMed]
64. Pflugmacher, S.; Wiegand, C.; Oberemm, A.; Beattie, K.A.; Krause, E.; Codd, G.A.; Steinberg, C.E. Identification of an enzymatically formed glutathione conjugate of the cyanobacterial hepatotoxin microcystin-LR: The first step of detoxication. *Biochim. Biophys. Acta* **1998**, *1425*, 527–533. [CrossRef]
65. Lankoff, A.; Bialczyk, J.; Dziga, D.; Carmichael, W.; Gradzka, I.; Lisowska, H.; Kuszewski, T.; Gozdz, S.; Piorun, I.; Wojcik, A. The repair of gamma-radiation-induced DNA damage is inhibited by microcystin-LR, the PP1 and PP2A phosphatase inhibitor. *Mutagenesis* **2006**, *21*, 83–90. [CrossRef] [PubMed]
66. Nesslany, F. The current limitations of in vitro genotoxicity testing and their relevance to the in vivo situation. *Food Chem. Toxicol.* **2017**, *106*, 609–615. [CrossRef]
67. De La Cruz, A.A.; Hiskia, A.; Kaloudis, T.; Chernoff, N.; Hill, D.; Antoniou, M.G.; He, X.; Loftin, K.; O'Shea, K.; Zhao, C.; et al. A review on cylindrospermopsin: The global occurrence, detection, toxicity and degradation of a potent cyanotoxin. *Environ. Sci. Process. Impacts* **2013**, *15*, 1979. [CrossRef] [PubMed]
68. Maron, D.M.; Ames, B.N. Revised methods for the Salmonella mutagenicity test. *Mutat. Res.* **1983**, *113*, 173–215. [CrossRef]
69. Maisanaba, S.; Prieto, A.I.; Puerto, M.; Gutiérrez-Praena, D.; Demir, E.; Marcos, R.; Cameán, A.M. *In vitro* genotoxicity testing of carvacrol and thymol using the micronucleus and mouse lymphoma assays. *Mutat. Res.* **2015**, *784*, 37–44. [CrossRef] [PubMed]
70. International Conferences on Harmonisation of Technical Requirements for Registration of Pharmaceuticals for Human Use. ICH Harmonised Tripartite Guideline. Guidance on Genotoxicity Testing and Data Interpretation for Pharmaceuticals Intended for Human Use. Available online: https://www.ich.org/fileadmin/Public_Web_Site/ICH_Products/Guidelines/Safety/S2_R1/Step4/S2R1_Step4.pdf (accessed on 1 April 2019).
71. Honma, M.; Hayashi, M.; Shimada, H.; Tanaka, N.; Wakuri, S.; Awogi, T.; Yamamoto, K.I.; Kodani, N.-U.; Nishi, Y.; Nakadate, M. Evaluation of the mouse lymphoma Tk assay (microwell method) as an alternative to the in vitro chromosomal aberration test. *Mutagenesis* **1999**, *14*, 5–22. [CrossRef] [PubMed]

72. Collins, A.R.; Azqueta, A. Chapter 4: Single-cell gel electrophoresis combined with lesion-specific enzymes to measure oxidative damage to DNA. In *Methods Cell Biology*; Elsevier: Amsterdam, The Netherlands, 2012; Volume 112, pp. 69–92.
73. LLana-Ruiz-Cabello, M.; Maisanaba, S.; Puerto, M.; Prieto, A.I.; Pichardo, S.; Jos, Á.; Cameán, A.M. Evaluation of the mutagenicity and genotoxic potential of carvacrol and thymol using the Ames Salmonella test and alkaline, Endo III-and FPG-modified comet assays with the human cell line Caco-2. *Food Chem. Toxicol.* **2014**, *72*, 122–128. [CrossRef] [PubMed]
74. Sambuy, Y.; De Angelis, I.; Ranaldi, G.; Scarino, M.; Stammati, A.; Zucco, F. The Caco-2 cell line as a model of the intestinal barrier: Influence of cell and culture-related factors on Caco-2 cell functional characteristics. *Cell Biol. Toxicol.* **2005**, *21*, 1–26. [CrossRef] [PubMed]
75. Puerto, M.; Pichardo, S.; Jos, Á.; Cameán, A.M. Comparison of the toxicity induced by microcystin-RR and microcystin-YR in differentiated and undifferentiated Caco-2 cells. *Toxicon* **2009**, *54*, 161–169. [CrossRef]
76. Puerto, M.; Pichardo, S.; Jos, Á.; Cameán, A.M. Microcystin-LR induces toxic effects in differentiated and undifferentiated Caco-2 cells. *Arch. Toxicol.* **2010**, *84*, 405–410. [CrossRef]
77. Gutiérrez-Praena, D.; Pichardo, S.; Jos, Á.; Moreno, F.J.; Cameán, A.M. Biochemical and pathological toxic effects induced by the cyanotoxin Cylindrospermopsin on the human cell line Caco-2. *Water Res.* **2012**, *46*, 1566–1575. [CrossRef]

toxins

MDPI

Article

Variations of Bacterial Community Composition and Functions in an Estuary Reservoir during Spring and Summer Alternation

Zheng Xu [1], Shu Harn Te [2], Cong Xu [1], Yiliang He [1,*] and Karina Yew-Hoong Gin [2,3]

[1] School of Environmental Science and Engineering, Shanghai Jiao Tong University, Shanghai 200240, China; xuzheng-2004@163.com (Z.X.); xucong90@sjtu.edu.cn (C.X.)

[2] NUS Environmental Research Institute (NERI), National University of Singapore, Singapore 138602, Singapore; eritsh@nus.edu.sg (S.H.T.); ceeginyh@nus.edu.sg (K.Y.-H.G.)

[3] Department of Civil and Environmental Engineering, National University of Singapore, Singapore 138602, Singapore

* Correspondence: ylhe@sjtu.edu.cn; Tel.: +86-21-5474-4008

Received: 18 July 2018; Accepted: 3 August 2018; Published: 6 August 2018

Abstract: In this study, we focused on the dynamics of bacterial community composition in a large reservoir in the Yangtze estuary during spring and summer seasons, especially the variations of functional mechanisms of microbial community during the seasonal alternation between spring and summer. Both 16S rRNA gene sequencing and shotgun metagenomic sequencing technology were used for these purposes. The results indicated that obvious variations of bacterial community structures were found at different sites. Particle-associated bacterial taxa exhibited higher abundance at the inlet site, which was closer to the Yangtze River with a high level of turbidity. In other sites, *Synechococcus*, as the most dominant cyanobacterial species, revealed high abundance driven by increased temperature. Moreover, some heterotrophic bacterial taxa revealed high abundance following the increased *Synechococcus* in summer, which indicated potential correlations about carbon source utilization between these microorganisms. In addition, the shotgun metagenomic data indicated during the period of seasonal alternation between spring and summer, the carbohydrate transport and metabolism, energy production and conversion, translation/ribosomal biogenesis, and cell wall/membrane/envelope biogenesis were significantly enhanced at the exit site. However, the course of cell cycle control/division was more active at the internal site.

Keywords: reservoir; Yangtze estuary; 16S rRNA gene sequencing; shotgun metagenomic sequencing; bacterial community; microbial metabolisms

Key Contribution: The technologies of high-throughput sequencing and whole metagenomic sequencing revealed that obvious variations of bacterial composition and function were found in an estuary reservoir during spring to summer transition.

1. Introduction

Estuary reservoirs, as important water sources for estuarine cities, are strongly influenced both by terrestrial and coastal environmental changes [1–5]. In the estuarine area, large accounts of organic matter originate (including N/P nutrients) from land and rivers, flowing through these systems and, finally, into oceans [2,6,7]. In addition, during some special seasons, the salt water invaded into the estuary because of the declined water levels of the river, which resulted in a high level of concentrations of salt ions in these areas [5,8]. Due to the unique geographical locations, the microbial community compositions within estuary reservoirs are very different from microbial community structures within lakes and oceans [9,10].

In estuary ecosystems, bacterial community plays important role in the microbial food web, such as recycling and consuming organic matters [9,11]. Research has indicated that the distinctly different distributions of particle-associated bacteria and free-living bacterial community in estuary areas, have been strongly affected by environmental factors such as turbidity and organic matters [11–13]. Although the microbial community composition in estuary aquatic ecosystems was widely studied and have got certain achievements in recent years, there is still a larger number of unclassified bacterial taxa and unknown ecological functions in estuary systems compared with terrestrial, inland lake, and ocean studies [14–17].

In Addition, Cyanobacteria as one of the most dominant members within the bacterial community should be paid more attention to in aquatic ecosystems, which could possible to form harmful cyanobacterial blooms when the environmental conditions became suitable in water bodies. Although the harmful cyanobacterial species (such as *Microcystis* and *Anabeana*) in freshwater lakes have been widely studied, the *Synechococcus* as one of the most dominant cyanobacterial species in estuarine and marine environments has been less studied. Especially, some strains of *Synechococcus* has been found to have toxicity effects on other marine organisms in recent years [18–20]. Related studies indicated that during the period of cyanobacterial proliferation, obvious variations of bacterial community composition were found in water bodies [21,22]. This implied that the functional mechanisms and ecological roles of the bacterial community changed in the process of cyanobacterial proliferation, which might correlate with nutrient utilization and spatial competition.

In this study, we utilized systematic methods that including high-throughput sequencing, molecular ecological network and metagenomics to reveal the composing characteristics of the bacterial community in the estuarine reservoir and identify the categories of dominate members within these complex bacterial communities in temporal and spatial scales. Additionally, we evaluated the effects of water environmental factors on bacterial community composition. Moreover, we explored the variations of functional metabolic mechanisms within the microbial community from later spring to early summer, which was the period of cyanobacterial proliferation.

2. Results

2.1. Physico-Chemical Parameters and Environmental Factor in QCS Reservoir

During the sampling period, water temperature varied from 15.3 to 29.1 °C, which increased rapidly from April to July and decreased gradually from July to September at all three sites (Figure 1A). Fluctuating pH changes were both found at all sites, which ranged from 7.8 to 9.3. In addition, the pH at both internal and exit sites were much higher than at the inlet site (Figure 1B). Although the electrical conductivity (EC) exhibited obviously decreased trends at all three sampling sites from April to September, the value of EC at the inlet site was much lower than other sites during spring (April–June) (Figure 1C). The turbidity both at internal and exit sites were relatively stable during the whole sampling period, which ranged from 6.83 NTU to 13.7 NTU. In contrast, the turbidity at the inlet site (34.9–125 NTU) was obviously higher than the other sites and remarkably increased from August to September (Figure 1D). The concentrations of ammonium nitrogen ($NH^{+}{}_4$-N), inorganic carbon (IC), dissolved oxygen (DO), and total nitrogen (TN) decreased obviously when water temperature increased (Figure 1E–H). Among these environmental factors, higher levels of $NH^{+}{}_4$-N, IC and DO were observed both at internal and exit sites. Especially, the concentration of DO was obviously higher at the internal site from April to July. In addition, both concentrations of TN and total phosphorus (TP) obviously decreased from the inlet to other sites, with the exception of TP concentration in July and August at the internal site (Figure 1H–I).

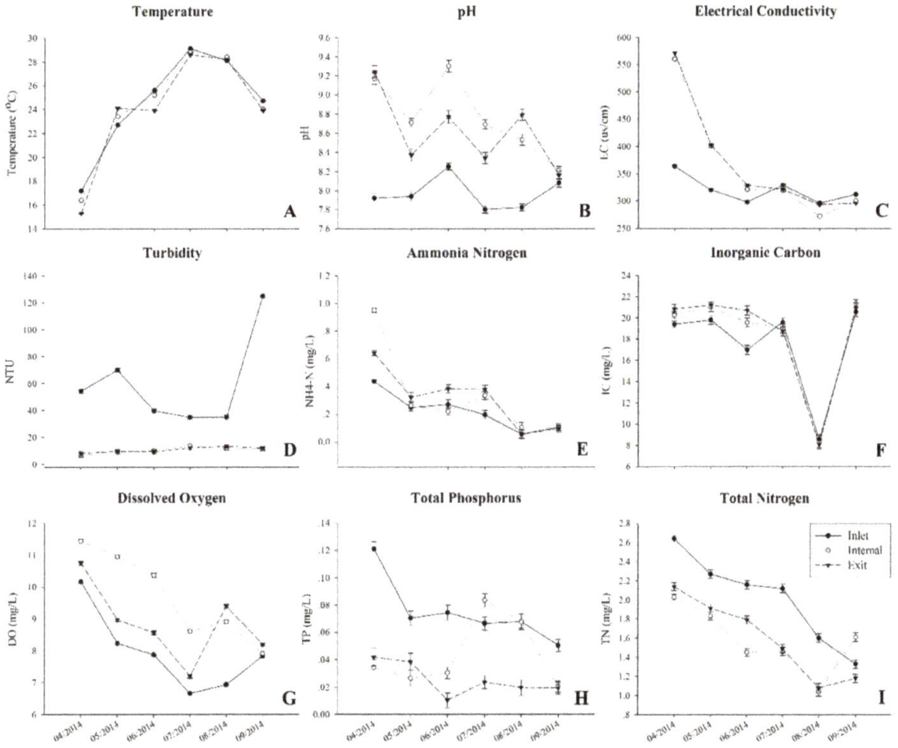

Figure 1. Water chemistry and environmental parameters. (**A**) Temperature, (**B**) pH, (**C**) electrical conductivity (EC), (**D**) turbidity (NTU), (**E**) ammonia nitrogen (NH_4-N^+), (**F**) inorganic carbon (IC), (**G**) dissolved oxygen (DO), (**H**) total phosphorus (TP), and (**I**) total nitrogen (TN).

2.2. The Variations of Chlorophyll-α Concentrations in the QCS Reservoir

In this study, the concentrations of chlorophyll-α from different algae exhibited distinct variation tendencies inside the reservoir (Figure 2). The cyanobacterial chlorophyll-α exhibited relative higher concentrations during July and August compared with other periods, and reached the maximum value at 10.8 μg/L at the exit site in August. While the chlorophyll-α of Chlorophyta only appeared higher concentration in April at the exit site (44.8 μg/L). In contrast, the chlorophyll-α of diatoms and dinoflagellates exhibited obviously higher concentrations from June to August, especially at the internal and exit sites with an average value of 25.1 μg/L.

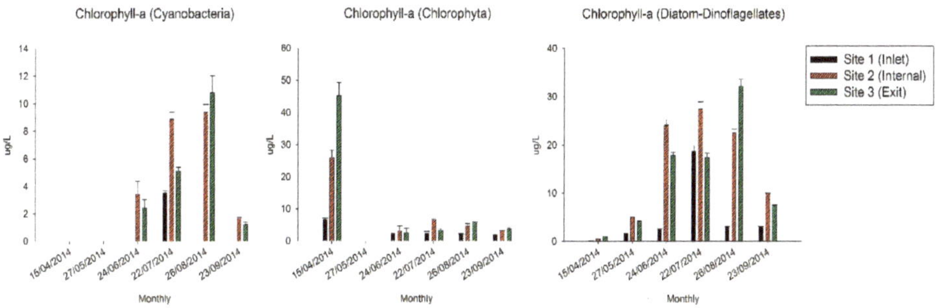

Figure 2. The concentrations of chlorophyll-α in the QCS Reservoir.

2.3. *Dynamic Analysis of Bacterial Community Composition based on the 16S rRNA Sequencing Data*

Based on bacterial community composition analysis assessed by sequencing of V4 region of the 16S rRNA gene, we identified a total of 5,132 OTUs based on 97% similarity during the whole sampling period. The most dominant bacterial phyla were Proteobacteria (31.3%), followed by Actinobacteria (24.8%), Cyanobacteria (10.8%), Bacteroidetes (10.4%), Planctomycetes (8.2%), Verrucomicrobia (5.4%), Chlorobi (2.2%), Gemmatimonadetes (1.9%), Acidobacteria (1.6%), and Chloroflexi (1.2%) at three sites across the whole sampling period at phylum level (Figure 3A). Overall, the variation trends of bacterial community composition at the internal and exit sites were quite similar, which were largely different from bacterial community composition at the inlet site.

For further classification, most of the proteobacterial OTUs classified as Alpha- and Betapeoteobacteria exhibited much higher abundance at the inlet site (16.5% and 16.4%, respectively) than other two sites (11.2% and 9.5% at the internal site, 10.9% and 8.9% at the exit site) (Figure 3B). In addition, the Gammaproteobacteria exhibited higher relative abundance at internal and exit sites (5.9% and 4.8%, respectively) than at the inlet site (3.7%). The relative abundance of both Acidimicrobiia and Actinobacteria, the most dominant actinobacterial OTUs, were relatively stable across the sampling period at the inlet site (with averages of 9.4% and 15%, respectively). In contrast, the fluctuation of Acidimicrobiia abundance was observed both at internal and exit sites (ranging from 4.3% to 12% at the internal site and from 4.4% to 13% at the exit site). However, the relative abundance of Actinobacteria (at the class level) was quite stable at the internal and exit sites (with an average of 15.6% at the internal site and 15% at the exit site). Synechococcophycideae, as the most abundant cyanobacterial OTUs always maintained at lower relative abundance during the sampling period at the inlet site (with an average of 3%), except in July (almost 30.5%). In contrast, the relative abundance of Synechococcophycideae was much higher at the internal and exit sites, especially from July to September (20.6% and 20.1%, respectively). The peaking value was appeared in July at both internal and exit sites (27.6% and 25%, respectively). As the most dominant taxa of the Bacteroidetes phylum, the Flavobacteriia exhibited the highest relative abundance in April at both internal and exit sites (20.4% and 9%, respectively), and maintained at higher relative abundance from April to June at these two sites (10.6% and 5.6%, respectively) compared with inlet site (1.7%). Additionally, the Sphingobacteriia as the second largest group of the Bacteroidetes phylum exhibited relatively higher abundance at internal and exit sites (2% and 2.1%, respectively) than the inlet site (0.9%) during the whole sampling period. In addition, the OPB56 that represented the most dominant Chlorobi OTUs had a higher relative abundance at the inlet site (3.3%) than the internal and exit sites (1.5% and 1.8%, respectively), especially in April and May.

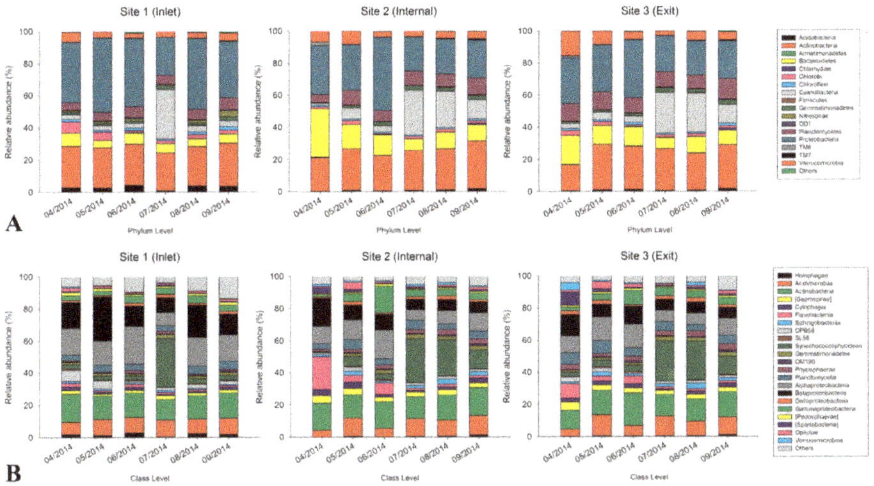

Figure 3. (**A**,**B**) Relative abundance of 16S rRNA bacterial OTUs across the whole sampling period ((**A**). phylum level, (**B**). class level).

The heat-map analysis of the bacterial OTUs with high relative abundance (1% of the total abundance of each sample) revealed that all samples were clustered into three groups (Figure 4). Group 1 was composed of samples from April to September at the inlet site except for July. The second group consisted of samples at both the internal and exit sites in May, August, and September, as well as samples in July at all three sites. Group 3 was mainly composed of samples at both internal and exit sites in April and June. Within these groups, bacterial OTUs including Methylophilaceae (OTU3238 and OTU19751), Holophagaceae (OTU18808), *Zymomonas mobilis* (OTU4718), Comamonadaceae (OTU12896), Rhodospirillaceae (OTU16090), and *Nitrospira* (OTU22889) revealed higher relative abundance in group 1 than other groups. Bacterial OTUs including KD8-87 (OTU5979), Cytophagaceae (OTU6344), Chitinophagaceae (OTU15974 and OTU10486), C111 (OTU20094 and OTU 14914) Sinobacteraceae (OTU7115), Comamonadaceae (OTU19343), Phycisphaerales (OTU5985), Sphingobacteriales (OTU7902), and *Luteolibacter* (OTU17847), Pirellulaceae (OTU1834) and PHOS-HD29 (OTU5225) exhibited higher relative abundance in group 2 than other samples. In addition, the relative abundance of *Opitutus* (OTU19459), *Planctomyces* (OTU2401), Gemmataceae (OTU22474), Sphingobacteriales (OTU3735), [Cerasicoccaceae] (OTU8322), and *Fluviicola* (OTU18054) revealed opposite trends withabundance of *Synechococcus* in group 2. In group 3, bacterial OTUs including Xanthomonadaceae (OTU3160 and OTU14940), *Rheinheimera* (OTU22391), *Flavobacterium* (OTU14035) and *Rhodobacter* (OTU23719) only revealed high abundance in June. In contrast, bacterial OTUs including *Flavobacterium* (OTU606 and OTU21726), SJA-4 (OTU7484), *Calciphila* (OTU7033), Chitinophagaceae (OTU8668), *Fluviicola* (OTU21307), *Luteolibacter* (OTU9323), and Verrucomicrobiaceae (OTU11682) exhibited high abundance only in April in group 3. In addition, *Synechococcus* (OTU1659), Pelagibacteraceae (OTU22095), C111 (OTU278), ACK-M1 (OTU7991), and Actinomycetales (OTU4207) revealed relatively higher abundance than other bacterial OTUs during the sampling period, and bacterial OTUs including *Limnohabitans* (OTU3412), ACK-M1 (OTU16592 and OTU21614) and Comamonadaceae (OTU17668) exhibited higher relative abundance both in group 1 and 3 compared with group 2.

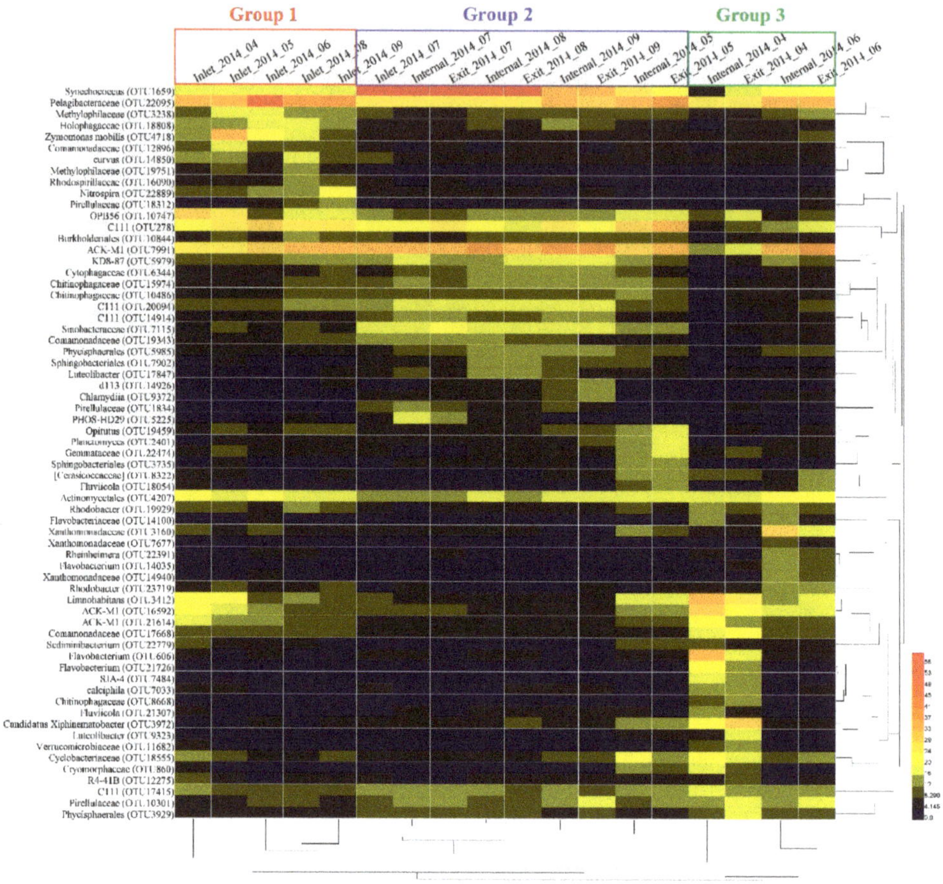

Figure 4. Heat-map analysis for the variations of dominant bacterial OTUs (1% of the total abundance of each sample) based on the Bray-Curtis similarity (OTU level).

2.4. Covariance Analysis of Bacterial Community Composition and Environmental Variables

Marginal test of biotic and abiotic factors in each site based on distance-based linear modelling (DistLM) indicated that the environmental factors exhibited obviously different effects on bacterial community composition between different sampling sites (Table 1). At the inlet site, only DO and temperature significantly affected the variations of bacterial community composition ($p < 0.05$). However, more environmental factors including NH_4^+-N, DO, EC, turbidity, temperature, K^+, Na^+, Mg^{2+}, Cl^-, and F^- exhibited significant effects on bacterial community composition at the internal site. In addition, TN, NH_4^+-N, EC, turbidity, temperature, K^+, Na^+, Ca^{2+}, Mg^{2+}, and Cl^- significantly affected the bacterial community composition at the exit site. Especially, turbidity extremely significantly affected the composition of the bacterial community at the exit site ($p < 0.01$).

Table 1. DistLM results of abundant bacterial community data against environmental variables (999 permutations).

Variables	Inlet		Internal		Exit	
	Pseudo-F	p	Pseudo-F	p	Pseudo-F	p
TC	0.751	0.933	0.731	0.655	1.528	0.059
TOC	0.734	0.943	0.673	0.828	0.673	0.971
IC	0.717	0.952	0.718	0.716	0.967	0.391
TN	1.352	0.059	1.669	0.090	1.923	**0.020**
TP	1.231	0.172	0.883	0.470	1.215	0.247
NH_4^+-N	1.368	0.052	2.437	**0.045**	1.658	**0.034**
pH	1.205	0.165	1.839	0.057	1.291	0.148
DO	1.426	**0.023**	2.119	**0.040**	1.262	0.193
EC	1.158	0.239	2.628	**0.029**	1.802	**0.021**
Turbidity	0.813	0.802	2.366	**0.018**	2.094	**0.003**
Temperature	1.426	**0.029**	2.417	**0.030**	1.738	**0.033**
Cl^-	1.164	0.199	2.674	**0.018**	1.797	**0.008**
SO_4^{2-}	0.730	0.917	0.986	0.381	1.209	0.205
F^-	1.337	0.089	2.357	**0.011**	1.441	0.134
Ca^{2+}	0.733	0.950	1.637	0.097	1.807	**0.033**
Mg^{2+}	0.970	0.528	2.524	**0.024**	1.849	**0.015**
Na^+	1.216	0.149	2.686	**0.019**	1.803	**0.010**
Al^{3+}	0.910	0.642	1.234	0.191	1.516	0.080
K^+	0.962	0.540	2.438	**0.023**	1.889	**0.012**
Si^{4+}	0.683	0.978	0.795	0.605	0.863	0.604
Chl-α	1.345	0.167	1.348	0.170	1.172	0.252

Bold: Significantly correlated with community structure at $p < 0.05$.

The alpha-diversity indices (including species richness, Pielou's evenness, and Shannon and Simpson indices) exhibited relatively consistent variation tendencies on a temporal scale and a marked decline appeared at all three sites in July (Table 2). In addition, although the total species were also decreased in July at all sites, the minimum values of total species at the internal and exit sites both appeared in April. On the spatial scale, the values of alpha-diversity indices at the inlet site were obviously higher than internal and exit sites.

Table 2. The diversity of microbial community composition between samples.

Sample	Total Species	Species Richness	Pielou's Evenness	Shannon	Simpson
Inlet_2014_04	1380	144.6	0.6997	5.059	0.9762
Inlet_2014_05	1349	141.4	0.6724	4.846	0.9708
Inlet_2014_06	1601	167.8	0.6644	4.903	0.9536
Inlet_2014_07	1131	118.5	0.6019	4.232	0.9084
Inlet_2014_08	1382	144.8	0.6969	5.039	0.9739
Inlet_2014_09	1744	182.8	0.6991	5.218	0.9718
Internal_2014_04	701	73.41	0.6637	4.349	0.9685
Internal_2014_05	1138	119.2	0.6798	4.784	0.9724
Internal_2014_06	1127	118.1	0.6946	4.881	0.9747
Internal_2014_07	1015	106.3	0.5991	4.148	0.9185
Internal_2014_08	1107	116	0.6271	4.395	0.9375
Internal_2014_09	1163	121.9	0.6591	4.652	0.9627
Exit_2014_04	990	103.7	0.7044	4.859	0.9806
Exit_2014_05	1128	118.2	0.6642	4.667	0.9661
Exit_2014_06	1251	131.1	0.7031	5.014	0.9743
Exit_2014_07	1133	118.7	0.6150	4.325	0.9294
Exit_2014_08	1173	122.9	0.6311	4.459	0.9361
Exit_2014_09	1265	132.6	0.6752	4.823	0.9672

The distance-based linear redundancy analysis (dbRDA) visualized the relative contribution of measured environmental variables on total bacterial community composition determined by 16S rRNA gene amplicon sequencing (Figure 5). The distributions of inlet samples were quite different from the samples at other sites. Most inlet samples (except July) clustered loosely and positively correlated with high turbidity and TN. However, samples in July at the inlet site exhibited positive correlation with temperature. In contrast, samples from July to September at both internal and exit sites clustered closely and positively correlated with chlorophyll-α, F^-, and high temperature. In addition, samples at internal and exit sites distributed widely in other months and positively correlated with high concentrations of K^+, Na^+, Mg^{2+}, Cl^-, $NH_4{}^+$-N, DO, EC, and pH in April.

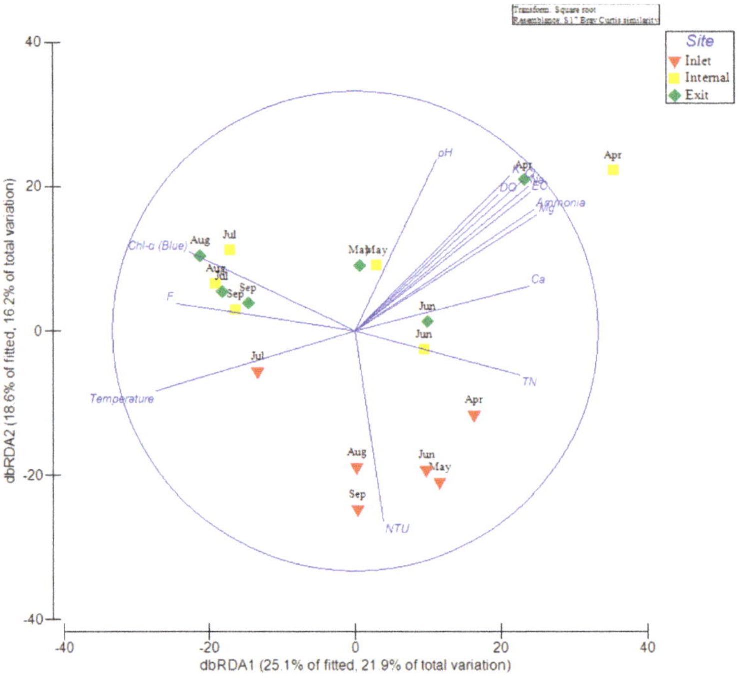

Figure 5. The distance-based linear redundancy analysis (dbRDA) reflecting the distribution of bacterial communities with environmental variables in an estuary reservoir.

2.5. Multivariate Analysis of Biotic and Abiotic Factors in the QCS Reservoir

A total of 89 measured variables include 65 16S rRNA OTUs contributed >1% to any samples and 24 environmental variables were shown in the single interconnected network. A total of 3916 tested correlations were calculated by using rcor.test in ltm package. During these correlations, only 605 ultimately considered significantly correlated with each other. The significant correlations were further used to construct a visual edge-weighted spring-embedded network, with *r* score as the edge-weight in the network (Figure 6).

By scrutinizing the distribution of biotic and abiotic parameters, it was apparent that the network exhibited a similar distribution trend with the result of dbRDA plot (Figure 5). Most bacterial OTUs and Environmental variables clustered into two obviously different groups (spring group and summer group). Within the edge-weighted spring-embedded network, betweenness is a much more significant indicator of essentiality than other topological parameters. Nodes with high betweenness centrality (large nodes) show high centrality—i.e., higher control over the network. Based on the topological characteristic analysis of nodes within the network, temperature was the only environmental variable with high betweenness centrality (>0.02) in summer group, meanwhile, nine bacterial OTUs including Phycisphaerales (Planctomycetes), Sphingobacteriales (Bacteroidetes), C111 (Actinobacteria), KD8-87 (Gemmatimonadetes), Chitinophagaceae (Bacteroidetes), Cytophagaceae (Bacteroidetes), ACK-M1 (Actinobacteria), and *Synechococcus* (Cyanobacteria) revealed high betweenness centrality (>0.02) within this group. In contrast, environmental variables, including pH, Ca^{2+}, NH_4^+-N, K^+, and Cl^- with high betweenness centrality (>0.02) during spring group, meanwhile, ten bacterial OTUs including *Flavobacterium* (Bacteroidetes), *Rhodobacter* (Alphaproteobacteria), ACK-M1 (Actinobacteria), *Sediminibacterium* (Betaproteobacteria), Cyclobacteriaceae (Bacteroidetes), Comamonadaceae (Betaproteobacteria), C111 (Actinobacteria), *Limnohabitans* (Betaproteobacteria), and Phycisphaerales (Planctomycetes) exhibited high betweenness centrality (>0.02) in spring groups. Although the biological network indicated these biotic/abiotic factors (bacteria/environmental factors) with high betweenness centrality might play important roles in network composition, there is less evidence to explain how these biotic/abiotic factors affected and controlled the whole network (such as their functions and roles in the ecosystem) due to technical restriction.

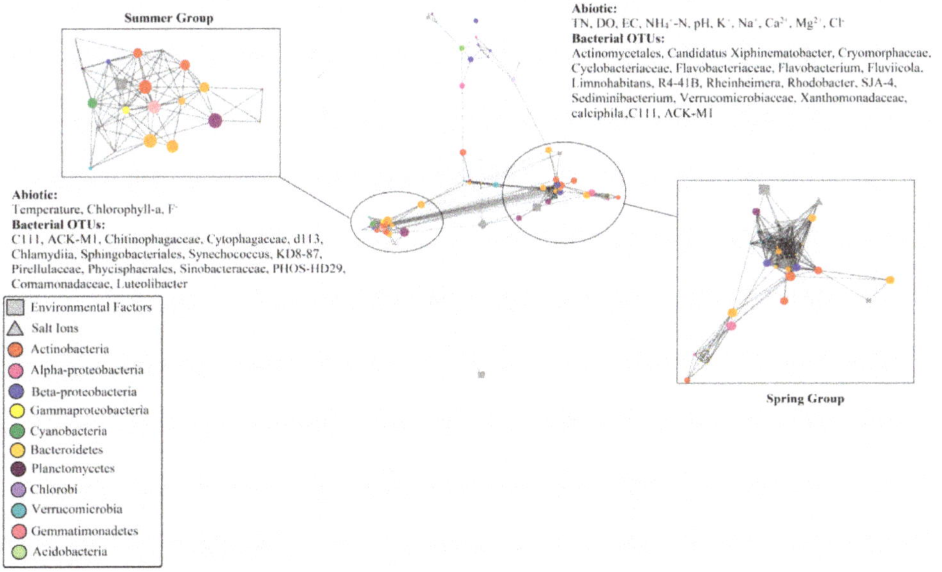

Figure 6. The edge-weighted spring-embedded network based on the Pearson correlation coefficient reflecting the significant correlations between biotic and abiotic factors (node size has reflected the value of betweenness centrality of the variables. Solid lines represent positive correlations and dashed lines represent negative correlation).

An organic correlation sub-network was constructed to visualize pair-wise correlations between the dominant *Synechococcus* (OTU1659) and other non-cyanobacterial OTUs (Figure 7). Environmental variables including temperature and Chlorophyll-α were positively associated with *Synechococcus* (OTU1659), and Ca^{2+} as the only salt ions was negatively correlated with *Synechococcus* (OTU1659) in our study. In addition, nine non-cyanobacterial OTUs including Sinobacteraceae (Gammaproteobacteria), C111 (Actinobacteria), KD8-87 (Gemmatimonadetes), Comamonadaceae (Betaproteobacteria), *Luteolibacter* (Verrucomicrobia), Sphingobacteriales (Bacteroidetes), Pirellulaceae (Planctomycetes), and PHOS-HD29 (Proteobacteria) revealedpositive correlations with *Synechococcus*

(OTU1659). In contrast, two non-cyanobacterial OTUs including *Rhodobacter* (Alphaproteobacteria) and Actinomycetales (Actinobacteria) negatively correlated with *Synechococcus* (OTU1659).

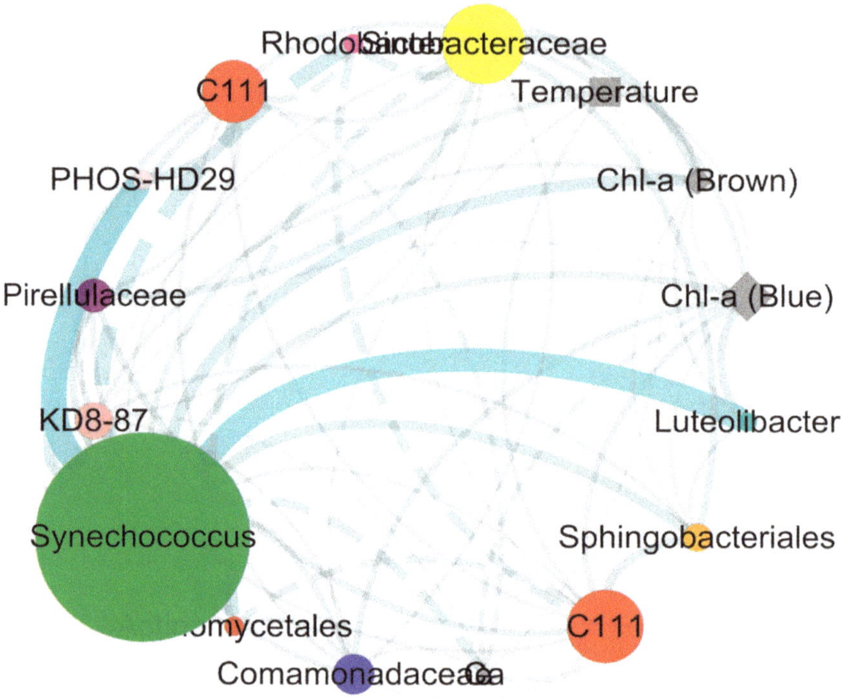

Figure 7. Organic correlation sub-network reflecting the pair-wise correlations between *Synechococcus*, other bacterial OTUsm and environmental factors (node size reflects the value of betweenness centrality of the variables, solid lines represent positive correlations, and dashed lines represent negative correlations).

2.6. Shotgun Metagenomic Analysis

Aiming to determine the functional mechanism variations of microbial community during the period of seasonal transition between spring and summer at different sites inside the reservoir, four samples (May–June at the internal site and June–July at the exit site) within this period were selected to assess the variations of enriched set of metabolic genes using the shotgun metagenomic sequencing technology. We obtained about 47.5 Gb of community shotgun metagenomic sequence data in total from four samples inside the reservoir, and a total of 2.76×10^2 million clean reads were generated from the metagenomic dataset of four samples (Table S3). The number of contigs ranged from 153,619 to 219,616 from scaffolds longer than 500 bp when the k-mer value set as 41 across all four samples (Tables S1 and S2). The statistical information including contigs_N50 and N90 length indicated we obtained a relatively high assembly efficiency of the contigs in our study.

Different from the 16S rRNA sequencing technology, the shotgun metagenomic sequencing technology could provide more information to explore the potential functional mechanisms within the microbial community. In this study, the eggNOGs database categories of non-redundant genes indicated that the microbial community inside the reservoir had a relatively high abundance of genes devoted to amino acid transport and metabolism, general function prediction (only), energy production and conversion, replication, recombination and repair, translation, ribosomal structure and biogenesis, cell wall/membrane/envelope biogenesis, inorganic ion transport and metabolism, carbohydrate transport and metabolism and posttranslational modification, protein turnover, chaperones (Gene abundance > 500,000). However, the category of unknown function still accounted for a large proportion of total gene abundance (gene abundance > 1,500,000). Furthermore, thirty COGs with high gene abundance annotated as sulfatase, ABC transporter, DNA polymerase, and other functions were also shown in this Figure.

The Statistical Analysis of Metagenomic Profiles (STAMP) on COGs categories between different sites revealed that NOG24668, NOG05037, COG0062, NOG00596, COG0809, COG0507, and COG1807 have significantly higher abundance at the exit site than at internal site, only NOG22510 were obviously higher at the internal site than at the exit site (Figure 8).

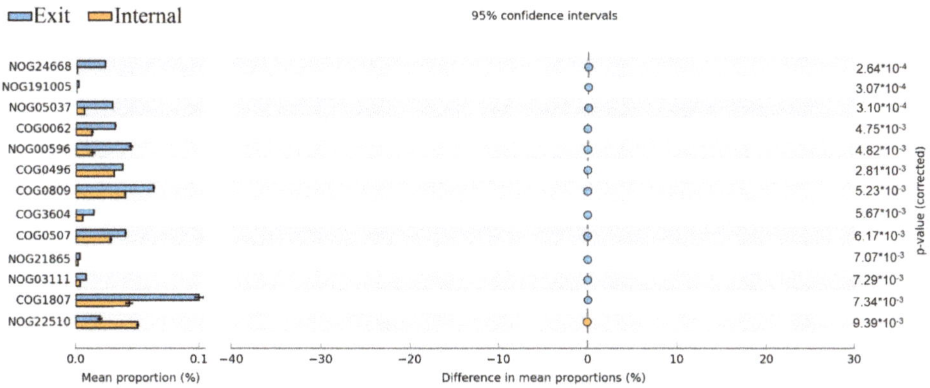

Figure 8. Proportion differences analysis of COG functional categories between internal and exit sites are represented in an extended error plot (the top 80 abundant COG functional categories were selected for analysis). Total mean proportions (%) in the COG categories are exhibited by the bar graph (**left column**); the upper bar graph (blue) represents the samples at the exit site, whereas the other bar graph (yellow) represents the samples at the internal site in each category. The coloured circles corresponding to the (**right column**) (blue and yellow) represent 95% confidence intervals calculated by Welch's *t*-test. COG functional categories were filtered by *p*-value (0.05) and effect size (0.04).

The non-redundant genes also were aligned against the Kyoto Encyclopaedia of Genes and Genomes (KEGG) database using BLAST, to visualize the differences of metabolic pathways within the microbial community. In our study, a total of 4050 KEGG categories were observed within four samples, which are involved in 376 KEGG pathways. Among these KEGG pathways, thirty-two KEGG pathways such as glycolysis/gluconeogenesis, TCA cycle, oxidative phosphorylation, purine/pyrimidine metabolism, carbon metabolism, biosynthesis of amino acids were the most dominant metabolic pathways with an obvious high abundance of all samples. In addition, more than 100 KEGG pathways resulted from STAMP analysis revealed the significant spatial differences between internal and exit sites. We further selected 27 KEGG pathways from these above and displayed in this paper (Figure 9). Among these KEGG pathways, the most notable KEGG pathway was K00525 (ribonucleoside-diphosphate reductase alpha chain), which exhibited much higher relative abundance in all samples, but significantly higher at the internal site.

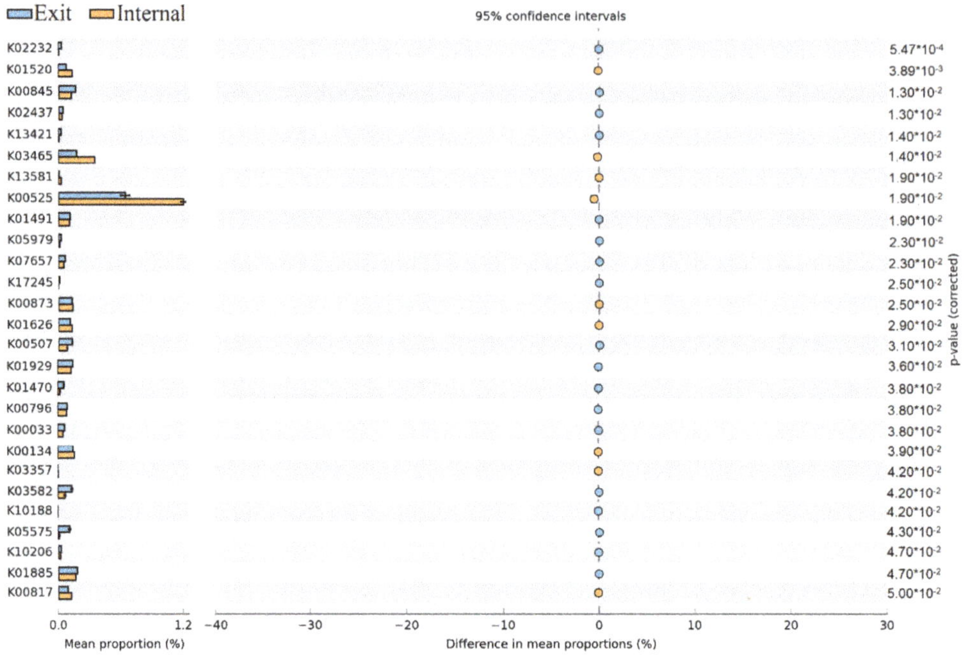

Figure 9. Proportion differences analysis of KEGG functional categories between internal and exit sites are represented in an extended error plot (the top 80 abundant KEGG functional categories were selected for analysis). Total mean proportions (%) in the KEGG categories are exhibited by the bar graph (**left column**); the upper bar graph (blue) represents the samples at the exit site, whereas the other bar graph (yellow) represents the samples at the internal site in each category. The coloured circles corresponding to the (**right column**) (blue and yellow) represent 95% confidence intervals calculated by Welch's *t*-test. KEGG functional categories were filtered by *p*-value (0.05) and effect size (0.04).

3. Discussion

3.1. Temporal and Spatial Dynamics of Microbial Community Composition in the QCS Reservoir

In this study, Illumina MiSeq (16S rRNA sequencing) technology was used to evaluate the microbial community diversity and composition spanning from end spring to summer in different sites of the reservoir. Based on these data, we further used shotgun metagenomic sequencing technology (Illumina HiSeq 4000 platform) to explore functional mechanism variations within the microbial community during the seasonal transition between spring and summer at different sites inside the reservoir.

The variations of microbial community composition indicated microorganisms had similar dominant community structure (at phylum level) at all three sites, but the relative abundance of these dominant bacterial phyla were obvious differences, especially at the inlet site. In addition, the alpha-diversity of bacterial community at the inlet site was also clearly higher than other two sites (Table 2). These changes were mainly due to the differences of water environmental conditions between different sites. At the inlet site, raw water from the Yangtze River runs into the reservoir. This means the aquatic ecological environment at the inlet site was linked with water quality parameters in the Yangtze River characterized for higher concentrations of nutrients (N, P) and turbidity, which is affected by seriously non-point pollution and soil erosion in upstream [23–25]. Thus, the inorganic nutrients and organic matter were not restrictive factors for microbial metabolisms at the inlet site.

However, the higher turbidity reduced the transparency in a surface water body and further limited the photosynthesis of photosynthetic microorganisms. Therefore, the DO became the restrictive environmental factor. These were coinciding with our experimental data that only DO and temperature were significantly affected the bacterial community composition at the inlet site ($p < 0.05$) (Table 1).

At the inlet site, Methylophilaceae (OTU3238 and OTU19751), *Zymomonas mobilis* (OTU4718), Comamonadaceae (OTU12896), *Limnohabitans curvus* (OTU14850), and Rhodospirillaceae (OTU16090) were representatives of the dominant Alpha- and Betaproteobacterial taxa exhibited higher relative abundance during the sampling period (Figure 4). Strains of family Methylophilaceae have the characteristic function of utilizing methanol/methylamine as the only energy and carbon source, were widely distributed in surface sediment of freshwater lakes [26,27]. Therefore, we assumed that the high abundant Methylophilaceae at the inlet site may be associated with high turbidity, which was derived from soil erosion upstream of the Yangtze River. *Limnohabitans curvus*, as the first described species of the family Limnohabitans, exhibited high relative abundance at the inlet site, which was reported as chemoorganotrophic, aerobic, and facultative anaerobe metabolic types [28]. In addition, these clades were also capable of assimilating glucose and types of small organic acids, excluding amino acids [28]. This implied a potential association between *Limnohabitans curvus* and the high concentration of total organic carbon (TOC) in raw water from the Yangtze River. Another dominant family, Comamonadaceae of Betaproteobacteria, was difficult to obtain more specific information to explain the high abundance at the inlet site, due to a large diversity of phylogeny and functions within this family [29]. The Alphaproteobacteria represented by family Rhodospirillaceae with high abundance at the inlet site have been considered with varying metabolic types, including photoheterotrophs, photoautotrophs, and chemoheterotrophs [30]. In contrast, internal and exit sites were midstream and downstream of the reservoir, respectively. The water flow velocity obviously declined, and have sufficient retention times for purification to increase the transparency of the water column in these areas. The water quality parameters were also indicated that the concentrations of TN, TP and turbidity were remarkably decreased at internal and exit sites than at the inlet site (Figure 1). To some extent, the higher transparency and lower nutrient level at these sites reduced the diversity of the bacterial community and increased the potential possibility of cyanobacterial proliferation in surface water. In addition, there were some potential associations between increased cyanobacterial abundance and reduced diversity of the bacterial community [21,22], which could partly explain the remarkable decline of alpha-diversity indices in July inside the reservoir. In addition, the dominant environmental factors were consistent at both internal and exit site (Table 1), which implied that the microbial community compositions were similar at two sites. The dbRDA plot showed that all July samples grouped together (Figure 5). Combined with the results of the heat-map (Figure 4) and network (Figure 6), we found most of the dominant bacterial taxa in July exhibited positive correlations with temperature. Thus, we speculated that the temperature was the key factor for the composition of the bacterial community in July at all three sites. Additionally, the increased water temperature further promoted some kinds of mesophile bacterial growth. In addition, the bacterial OTUs were strongly connected (negatively correlated) between spring and summer groups within the network (Figure 6). To some extent, this implied that the dynamics and continuity of bacterial community composition varied seasonally inside the reservoir, although some variations in the short-term (days or one week) might be ignored in our study. However, high-frequency sampling in the short-term would perform inside the reservoir, which could further validate these conclusions. While we also found few bacterial OTUs were both excluded from summer and spring groups in the network, these bacterial OTUs exhibited relatively lower abundance inside the reservoir and were less connected with other biotic/abiotic factors, which implied little dependence of these bacterial OTUs on other biotic/abiotic factors.

The *Synechococcus* was the most common Cyanobacteria in coastal areas, which was found to have high abundance in the Yangtze estuary during the summer season in history [9,31]. Early studies have shown that the counterparts of *Synechococcus* in marine ecosystems were found to have the capability to utilize nitrate, ammonia, or urea as nitrogen sources [32], but we are still unclear whether the

Synechococcus in the estuarine ecosystem has a similar capacity or not. The molecular ecological network further indicated that some bacterial OTUs positively correlated with the increased *Synechococcus*, which implied that co-occurrence correlations probably existed between these bacterial taxa and *Synechococcus* (Figure 7). Among these bacterial OTUs, the C111 of actinobacterial phylum were found to have strong connections with Synechococcaceae in a previous study [33], which indicated that the C111 clades might depend on the carbon source released by these cyanobacterial species. Additionally, some other bacterial OTUs of Betaproteobacteria and Bacteroidetes phyla were also found to have similar functional relationships with the dominant *Synechococcus* (Figure 7). The results were consistent with previous study that these bacterial clades have a similar tendency to increased cyanobacterial abundance, and assimilated dissolved organic matters derived from cyanobacterial cell metabolism as their carbon sources [34].

3.2. The Variations of Ecological Functions within the Microbial Community during the Period of Later Spring/Early Summer

The results of dbRDA plot demonstrated that the microbial community composition was obviously different between samples in later spring and early summer both at internal and exit site, which implied the potential ecological functions of the dominant microbial community also changed obviously from later spring to early summer at these sites (Figure 5). Subsequently, the functional annotations through eggNOG were shown that the relative abundance of some COG/NOGs taxonomies in critical metabolic reactions was significantly enhanced at the exit site ($p < 0.05$). These COG/NOGs include COG0062, NOG00596, COG0809 and COG1807 (Figure 8). The COG0062 was annotated as ADP-dependent NADHX epimerase, which played important roles in the course of carbohydrate transport and metabolism. NOG00596 was annotated as AMP-binding protein, which played key roles in energy production and conversion. COG0809 and COG1807 were both annotated as glycosyl transferase, which played important roles in translation/ribosomal biogenesis and cell wall/membrane/envelope biogenesis, respectively. These results indicated that the activities of key enzymes involved in carbohydrate transport and metabolism, energy production and conversion, translation/ribosomal biogenesis and cell wall/membrane/envelope biogenesis were significantly enhanced at the exit site. In contrast, only NOG22510 exhibited higher relative abundance at internal site than exit site (Figure 8). The NOG22510 was further annotated as TGFb_propeptide, which was correlated with Beta binding protein and was a main factor controlling cell cycle control/division. Thus, we speculated that the course of cell cycle control/division was more active at internal site than exit site. It was notable that in STAMP analysis, K00525 exhibited significantly higher abundance at internal site than exit site (Figure 9). K00525 was annotated as ribonucleoside-diphosphate reductase, mainly involved in the courses of Purine metabolism (ko00230) and Pyrimidine metabolism (ko00240). These reactions were mainly provided the raw material for DNA synthesis. This result was consistent with the conclusion of COG/NOGs variation analysis that the course of cell cycle control/division was enhanced at an internal site. In addition, it's important to note that some changes of metabolisms and functions of cells within the bacterial community in short-term (days or one week) might be ignored due to the monthly sampled intervals in our study. Previous study indicated that although the bacterial community composition retained relatively stable over weeks or a month, obvious dynamic of bacterial community composition were observed within short-term (days or one week) [17,35]. This could result in the sharp shifts of potential cell metabolisms within the microbial community during a short time, which would be missing in our study.

Combined with variations of environmental factors, we can found that the dissolved oxygen and water pH at the internal site was much higher than at the exit site from May to July (Figure 1). We speculated that the respirations of microbial community at the exit site were much stronger than internal site, which could deplete more dissolved oxygen and accumulate more CO_2 in the water, which resulted in a lower concentration of dissolved oxygen and pH. Besides, the concentration of TP has obviously decreased at the exit site after June compared with the internal site (Figure 1).

Phosphorus was an essential nutrient element for the bacterial community in aquatic ecosystems, which played important roles in cell metabolisms and cell structures [36,37]. Early studies indicated that the bacteria have higher cellular requirements for phosphorus relative to carbon in freshwater lakes [38]. Therefore, we speculated that the obviously decreased of TP concentration at the exit site mainly correlated with the increased metabolic activity of carbohydrate transport and metabolism, energy production and conversion, translation/ribosomal biogenesis and cell wall/membrane/envelope biogenesis in this area.

In this study, we used both 16S rRNA sequencing and shotgun metagenomic sequencing technology to detect the diversity and functions of the bacterial community in samples at internal and exit sites during the period of later spring and early summer. The majority bacterial community structure characterized by the shotgun metagenomic sequencing approach was quite similar to the results based on the 16S rRNA sequencing technology. However, there are a few discrepancies in the classification of some individual bacterial taxa by using these two approaches. For example, the relative abundance of *Synechococcus* was obviously higher in June at the exit site by using shotgun metagenomic sequencing technology than the 16S rRNA sequencing technology. These were likely caused by different sequencing procedures between shotgun metagenomic sequencing and 16S rRNA sequencing technology. As such, the 16S rRNA targeted sequencing included extra PCR steps, and other reasons including primer bias or suboptimal PCR conditions in the process [39–41]. In addition, the eggNOG database could provide accurate clusters of orthologous groups' information on proteins, but a great amount of gene was classified as unknown functions. These together indicated that although the shotgun metagenomic sequencing technology could reflect the functional characteristics of microbial community to some extent, there is still a certain gap between the real functions of microorganisms and environments. Furthermore, some specific metabolic pathways were still not clear, which need to be further improved and perfected.

4. Conclusions

To fully understand the dynamics of bacterial community composition during spring and summer in a large estuary reservoir, the 16S rRNA sequencing technology was used to assess characteristics of the bacterial community in different sites monthly. Moreover, the shotgun metagenomic sequencing technology was used to further detect the variations of potential functional mechanisms within the microbial community during the seasonal transition from later spring to early summer. The 16S rRNA sequencing data indicated that obvious differences of bacterial community composition at different sites inside the reservoir. Particle-associated bacterial taxa exhibited obviously higher abundance at the inlet site than at two other sites. In contrast, heterotrophic bacterial taxa exhibited higher abundance with increased *Synechococcus* at internal and exit sites during summer. Correlation analysis indicated temperature was the major factor contributing to the increase of the abundance of *Synechococcus*. The shotgun metagenomic sequencing data indicated that the carbohydrate transport and metabolism, energy production and conversion, translation/ribosomal biogenesis, and cell wall/membrane/envelope biogenesis were significantly enhanced at the exit site. However, the course of cell cycle control/division was more active at the internal site.

5. Materials and Methods

5.1. Sampling Sites and In Situ Measurements

QCS Reservoir is the largest estuary reservoir in China located at the Yangtze estuary area near Shanghai (Figure 10). The reservoir covers a total catchment area of 66.27 km^2, with a depth ranged from 2.5 to 13.5 m. Its main purpose is compensating for drinking water shortage in Shanghai, which inputs high turbidity water from the Yangtze River estuary and outputs clean water to water plants after the self-purification in the reservoir [2,8,42]. During our study, we set three sampling sites along the reservoir. The raw water entered the reservoir from the inlet site. The internal and exit sites represented the midstream and downstream of the reservoir, respectively (Figure 10). All water samples were collected at a depth of 0.5 m below the surface monthly from April to September 2014, which is the warm seasons from spring to summer with a high risk of cyanobacterial bloom [9,31]. Water temperature and dissolved oxygen (DO) were detected in situ using multi-parameter water quality analyser (Multi3410, WTW Company, Weilheim, Germany).

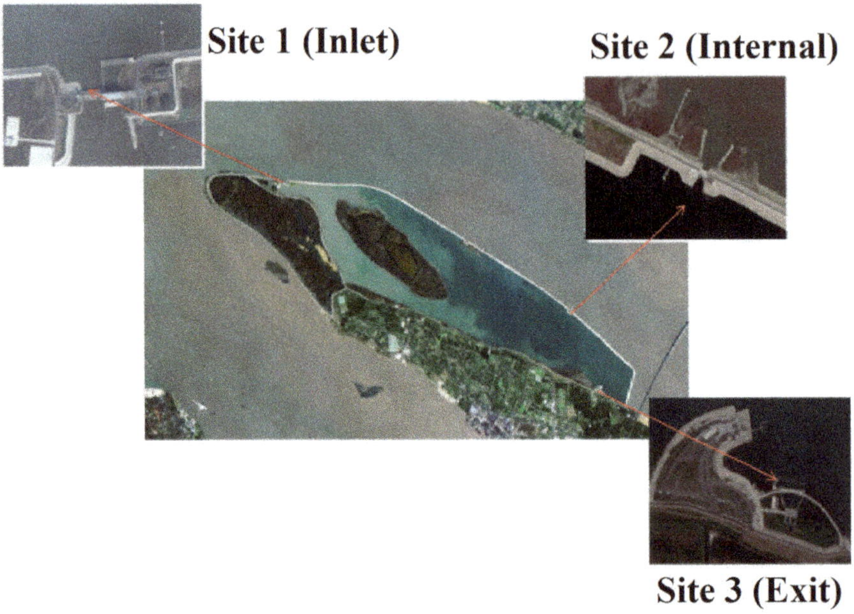

Figure 10. Aerial schematic of the Qingcaosha (QCS) Reservoir and annotated sampling locations (Site 1 (Inlet), Site 2 (Internal), and Site 3 (Exit)).

5.2. *Physic-Chemical Parameters and Environmental Factors*

Physico-chemical parameters and environmental factors include pH, electrical conductivity (EC), turbidity, total phosphorus (TP), and ammonium nitrogen (NH_4^+-N) were analysed according to water and wastewater monitoring analysis standard methods. Total carbon (TC), total nitrogen (TN), total organic carbon (TOC), and inorganic carbon (IC) were detected by using a Multi N/C 3100 Analyser (Jena, Germany). The concentrations of chlorophyll-α (Chl-α), which represented phytoplankton biomass, were measured using PHYTO-PAM phytoplankton analyser (Waltz, Germany) [43]. The PHYTO-PAM phytoplankton analyser could distinguish different types of phytoplankton, like chlorophyta, diatoms, and cyanobacteria, based on the specific fluorescence excitation properties of differently pigmented phytoplankton groups and exhibited high detection precision. After all water samples were filtrated through 0.45 μm Cellulose Acetate filter membranes, the concentrations of K^+, Na^+, Ca^{2+}, Mg^{2+}, Al^{3+}, and Si^{4+} ions were detected by inductively-coupled plasma (ICP) spectroscopy. The F^-, Cl^-, and SO_4^{2-} ions contents were detected using a Metrohm 830 ion chromatographer [8].

5.3. *DNA Extraction*

A total of 500 mL volume water samples at each site were filtrated through 0.22 μm cellulose acetate filter membranes immediately on receipt at Shanghai Jiaotong University (SJTU, Shanghai, China). Total DNA was extracted directly from the same amount of membranes using an E.Z.N.A.

Water DNA Kit (Omega, Irving, TX, USA) in according to the manufacturer's specifications. To ensure the DNA samples were adequate for metagenomic analysis, we conducted six replicates for DNA extraction per water sample using the same DNA extraction approach described previously.

5.4. The 16S rRNA Gene Sequencing via PCR Amplification

To determine the diversity and variation of bacterial community composition in different sites of the reservoir, we used PCR amplification for each water sample with the 515F/806R primer set which could amplify the V4 region of the 16S rRNA gene. This primer set exhibited lower biases and more accurate taxonomic and phylogenetic information for individual bacterial taxa [44]. The PCR amplifications were performed in 25 μL reaction mixtures containing 5.0 μL 5* Q5 Reaction Buffer, 5.0 μL 5*Q5 GC high Enhancer, 2.5 mM of dNTPs 2.0 μL, 1.0 μL of forward and reverse primers (10 μM each), 0.25 μL of Q5 DNA Polymerase (5 U/μL), and 1 μL DNA template (20 ng/μL each). The following PCR cycling processes included an initial denaturation at 98 °C for 5 min, then followed by 27 cycles of denaturation at 98 °C for 30 s, annealing at 50 °C for 30 s, extension at 72 °C for 30 s, and a final elongation at 72 °C for 5 min. The 16S rRNA PCR products were then further purified using MinElute PCR Purification Kit (Qiagen, Gmbh, Germany).

After purification, paired-end amplicon sequencing (2 × 150 bp) were sent to Personal Biotechnology Co., Ltd. (Shanghai, China) for Illumina sequencing. Raw data were processed according to procedures described previously [45,46], using the Quantitative Insights into Microbial Ecology (QIIME) pipeline (version 1.7.0, http://qiime.org/) for quality control. Uchime was implemented in Mothur (version 1.31.2, http://www.mothur.org/) to identify and remove chimeric sequences [47,48]. At this stage, sequences less than 150 bp in length, which means quality less than 20, and sequences containing Ns and any ambiguous bases pairs were eliminated from pair-end sequence reads. Sequences were subsampled at a level of 21,869 reads for each sample before the further analysis. Purified sequences were binned into operational taxonomic units (OTUs) based on a 97% identity threshold, while the longest sequence of each OTU was selected as the representative sequence for that OTU based on UCLUST algorithm using QIIME [49]. The taxonomic identity of OTUs was aligned and compared with Ribosomal Database Project classifier (Release 11.1, http://rdp.cme.msu.edu/), SILVA database (Release 119, http://www.arb-silva.de), and the Greengenes database (97% taxonomy) (Release 13.8, http://greengenes.secondgenome.com/), which were used for taxonomy assignment of bacteria and archaea [50–52]. The raw sequencing datasets were available from NCBI Sequence Read Achieve under BioProject PRJNA397386.

5.5. Statistical Analysis of the 16S rRNA Sequencing Data

To better understand the dynamic distributions of dominant bacterial OTUs, we selected the bacterial OTUs which were detected at least 20% samples and contributed to at least 1% of the total abundance of each sample. The relative abundance of these OTUs was further transformed by square root to reduce the disturbance of highly abundant OTUs in the analysis progress. A heat-map was constructed for cluster analysis of the distributions of these bacterial OTUs based on Bray-Curtis similarity at the genus level.

Distance-based linear models (DistLM) were created to model and evaluate the contribution of each measured environmental variable on variations of microbial community composition by using PRIMER v6 and PERMANOVA+ (PRIMER-E Ltd., Plymouth, UK). Alpha-diversity parameters including total species, species richness, Pielou's evenness, Shannon and Simpson indices between samples were calculated using PRIMER v6. Furthermore, Distance-based redundancy analysis (dbRDA) was implemented to assess the correlations between environmental factors and distributions of microbial community in spatial and temporal scales. The degree of paired correlations between each biotic and abiotic factors across the whole sampling period was calculated using Pearson's correlation coefficient (r). Highly abundant OTUs, which were observed at least four samples (>20% samples) and contributed at least 1% to any given samples were selected. All original abundance values of these OTUs were retained without any alteration. Both Pearson's correlation coefficient (*r*) and *p*-value were calculated pairwise based on a rcor.test algorithm by using ltm package in R (version 3.2.0) for each OTU. During the operational processes, the *p*-value was generated with each counterpart correlation and the false discovery rate was constantly kept below 5% based on the Benjamini-Hochberg procedure [53]. Based on these significant correlations, a visualized edge-weighted spring-embedded network was generated by using Cytoscape package (version 3.2.1), which was according to *r*-value as the edge-weighted of the network. Within the network, relevant topological and node/edge metrics including betweenness centrality was also enumerated through the network analysis plug-in [54].

5.6. Shotgun Metagenomic Analyses

Shotgun metagenomic sequencing was used for the same DNA extracts of four selected samples (May–June at the internal site and June–July at the exit site) inside the reservoir. The genome DNA was mechanically sheared into ~300 bp fragments using an M220 Focused-ultrasonicator^{TM} (Covaris Inc., Woburn, MA, USA). Meanwhile, paired-end library (2 × 150 bp) was constructed. After the procedures of DNA templates enrichment and bridge PCR amplification, the paired-end reads (2 × 150 bp) were sequenced by Illumina HiSeq 4000 at Majorbio Bio-Pharm Technology Co., Ltd. (Shanghai, China) using Truseq SBS Kit v3-HS following the standard protocol (www.illumina.com). All the raw metagenomic datasets have been submitted into NCBI Sequence Read Achieve under accession BioProject PRJNA393607.

5.7. Sequence Quality Control and Assembly

In order to improve the quality and reliability of subsequent analysis, Seqprep (https://github.com/jstjohn/SeqPrep) software was used for quality control. Sickle (https://github.com/najoshi/sickle) was used to remove reads of which the length is less than 50 bp, mean Quality is less than 20 and contain N [55]. The clean reads were assembled using SOAPdenovo (Version 1.06, http://soap.genomics.org.cn/) based on De-Brujin graph with a range of k-mers (39–47). The length of scaffolds over than 500 bp was chosen for further analysis. Based on the quality and quantity of the scaffolds assembly, the maximum number of the scaffold and the peak value of N50 and N90 were obtained when the k-mer value was set at 41. New contigs were extracted when the scaffolds were broken from gaps inside. Then, the contigs with length over 500 bp were further used for prediction and annotation. The statistics of assembly results can be found in Table S1.

5.8. Gene Prediction, Taxonomy, and Functional Annotation

Open reading frames (ORFs) (Table S2) of contigs in each sample were predicted using the MetaGene software (http://metagene.cb.k.u-tokyo.ac.jp/). The ORFs with length over 500 bp were extracted and translated to amino acid sequences. In order to better understand the commonness and difference between samples, the dynamic changes of abundance of microorganisms (or genes) were compared. Moreover, the non-redundant gene catalogue was constructed using CD-HIT software (http://www.bioinformatics.org/cd-hit/) (Parameters: 95% identity, 90% coverage), then the longest genes of each cluster were chosen as representative sequences. High-quality reads were aligned to the Non-redundant gene catalogue (95% identity) using SOAPaligner software (http://soap.genomics.org.cn/) and the abundance of each Non-redundant genes was counted for each sample. Non-redundant gene catalogue was aligned against eggNOG database (cut-off: *e* value < 1×10^{-5}) by BLASTP (BLAST Version 2.2.28+, http://blast.ncbi.nlm.nih.gov/Blast.cgi) for Clusters of orthologous groups (COGs) of proteins assignment and the Kyoto Encyclopedia of Genes and Genome (KEGG) database (cut-off: *e* value < 1×10^{-5}) by BLAST (BLAST Version 2.2.28+, http://blast.ncbi.nlm.nih.gov/Blast.cgi). The catalogue was also assigned KEGG functional annotation by KOBAS 2.0 (KEGG Orthology Based Annotation System, http://kobas.cbi.pku.edu.cn/home.do). The pairwise statistical comparative analyses of COG and KEGG functional classification between samples were realized by STAMP software (http://kiwi.cs.dal.ca/Software/STAMP). The significance of the results was evaluated based on the Welch's-test. The COGs and KEGG categories that were larger than 1% of total abundance in each sample were selected and then calculated using PRIMER v6 and PERMANOVA+ (PRIMER-E Ltd., Plymouth, UK). Further, the independent sample *t*-test was calculated to compare the values of these COGs and KEGG categories between site 2 and 3 using SPSS software (SPSS v22. Inc., Armonk, NY, USA).

Supplementary Materials: The following are available online at http://www.mdpi.com/2072-6651/10/8/315/s1, Table S1: Statistics of assembly result, Table S2: Statistics of predicted ORF, Table S3: Statistics of clean data.

Author Contributions: Author Contributions: Conceptualization, Y.H. and K.Y.-H.G.; Methodology, Z.X.; Software, Z.X.; Validation, Z.X.; Formal Analysis, Z.X.; Investigation, Z.X.; Resources, Z.X.; Data Curation, Z.X.; Writing-Original Draft Preparation, Z.X.; Writing-Review & Editing, Z.X. and C.X.; Visualization, Z.X. and S.H.T.; Supervision, Y.H. and K.Y.-H.G.; Project Administration, Y.H.; Funding Acquisition, Y.H.

Funding: This research received no external funding.

Acknowledgments: This research grant is supported by the joint program between Shanghai Jiaotong University (SJTU) and National University of Singapore (NUS), and we are also grateful to the Campus for Research Excellence And Technological Enterprise (CREATE) programme under the joint program Energy and Environmental Sustainability Solutions for Megacities (E2S2) between Shanghai Jiaotong University (SJTU) and the National University of Singapore (NUS), also Singapore National Research Foundation (NRF) under its Environmental and Water Technologies Strategic Research Programme and administered by PUB, the Singapore's National Water Agency (Grant number: 1102-IRIS-14-02).

Conflicts of Interest: The authors declare no conflict of interest.

References

1. Shu, H.T.; Gin, Y.H. The dynamics of cyanobacteria and microcystin production in a tropical reservoir of singapore. *Harmful Algae* **2011**, *10*, 319–329.
2. Jin, X.; He, Y.; Kirumba, G.; Hassan, Y.; Li, J. Phosphorus fractions and phosphate sorption-release characteristics of the sediment in the yangtze river estuary reservoir. *Ecol. Eng.* **2013**, *55*, 62–66. [CrossRef]
3. Dong, Y.L.; Owens, M.S.; Doherty, M.; Eggleston, E.M.; Hewson, I.; Crump, B.C.; Cornwell, J.C. The effects of oxygen transition on community respiration and potential chemoautotrophic production in a seasonally stratified anoxic estuary. *Estuaries Coasts* **2014**, *38*, 1–14.
4. Yong, H.J.; Yang, J.S.; Park, K. Changes in water quality after the construction of an estuary dam in the geum river estuary dam system, korea. *J. Coast. Res.* **2014**, *30*, 1278–1286.
5. Chen, W.; Chen, K.; Kuang, C.; Zhu, D.Z.; He, L.; Mao, X.; Liang, H.; Song, H. Influence of sea level rise on saline water intrusion in the yangtze river estuary, china. *Appl. Ocean Res.* **2016**, *54*, 12–25. [CrossRef]
6. Canuel, E.A.; Hardison, A.K. Sources, ages, and alteration of organic matter in estuaries. *Ann. Rev. Mar. Sci.* **2016**, *8*, 409–434. [CrossRef] [PubMed]
7. Osburn, C.L.; Boyd, T.J.; Montgomery, M.T.; Bianchi, T.S.; Coffin, R.B.; Paerl, H.W. Optical proxies for terrestrial dissolved organic matter in estuaries and coastal waters. *Front. Mar. Sci.* **2016**, *2*. [CrossRef]
8. Jin, X.; He, Y.; Zhang, B.; Hassan, Y.; George, K. Impact of sulfate and chloride on sediment phosphorus release in the yangtze estuary reservoir, china. *Water Sci. Technol. A J. Int. Assoc. Water Poll. Res.* **2013**, *67*, 1748–1756. [CrossRef] [PubMed]
9. Sun, Z.; Li, G.; Wang, C.; Jing, Y.; Zhu, Y.; Zhang, S.; Liu, Y. Community dynamics of prokaryotic and eukaryotic microbes in an estuary reservoir. *Sci. Rep.* **2014**, *4*, 6966. [CrossRef] [PubMed]
10. Lee, S.Y.; Eom, Y.B. Analysis of microbial composition associated with freshwater and seawater. *Biomed. Sci. Lett.* **2016**, *22*, 150–159. [CrossRef]
11. Crump, B.C.; Baross, J.A.; Simenstad, C.A. Dominance of particle-attached bacteria in the columbia river estuary, USA. *Aquat. Microb. Ecol.* **1998**, *14*, 7–18. [CrossRef]
12. Angelika, R.; Herlemann, D.P.R.; Klaus, J.; Hans-Peter, G. Particle-associated differ from free-living bacteria in surface waters of the baltic sea. *Front. Microbiol.* **2015**, *6*, 1297.
13. Yung, C.M.; Ward, C.S.; Davis, K.M.; Johnson, Z.I.; Hunt, D.E. Insensitivity of diverse and temporally variable particle-associated microbial communities to bulk seawater environmental parameters. *Appl. Environ. Microbiol.* **2016**, *82*, 3431. [CrossRef] [PubMed]

14. Ochsenreiter, T.; Selezi, D.; Quaiser, A.; Bonch-Osmolovskaya, L.; Schleper, C. Diversity and abundance of crenarchaeota in terrestrial habitats studied by 16S rna surveys and real time pcr. *Environ. Microbiol.* **2010**, *5*, 787–797. [CrossRef]
15. Humbert, S.; Tarnawski, S.; Fromin, N.; Mallet, M.P.; Aragno, M.; Zopfi, J. Molecular detection of anammox bacteria in terrestrial ecosystems: Distribution and diversity. *J. Emultidiscip. J. Microb. Ecol.* **2010**, *4*, 450–454. [CrossRef] [PubMed]
16. Oh, S.; Caroquintero, A.; Tsementzi, D.; Deleonrodriguez, N.; Luo, C.; Poretsky, R.; Konstantinidis, K.T. Metagenomic insights into the evolution, function, and complexity of the planktonic microbial community of lake lanier, a temperate freshwater ecosystem. *Appl. Environ. Microbiol.* **2011**, *77*, 6000–6011. [CrossRef] [PubMed]
17. Lindh, M.V.; Sjöstedt, J.; Andersson, A.F.; Baltar, F.; Hugerth, L.W.; Lundin, D.; Muthusamy, S.; Legrand, C.; Pinhassi, J. Disentangling seasonal bacterioplankton population dynamics by high-frequency sampling. *Environ. Microbiol.* **2015**, *17*, 2459–2476. [CrossRef] [PubMed]
18. Frazão, B.; Martins, R.; Vasconcelos, V. Are known cyanotoxins involved in the toxicity of picoplanktonic and filamentous north atlantic marine cyanobacteria? *Mar. Drugs* **2010**, *8*, 1908–1919. [CrossRef] [PubMed]
19. Gantar, M.; Sekar, R.; Richardson, L.L. Cyanotoxins from black band disease of corals and from other coral reef environments. *Microb. Ecol.* **2009**, *58*, 856–864. [CrossRef] [PubMed]
20. Martins, R.; Fernandez, N.; Beiras, R.; Vasconcelos, V. Toxicity assessment of crude and partially purified extracts of marine synechocystis and synechococcus cyanobacterial strains in marine invertebrates. *Toxicon* **2007**, *50*, 791–799. [CrossRef] [PubMed]
21. Woodhouse, J.N.; Kinsela, A.S.; Collins, R.N.; Bowling, L.C.; Honeyman, G.L.; Holliday, J.K.; Neilan, B.A. Microbial communities reflect temporal changes in cyanobacterial composition in a shallow ephemeral freshwater lake. *ISME J.* **2016**, *10*, 1337–1351. [CrossRef] [PubMed]
22. Xu, Z.; Woodhouse, J.N.; Te, S.H.; Yew-Hoong, G.K.; He, Y.; Xu, C.; Chen, L. Seasonal variation in the bacterial community composition of a large estuarine reservoir and response to cyanobacterial proliferation. *Chemosphere* **2018**, *202*, 576–585. [CrossRef] [PubMed]
23. Bao, L.J.; Maruya, K.A.; Snyder, S.A.; Zeng, E.Y. China's water pollution by persistent organic pollutants. *Environ. Pollut.* **2012**, *163*, 100–108. [CrossRef] [PubMed]
24. Zhang, L.; Liang, D.; Ren, L.; Shi, S.; Li, Z.; Zhang, T.; Huang, Y. Concentration and source identification of polycyclic aromatic hydrocarbons and phthalic acid esters in the surface water of the yangtze river delta, china. *J. Environ. Sci. (China)* **2012**, *24*, 335–342. [CrossRef]
25. Floehr, T.; Xiao, H.; Scholz-Starke, B.; Wu, L.; Hou, J.; Yin, D.; Zhang, X.; Ji, R.; Yuan, X.; Ottermanns, R. Solution by dilution?—A review on the pollution status of the yangtze river. *Environ. Sci. Pollut. Res.* **2013**, *20*, 6934–6971. [CrossRef] [PubMed]
26. Beck, D.A.C.; Kalyuzhnaya, M.G.; Malfatti, S.; Tringe, S.G.; Rio, T.G.D.; Ivanova, N.; Lidstrom, M.E.; Chistoserdova, L. A metagenomic insight into freshwater methane-utilizing communities and evidence for cooperation between the methylococcaceae and the methylophilaceae. *PeerJ* **2013**, *1*, e23. [CrossRef] [PubMed]
27. Garrity, G.M.; Bell, J.A.; Lilburn, T. *Methylophilaceae fam. Nov*; John Wiley & Sons, Ltd.: Hoboken, NJ, USA, 2015.
28. Hahn, M.W.; Kasalický, V.; Jezbera, J.; Brandt, U.; Jezberová, J.; Simek, K. *Limnohabitans curvus* gen. Nov., sp. Nov., a planktonic bacterium isolated from a freshwater lake. *Int. J. Syst. Evol. Microbiol.* **2010**, *60*, 1358–1365. [CrossRef] [PubMed]
29. Willems, A. *The Family Comamonadaceae*; Springer: Heidelberg/Berlin, Germany, 2014; pp. 777–851.
30. Pujalte, M.J.; Lucena, T.; Ruvira, M.A.; Arahal, D.R.; Macián, M.C. *The Family Rhodobacteraceae*; Springer: Heidelberg/Berlin, Germany, 2014; pp. 439–512.
31. Huang, Z.; Xie, B.; Yuan, Q.; Xu, W.; Lu, J. Microbial community study in newly established qingcaosha reservoir of shanghai, china. *Appl. Microbiol. Biotechnol.* **2014**, *98*, 9849–9858. [CrossRef] [PubMed]
32. Collier, J.L.; Lovindeer, R.; Xi, Y.; Radway, J.C.; Armstrong, R.A. Differences in growth and physiology of marine synechococcus (cyanobacteria) on nitrate versus ammonium are not determined solely by nitrogen source redox state1. *J. Phycol.* **2012**, *48*, 106–116. [CrossRef] [PubMed]
33. Li, J.; Zhang, J.; Liu, L.; Fan, Y.; Li, L.; Yang, Y.; Lu, Z.; Zhang, X. Annual periodicity in planktonic bacterial and archaeal community composition of eutrophic lake taihu. *Sci. Rep.* **2015**, *5*, 15488. [CrossRef] [PubMed]

34. Newton, R.J.; Jones, S.E.; Eiler, A.; Mcmahon, K.D.; Bertilsson, S. A guide to the natural history of freshwater lake bacteria. *Microbiol. Mol. Biol. Rev.* **2011**, *75*, 14–49. [CrossRef] [PubMed]
35. Needham, D.M.; Chow, C.E.T.; Cram, J.A.; Sachdeva, R.; Parada, A.; Fuhrman, J.A. Short-term observations of marine bacterial and viral communities: Patterns, connections and resilience. *ISME J.* **2013**, *7*, 1274–1285. [CrossRef] [PubMed]
36. Currie, D.J.; Kalff, J. The relative importance of bacterioplankton and phytoplankton in phosphorus uptake in freshwater. *Limnol. Oceanogr.* **1984**, *29*, 311–321. [CrossRef]
37. Vadstein, O. Growth and phosporus status of limnetic phytoplankton and bacteria. *Limnol. Oceanogr.* **1988**, *33*, 489–503. [CrossRef]
38. Smith, E.M.; Prairie, Y.T. Bacterial metabolism and growth efficiency in lakes: The importance of phosphorus availability. *Limnol. Oceanogr.* **2004**, *49*, 137–147. [CrossRef]
39. Fierer, N.; Leff, J.W.; Adams, B.J.; Nielsen, U.N.; Bates, S.T.; Lauber, C.L.; Owens, S.; Gilbert, J.A.; Wall, D.H.; Caporaso, J.G. Cross-biome metagenomic analyses of soil microbial communities and their functional attributes. *Proc. Natl. Acad. Sci. USA* **2012**, *109*, 21390–21395. [CrossRef] [PubMed]
40. Weiss, S.; Treuren, W.V.; Lozupone, C.; Faust, K.; Friedman, J.; Ye, D.; Li, C.X.; Xu, Z.Z.; Ursell, L.; Alm, E.J. Correlation detection strategies in microbial data sets vary widely in sensitivity and precision. *ISME J.* **2016**, *10*, 1669–1681. [CrossRef] [PubMed]
41. Reshef, D.N.; Reshef, Y.A.; Finucane, H.K.; Grossman, S.R.; Mcvean, G.; Turnbaugh, P.J.; Lander, E.S.; Mitzenmacher, M.; Sabeti, P.C. Detecting novel associations in large data sets. *Science* **2011**, *334*, 1518–1524. [CrossRef] [PubMed]
42. Ou, H.S.; Wei, C.H.; Deng, Y.; Gao, N.Y. Principal component analysis to assess the composition and fate of impurities in a large river-embedded reservoir: Qingcaosha reservoir. *Environ. Sci. Process. Impacts* **2013**, *15*, 1613–1621. [CrossRef] [PubMed]
43. Gera, A.; Alcoverro, T.; Mascarã, O.; PãRez, M.; Romero, J. Exploring the utility of posidonia oceanica chlorophyll fluorescence as an indicator of water quality within the european water framework directive. *Environ. Monit. Assess.* **2012**, *184*, 3675–3686. [CrossRef] [PubMed]
44. Caporaso, J.G.; Lauber, C.L.; Walters, W.A.; Berglyons, D.; Huntley, J.; Fierer, N.; Owens, S.M.; Betley, J.; Fraser, L.; Bauer, M. Ultra-high-throughput microbial community analysis on the illumina hiseq and miseq platforms. *ISME J. Multidiscip. J. Microb. Ecol.* **2012**, *6*, 1621–1624. [CrossRef] [PubMed]
45. Ma, J.; Wang, Z.; Li, H.; Park, H.D.; Wu, Z. Metagenomes reveal microbial structures, functional potentials, and biofouling-related genes in a membrane bioreactor. *Appl. Microbiol. Biotechnol.* **2016**, *100*, 1–13. [CrossRef] [PubMed]
46. Thomas, T.; Gilbert, J.; Meyer, F. Metagenomics—A guide from sampling to data analysis. *Microbial. Inf. Exp.* **2012**, *2*, 3. [CrossRef] [PubMed]
47. Schloss, P.D.; Westcott, S.L.; Ryabin, T.; Hall, J.R.; Hartmann, M.; Hollister, E.B.; Lesniewski, R.A.; Oakley, B.B.; Parks, D.H.; Robinson, C.J. Introducing mothur: Open-source, platform-independent, community-supported software for describing and comparing microbial communities. *Appl. Environ. Microbiol.* **2009**, *75*, 7537–7541. [CrossRef] [PubMed]
48. Edgar, R.C.; Haas, B.J.; Clemente, J.C.; Quince, C.; Knight, R. Uchime improves sensitivity and speed of chimera detection. *Bioinformatics* **2011**, *27*, 2194–2200. [CrossRef] [PubMed]
49. Edgar, R.C. Search and clustering orders of magnitude faster than blast. *Bioinformatics* **2010**, *26*, 2460–2461. [CrossRef] [PubMed]
50. Desantis, T.Z.; Hugenholtz, P.; Larsen, N.; Rojas, M.; Brodie, E.L.; Keller, K.; Huber, T.; Dalevi, D.; Hu, P.; Andersen, G.L. Greengenes, a chimera-checked 16s rrna gene database and workbench compatible with arb. *Appl. Environ. Microbiol.* **2006**, *72*, 5069–5072. [CrossRef] [PubMed]
51. Cole, J.R.; Wang, Q.; Cardenas, E.; Fish, J.; Chai, B.; Farris, R.J.; Kulam-Syed-Mohideen, A.S.; Mcgarrell, D.M.; Marsh, T.; Garrity, G.M. The ribosomal database project: Improved alignments and new tools for rrna analysis. *Nucleic Acids Res.* **2009**, *37*, D141–D145. [CrossRef] [PubMed]
52. Quast, C.; Pruesse, E.; Yilmaz, P.; Gerken, J.; Schweer, T.; Yarza, P.; Peplies, J.; Glöckner, F.O. The silva ribosomal rna gene database project: Improved data processing and web-based tools. *Nucleic Acids Res.* **2013**, *41*, 590–596. [CrossRef] [PubMed]
53. Benjamini, Y.; Hochberg, Y. Controlling the false discovery rate—A practical and powerful approach to multiple testing. *J. R. Stat. Soc.* **1995**, *57*, 289–300.

54. Assenov, Y.; Schelhorn, S.E.; Lengauer, T.; Albrecht, M. Computing topological parameters of biological networks. *Bioinformatics* **2008**, *24*, 282–284. [CrossRef] [PubMed]
55. Albertsen, M.; Hansen, L.B.S.; Saunders, A.M.; Nielsen, P.H.; Nielsen, K.L. A metagenome of a full-scale microbial community carrying out enhanced biological phosphorus removal. *ISME J.* **2012**, *6*, 1094–1106. [CrossRef] [PubMed]

toxins

MDPI

Article

Differential Toxicity of Cyanobacteria Isolated from Marine Sponges towards Echinoderms and Crustaceans

Ana Regueiras [1,2], Sandra Pereira [1], Maria Sofia Costa [1,3] and Vitor Vasconcelos [1,2,*]

1 CIIMAR/CIMAR, Blue Biotechnology and Ecotoxicology—Centre of Environmental and Marine Research, University of Porto, Terminal de Cruzeiros do Porto de Leixões, Avenida General Norton de Matos, S/N, Matosinhos 4450-208, Portugal; anaregueiras@gmail.com (A.R.); sandra.c.pereira28@gmail.com (S.P.); marysofs@gmail.com (M.S.C.)

2 Department of Biology, Sciences Faculty, University of Porto, Rua do Campo Alegre, Porto 4169-007, Portugal

3 Faculty of Pharmaceutical Sciences, University of Iceland, Hagi, Hofsvallagata 53, Reykjavik 107, Iceland

* Correspondence: vmvascon@fc.up.pt; Tel.: +351-220-402-738

Received: 28 May 2018; Accepted: 16 July 2018; Published: 18 July 2018

Abstract: Marine sponges and cyanobacteria have a long history of co-evolution, with documented genome adaptations in cyanobionts. Both organisms are known to produce a wide variety of natural compounds, with only scarce information about novel natural compounds produced by cyanobionts. In the present study, we aimed to address their toxicological potential, isolating cyanobacteria ($n = 12$) from different sponge species from the coast of Portugal (mainland, Azores, and Madeira Islands). After large-scale growth, we obtained both organic and aqueous extracts to perform a series of ecologically-relevant bioassays. In the acute toxicity assay, using nauplii of *Artemia salina*, only organic extracts showed lethality, especially in picocyanobacterial strains. In the bioassay with *Paracentrotus lividus*, both organic and aqueous extracts produced embryogenic toxicity (respectively 58% and 36%), pointing to the presence of compounds that interfere with growth factors on cells. No development of pluteus larvae was observed for the organic extract of the strain *Chroococcales* 6MA13ti, indicating the presence of compounds that affect skeleton formation. In the hemolytic assay, none of the extracts induced red blood cells lysis. Organic extracts, especially from picoplanktonic strains, proved to be the most promising for future bioassay-guided fractionation and compounds isolation. This approach allows us to classify the compounds extracted from the cyanobacteria into effect categories and bioactivity profiles.

Keywords: marine cyanobacteria; cyanotoxins; marine sponges; secondary metabolites; marine natural compounds; bioassays; *Artemia salina*; *Paracentrotus lividus*; hemolytic essay

Key Contribution: Marine sponges were used as a source for harvesting cyanobacteria. Being adapt to life inside sponges; these cyanobacteria can prove to have novel compounds produced from their secondary metabolism.

1. Introduction

Cyanobacteria are photosynthetic prokaryotes, with a high morphological, physiological, and metabolic diversity, with fossil records dating back to 3.5 billion years ago [1]. Secondary metabolite production was essential for their survival allowing for adaptation to several environmental conditions such as variations in temperature, pH, salinity, UV radiation, etc.

Climate change and eutrophication increased the occurrence and frequency of cyanobacterial blooms in water bodies [2], posing human and animals' health risks due to toxin production.

Apart from toxin production, these secondary metabolites have also been shown to be a source of compounds of interest in different industries, such as pharmaceutical, cosmetics, agriculture, energy, etc. In the last decade alone, estimations point to more than 400 new natural compounds extracted from marine cyanobacteria [3]. Coastal water blooms pose another health risk concerning cyanobacterial toxins, as many of them are able to accumulate in both vertebrates and invertebrates [4].

Assessing marine cyanobacterial diversity on the Portuguese coast has already been the focus of various studies (e.g., [5,6]), with *Cyanobium*, *Leptolyngbya* and *Pseudanabaena* as the most abundant genera among isolates [6]. Isolated strains from the coast of Portugal were found to be a source of bioactive compounds, both with toxicological and/or pharmaceutical interest [2,7–13]. Also, Brito et al. [14] evaluated the potential to produce secondary metabolites for some strains through molecular methods.

In marine environments, cyanobacteria are known to form associations with a variety of invertebrates, such as sponges (Phylum Porifera). Sponges are filter-feeders, capable of filtering thousands of liters of water per day. During this process, some filtered microorganisms can become part of the sponge microbiota. Sponge microbiota diversity can reach up to 4 orders of magnitude, when compared to the one from water column [15]. In temperate ecosystems, it is estimated that 45–60% of sponges have cyanobacterial symbionts (cyanobionts) [16], and are able to cover up to 50% of the sponge cell volume [17]. As they are able to concentrate microorganisms, sponges can be used as a source for cyanobacteria harvesting as already stated by Regueiras, et al. [18]. Sponges are a huge source of bioactive compounds [19], most of them known to be produced by their symbiotic microorganisms [15]. Actinobacteria, Cyanobacteria, Firmicutes, and Proteobacteria (alpha and gamma classes) are the main phyla producing secondary metabolites in sponges [20].

Both coccoid and filamentous cyanobacteria have been described in sponges. Recently, Konstantinou, et al. [21] made a review on the diversity of both sponge species harboring cyanobacteria, and cyanobacterial diversity. In Portugal, *Xenococcus*-like and *Acaryochloris* sp. were reported from the intertidal marine sponge *Hymeniacidon perlevis* [22,23]. Regueiras, et al. [18] were also able to identify cyanobacteria belonging to the genera *Synechococcus*, *Cyanobium*, *Synechocystis*, *Nodosilinea*, *Pseudanabaena*, *Phormidesmis*, *Acaryochloris*, and *Prochlorococcus* associated with the same marine sponge.

Due to a long evolutionary history of both cyanobacteria and marine sponges, co-evolution has already been documented, with some cyanobacteria being passed to new sponge generations through vertical transmission (from sponge to offspring through reproductive cells) [24]. The study of genomes from the symbiotic cyanobacteria "Ca. *Synechococcus spongiarum*" and its comparison with the genome of free-living ones, found adaptations to life inside sponges and the presence of different adaptations in different phylotypes [25,26]. These adaptations may also lead to the production of novel and unique natural compounds.

Bioassay-guided fractionation is a successful strategy in the isolation and discovery of novel compounds [27–31]. To address toxin production, several assays can be used. The use of the brine shrimp *Artemia salina*, has ecological relevance in marine ecosystems, as these organisms are a representation of the zooplankton community and vital on the ecology of seashores [11]. For preliminary toxicity assessment, the brine shrimp lethality assay is a standardized bioassay in marine and aquatic research [32]. For embryogenesis studies, the use of echinoids, such as the sea urchin *Paracentrotus lividus*, is very common. They occupy an important phylogenetic position (deuterostomes) when compared to other invertebrates {Lopes, 2010 #555227}. *P. lividus* are also common among the Portuguese seashore and key elements on their habitats [11], capable of producing a great amount of eggs feasible to be fertilized in seawater, and to develop optically clear embryos [33]. Apart from these common assays, less is known on hemolytic toxins from cyanobacteria. Cyanobacterial toxins are able to accumulate in marine vertebrate and invertebrates [34,35], posing risks for mammals, showing the importance of the use of such assays.

The present study aims to do a preliminary assessment on the cyanotoxin potential of marine cyanobacteria isolated from marine sponges. Most studies isolate marine cyanobacteria through filtration of large volumes of water, or by scratching coastal surfaces. In the present study, we aimed to isolate cyanobacteria from marine sponges off the coast of Portugal, as they are able to concentrate microorganisms, allowing them to obtain some cyanobacteria that can be present in seawater in amounts under detection. We intend to evaluate the toxic effects of organic (lipophilic) and aqueous (hydrophilic) crude extracts towards the nauplii of the brine shrimp *A. salina* and embryos of the sea urchin *P. lividus*, and their hemolytic activity. These assays will be useful to evaluate cyanobacterial potential to produce compounds with relevant bioactivity profiles to be further investigated and possibly identified in the future. This approach allow us to classify the compounds extracted from the cyanobacteria into effect categories and bioactivity profiles.

2. Results

2.1. Acute Toxicity Assay Using Nauplii of Artemia Salina

Aqueous extracts, containing the hydrophilic compounds from the cyanobacterial strains, did not exhibit statistically significant differences against control, in the bioassay to assess mortality in *Artemia salina* nauplii (Figure 1). However, for the organic extracts, toxicity was found after 48 h of exposure. Cyanobacterial strains *Synechoccocus* sp. LEGE11381 ($F = 68.80$, $p < 0.000$), *Synechocystis* sp. 44B13pa ($F = 21.82$, $p < 0.048$), unidentified filamentous *Synechococcales* LEGE11384 ($F = 24.74$, $p < 0.018$), *Chroococales* 6MA13ti ($F = 86.73$, $p < 0.000$), and *Cyanobium* sp. LEGE10375 ($F = 43.50$, $p < 0.000$) presented statistically significant differences when compared against the negative control.

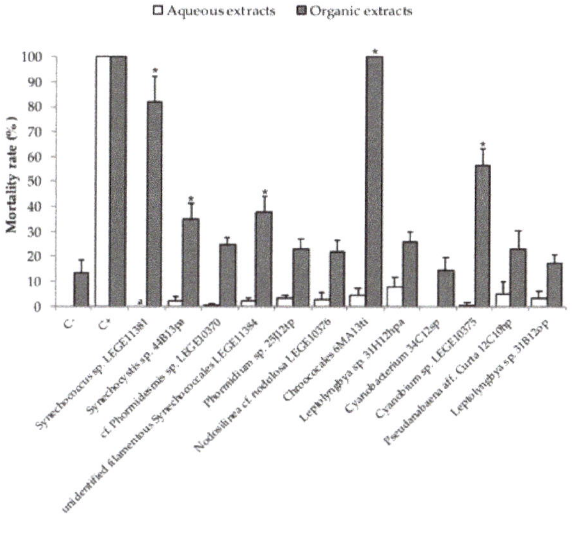

Figure 1. Mortality rate (%) for the *Artemia salina* bioassay, after 48 h of exposure, for the organic and aqueous extracts. Controls used included filtered seawater with 0.1% DMSO for negative control and potassium dichromate (8 μg/mL) for positive control. a Assay not performed; * Statistically significant differences between extract and control.

2.2. Embryo—Larval Acute Toxicity Assay with Paracentrotus Lividus

The toxicity of the cyanobacterial extracts in the bioassay with *P. lividus* was determined by analysis of the embryogenic success, i.e., the ability of the fertilized egg to reach the stage of pluteus larvae, and through growth of pluteus larvae (Figure 2). Development arrest indicates that no normal pluteus larvae were produced. The results gathered after 48 h of incubation with cyanobacterial extracts revealed that in the control, 67.5 ± 6.1% of the sea urchin fertilized eggs developed to normal pluteus larvae, with an average length of 330.0 ± 18.8 μm. Figure 3 shows significant difference in the embryogenic development, at $p < 0.05$, for the organic extract of the following strains: *Synechococcus* sp. LEGE11381 ($F = -62.78$, $p < 0.000$), *Synechocystis* sp. 44B13pa ($F = -41.80$, $p < 0.000$), unidentified filamentous *Synechococcales* LEGE11384 ($F = -36.05$, $p < 0.000$), *Phormidium* sp. 25J12tp ($F = -27.22$, $p < 0.010$), *Leptolyngbya* sp. 31H12hpa ($F = 67.48$, $p < 0.048$), and *Cyanobium* sp. LEGE10375 ($F = -52.38$, $p < 0.000$). The organic extract of the strain *Chroococcales* 6MA13ti caused development arrest with none of the larvae reaching the stage of viable pluteus. Amongst the aqueous extracts, unidentified filamentous *Synechococcales* LEGE11384 ($F = -41.75$, $p < 0.001$), *Phormidium* sp. 25J12tp ($F = -28.75$, $p < 0.033$), *Chroococcales* 6MA13pi ($F = -30.00$, $p < 0.024$), and *Cyanobacterium* 34C12sp ($F = -39.25$, $p < 0.002$) strains presented significant embryogenic effect. Regarding the results from the positive control, only embryos on gastrula stage were found.

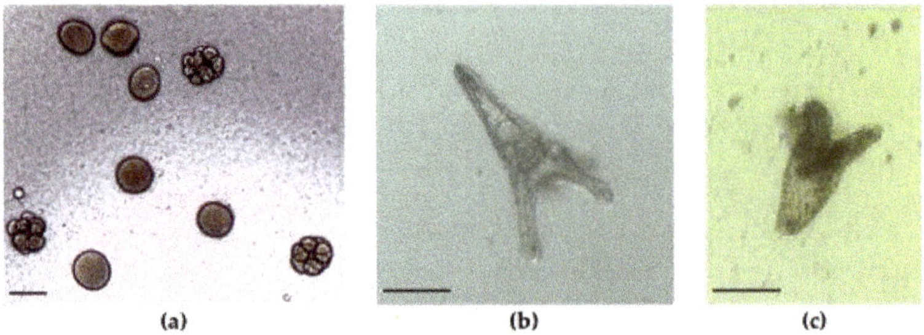

Figure 2. Effects of marine cyanobacterial extracts on embryogenesis of the sea urchin *Paracentrotus lividus*. (**a**) Fertilized sea urchin eggs; (**b**) Normal pluteus larvae resulting from control treatment and (**c**) Abnormally developed larvae resulting from treatments with cyanobacterial extracts. Scale bar: 100 μm.

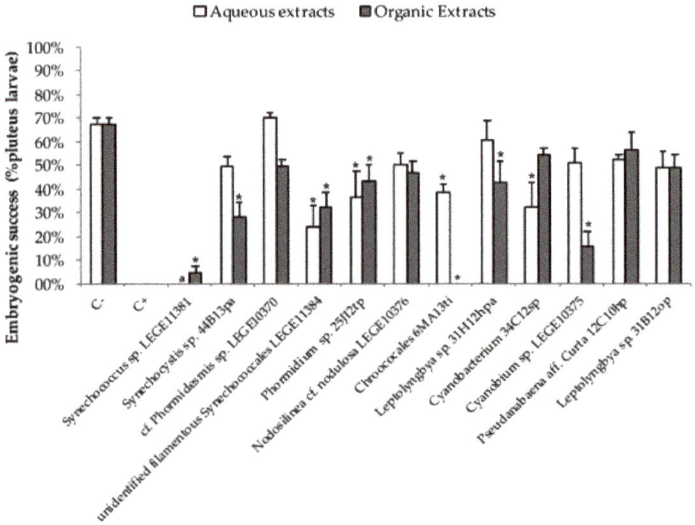

Figure 3. Percentage of pluteus larvae developed (embryogenic success) after exposure to aqueous and organic extracts of the cyanobacterial strains. For the controls, filtered seawater was used with 0.1% DMSO (negative) and potassium dichromate at 4 μg/mL (positive). a Assay not performed; * Statistically significant differences between extract and control.

Regarding larval growth data, no significant changes in larval length was observed in the aqueous extracts at $p < 0.05$ [F (11, 36) = 1.039, $p < 0.434$)] (Figure 4). However, differences in larval length were found in organic extracts. These differences were more significant in *Synechococcus* sp. LEGE11381 (246.2 ± 11.5 µm, $p < 0.001$) and *Cyanobium* sp. LEGE10375 (325.7 ± 9.7 µm, $p < 0.000$).

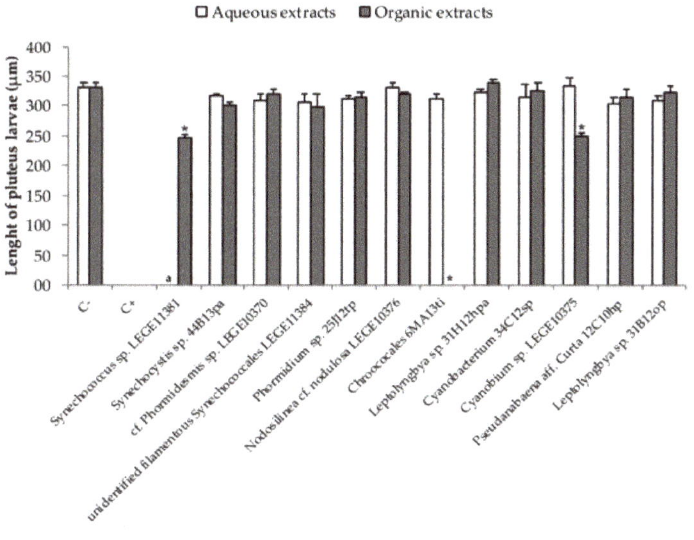

Figure 4. Larval growth from the organic extracts of the cyanobacterial strains. For the controls, filtered seawater was used with 0.1% DMSO (negative) and potassium dichromate at 4 µg/mL (positive). a Assay not performed; * Statistically significant differences between extract and control.

2.3. Hemolytic Assay

The hemolytic activity registered during the assay was below 10%, with the highest value obtained being 7% of activity by the strain *Chroococcales* 6MA13ti, in the organic extract. All strains and extracts did not present significant interference with the hemoglobin content.

3. Discussion

To date, most studies exploring the bioactivity of marine cyanobacteria have been focusing on free-living forms. Cyanobacteria can live in association with a variety of marine invertebrates, such as sponges, for example, and it is known that cyanobacteria can affect the biosynthesis of compounds from the host [36] and that symbionts have specific adaptations in their genome [25,26]. The biological potential of associated and/or symbiotic cyanobacteria is still mostly unexplored. In the present study, twelve marine cyanobacterial strains were isolated from sponges of the Portuguese coast. Aqueous and organic crude extracts of the isolated cyanobacterial strains were submitted to ecologically-relevant bioassays in order to do a preliminary assessment on the production of secondary metabolites with relevant bioactivity profiles.

Artemia spp. is known for its ability to adapt to different environmental conditions, making it a crucial test organism in ecotoxicology [37]. Results from the bioassay with the brine shrimp *Artemia salina* nauplii did not demonstrate acute toxicity with exposure to the aqueous extracts of the tested cyanobacterial strains. The organic extracts of *Synechococcus* sp. LEGE11381, *Synechocystis* sp. 44B13pa, unidentified filamentous *Synechococcales* LEGE11384, *Chroococcales* 6MA13ti, and *Cyanobium* sp. LEGE10375 cyanobacterial strains proved to be the most toxic to this crustacean species. In contrast with our results, most previous studies with cyanobacteria from the coast of Portugal found aqueous extracts to be more toxic. For example, Leão, et al. [10] reported lethality towards *A. salina*, in aqueous extracts in free-living forms from *Nodosilinea*, *Leptolyngbya*, and *Pseudanabaena* genera strains. Also, Frazão, et al. [2] found aqueous extracts of the genera *Cyanobium*, *Synechococcus*, *Leptolyngbya*, *Oscillatoria*, and *Phormidium* more toxic than organic ones. In brackish waters Lopes, et al. [33] also found aqueous extracts more toxic, and organic extracts did not induce more than 7% of mortality on *A. salina*. Organic extracts in our work showed a higher toxicity towards *A. salina*, leading to an assumption that cyanobacteria associated with marine sponges may produce different metabolites from the ones present in free-living forms of cyanobacteria, and therefore, their toxicological and pharmaceutical potential should be further investigated. The higher values of mortality here observed were all in picocyanobacterial strains. Costa, et al. [9] already reported the potential of these cyanobacteria as a source for novel metabolites. In the present work, toxicity was only found after 48 h.

In the bioassay with sea urchin *Paracentrotus lividus*, embryogenic toxicity occurred in 58% of the organic extracts and in 36% of the aqueous extracts tested. The unidentified filamentous *Synechococcales* LEGE11384, *Phormidium* sp. 25J12tp, *Chroococcales* 6MA13ti cyanobacterial strains demonstrated embryogenic toxicity in both extracts, which may lead us to infer that, for the same cyanobacterial strain, chemically different bioactive compounds are produced, having the same effect on the embryogenic activity of the sea urchin. Although the *Synechocystis* sp. 44B13pa, unidentified filamentous *Synechococcales* LEGE11384, *Phormidium* sp. 25J12tp, *Leptolyngbya* sp. 31H12hpa, *Chroococcales* 6MA13pi and *Cyanobacterium* 34C12sp cyanobacterial strains have demonstrated to be embryotoxic, no alteration on larval length was observed. This may suggest that the toxicity showed by these cyanobacterial strains only affected the early life stages of the sea urchin embryos development, providing strong evidence for the presence of compounds that interfere with growth factors on cells [11]. The organic extracts of *Synechococcus* sp. LEGE11381 and *Cyanobium* sp. LEGE10375 exhibited interference with the embryogenic development and also with the larval growth. From all the extracts tested, the organic extract from *Chroococcales* 6MA13ti seemed to have the most potent effect on *P. lividus* larvae, since it did not allow a normal development of any pluteus larvae. *Cyanobium* sp. organic extracts have already been shown to decrease *P. lividus* larvae length [9]. Lopes, et al. [33] found organic extracts from brackish waters to be more toxic to *P. lividus*, which is in accordance to our results. The inhibition of larval morphogenesis, here observed, point to the presence of compounds that affect skeleton formation.

Although hemolytic activity has already been documented in strains of *Synechocystis* [38], *Anabaena* [39], *Synechococcus* and *Leptolyngbya* [40], our results showed that in neither organic nor aqueous extracts analyzed, the lysis of the red mammalian blood cells was induced.

The present study aimed to assess a preliminary cyanotoxicological potential from twelve marine cyanobacteria isolated from the sponges of the Portuguese coast. Eight extracts from cyanobacterial strains have shown a promising potential on the performed ecologically-relevant bioassays (*Synechococcus* sp. LEGE11381, *Synechocystis* sp. 44B13pa; Unidentified filamentous *Synechococcales* LEGE11384; *Phormidium* sp. 25J12tp; *Chroococcales* 6MA13ti; *Leptolyngbya* sp. 31H12hpa; Cyanobacterium 34C12sp; *Cyanobium* sp. LEGE10375). Furthermore, the concentrations of the extracts here used (30 μg mL^{-1}) are an ecologically relevant concentration. This emphasizes the premise that sponges can harbor microorganisms with toxicological interest and that these invertebrates can and should be used in order to isolate new cyanobacteria. The extracts with the most promising bioactivity should be further fractionated to identify with more detail the bioactive compounds. Chemical elucidation should be performed once the purest compounds are achieved.

4. Materials and Methods

4.1. Cyanobacterial Strains Selection and Biomass Production

Cyanobacterial strains used in this study were previously isolated from marine sponges. Marine sponges were collected both from seashore rocks and by scuba diving. A small fraction of sponge tissue was collected in flaks with ambient seawater. Figure 5 shows sampling locations, being all intertidal sites, with exception from the one in Madeira Island, Caniçal (sponges collected through scuba diving). When collected from intertidal areas, beaches were chosen with a combination of sand and rocks. Sponges substratum were rocks or sand. Preparation of sponge samples and cyanobacterial isolation and characterization was done according to Regueiras, et al. [18]. Summarizing, sponges were cleaned of debris and 1 mm of the sponge surface was discarded, using a sterile razor to avoid cultivation of superficial bacteria. Small fragments of the sponge body (<0.5 cm^3) were placed in 2 different culture media, Z8 liquid media [41], supplemented with 30 g L^{-1} of NaCl and MN liquid medium [42]. Both culture media were supplemented with vitamin B12 and cyclohexamide [42]. After growth, through micromanipulation techniques, as described by Rippka [42], a single cell or filament of cyanobacteria were transfer to new liquid medium, until achievement of unicyanobacterial, non-axenic cultures.

Figure 5. Sampling locations. Two sampling locations were in Portugal mainland: Memória (N 41°13′52.27″, W 8°43′18.34″) and Porto Côvo (N 37°52′3.04″, W 8°47′37.19″). One was in Madeira Island: Caniçal (N 32°44′20.08″, W 16°44′17.55″) and the other in São Miguel Island, Azores: São Roque (N 37°45′15,35″, W 25°38′31.60″).

The selection of cyanobacterial strains was based on growth performance rates and cyanobacterial diversity. Morphological identification followed the criteria of Komárek and Anagnostinis [43–45], the Bergey's manual of systematic bacteriology [46] and Komárek, et al. [47]. Strains are deposited in the LEGE Culture Collection (Ramos et al., 2018). The twelve strains selected (Table 1) were cultured and up-scaled under laboratory conditions at 25 °C, light/dark cycle of 14/10 h and light intensity of approximately 25 × 10^{-6} E/$m^{-2}s^{-1}$. After 60 to 90 days of growth, the cyanobacterial biomass produced was collected (through centrifugation or filtration with a 20 µm pore net), frozen at −20 °C and freeze dried. Lyophilized material was kept at −20 °C.

Table 1. Cyanobacterial strains selected for the present study, with information about the marine sponge it was isolated from and collection site.

Cyanobacterial Strain	Sponge Species	Collection Site
Synechococcus sp. LEGE11381	*Polymastia* sp.	Memória
Synechocystis sp. 44B13pa	*Polymastia agglutinans*	São Roque, Azores
cf. *Phormidesmis* sp. LEGE10370	*Hymeniacidon perlevis*	Memória
Unidentified filamentous *Synechococcales* LEGE11384	*Phorbas plumosus*	Memória
Phormidium sp. 25J12tp	*Tedania pilarriosae*	Memória
Nodosilinea cf. *nodulosa* LEGE10376	*Hymeniacidon perlevis*	Porto Côvo
Chroococcales 6MA13ti	*Tedania ignis*	São Roque, Azores
Leptolyngbya sp. 31H12hpa	*Halichondria panicea*	Memória
Cyanobacterium 34C12sp	Unidentified sponge	Caniçal, Madeira
Cyanobium sp. LEGE10375	*Hymeniacidon perlevis*	Memória
Pseudanabaena aff. *curta* 12C10hp	*Hymeniacidon perlevis*	Memória
Leptolyngbya sp. 31B12op	*Ophlitaspongia papila*	Memória

4.2. Preparation of Cyanobacterial Extracts

The freeze dried biomass from each cyanobacterial strain was repeatedly extracted with a warm (<40 °C) mixture of dichloromethane and methanol (CH_2Cl_2:MeOH) (2:1) (P.A. Sigma, St Louis, MO, USA). Afterwards, the solvents were removed in vacuo and/or under a N_2 stream. Following the organic extraction, the remaining biomass was subjected to aqueous extraction (ultra-pure water), decanted, and centrifuged at 4600 rpm for 15 min. The resulting supernatant was freeze-dried, weighed, and stored at −20 °C. Just before the tests, organic extracts were dissolved (30 mg mL^{-1}) in dimethyl-sulfoxide (DMSO) and aqueous extracts in ultra-pure water.

4.3. Bioassays

4.3.1. Acute Toxiciyy Assay Using Nauplii of *Artemia Salina*

In the acute toxicity assay, the nauplii of the crustacean *Artemia salina* were used. The dried cysts (JBL Novotemia, Germany) hatched after 48 h in 35 g/L filtered seawater, at 25 °C, under conditions of continuous illumination and aeration. Toxicity was screened in a 96-well polystyrene plate, with 10–15 nauplii per well and 200 µL of organic or aqueous extract. Filtered seawater with 0.1% DMSO was used as negative control, and potassium dichromate at a concentration of 8 µg/mL as positive control. Four replicates were made for each treatment. The plates were covered with Parafilm to prevent water loss and then incubated at 25 °C, for 48 h in darkness. Dead larvae were counted in each well on an inverted microscope at 24 h and 48 h. Before determining the total number of larvae, organisms were fixed with a few drops of Lugol's solution. Mortality was calculated through percentage as described by Martins, et al. [11].

4.3.2. Embryo-Larval Acute Toxicity Assay with *Paracentrotus Lividus*

For the embryo-larval acute toxicity assay, sea urchins *Paracentrotus lividus* were captured in the intertidal rocky shore, during low tide in Praia da Memória, Matosinhos, Portugal and immediately transported to the laboratory, in natural sea water and under refrigeration. The protocol employed was the one described by Fernández and Beiras [48]. Briefly, a couple of specimens were dissected, and gametes were collected with a pipette directly from the gonads. The optimal condition from gametes (spherical eggs and mobile sperm) was granted through careful observation under the optical microscope. Eggs were transferred to a 100 mL measuring cylinder containing natural seawater filtered through a 0.45 µm pore filter. A few microliters of sperm were added to the eggs suspension and then carefully stirred to allow fertilization. Fertilized eggs were counted in four 10 µL aliquots in order to determine the fertilization success and egg density. In a 24-well plate, a concentration of 20 fertilized eggs per mL of solution were exposed to organic and aqueous extracts, during 48 h at 20 °C, in darkness. Test solutions consisted of 2.5 mL of each cyanobacterial extract; two negative controls were used, one with only filtered seawater and the other with 0.1% DMSO; as positive control was used potassium dichromate in a concentration of 4 μg/mL. Four replicates were made for each treatment. After 48 h of incubation, the solutions were fixed with 40% formalin. Results were evaluated through percentage of pluteus larvae (embryogenic success) and larval length (larval growth) [11].

4.3.3. Hemolytic Assay

For the hemolytic assay, mice blood, stabilized with heparin, was provided by IBMC Bioterium, from healthy specimens without need to sacrifice the animals. The protocol used was an adaptation of the ones described by Rangel, et al. [49] and Slowing, et al. [50]. Summarizing, the erythrocytes solution was diluted with 30 volumes of a saline solution (0.85% NaCl with 10 mM $CaCl_2$) and centrifuged at 1100 *g* for 5 min, discarding the supernatant and then washed three times with the same solution followed by centrifugations (1100 *g* for 5 min). After the final wash, the cells were diluted to a final concentration of 1% in sterile PBS solution. The assay was performed with 100 μL of each extract mixed with equal volume of erythrocytes suspension, using three replicates per treatment. For the negative and positive controls were used PBS and 0.1% Triton100, respectively. Eppendorfs with the mixtures were incubated for 2 h, at a temperature of 37 °C, with slow agitation. After that period, the mixtures were centrifuged at 4000 *g* for 1 min at 4 °C. The supernatants were transferred to a 96 well plate. Hemoglobin content was evaluated spectrophotometrically at 540 nm [49].

$$Hemolytic\ activity = \frac{Abs_{sample} - Abs_{negative\ control}}{Abs_{positive\ control} - Abs_{negative\ control}} \times 100\% \quad (1)$$

4.3.4. Analysis

Data collected during the bioassays were analyzed using a one-way analysis of variance (ANOVA), followed by a multi-comparisons Dunnett test ($p < 0.05$). The software IBM SPSS Statistics 24 (Version 24.0.0.0 edition 64-bit, IBM Corporation, New York, NY, USA, 2016) was used for statistical analysis.

Author Contributions: A.R. and V.V. conceived the conceptualization. A.R., S.P. and M.S.C. did the experimental work. Analysis of the data was done by A.R. and S.P. as well as the writing of the original draft. The review and editing of the writing was done by A.R. and V.V.

Funding: This work was financed by UID/Multi/04423/2013 and by the Structured Program of R&D&I INNOVMAR—Innovation and Sustainability in the Management and Exploitation of Marine Resources (reference NORTE-01-0145-FEDER-000035, Research Line NOVELMAR), funded by the Northern Regional Operational Program (NORTE2020) through the European Regional Development Fund (ERDF).and by the grants PTDC/MAR/099642/2008, PhD grants SFRH/BD/73033/2010 and the Fellowship grant BI/PTDC/MAR/099642/2008/2011-030.

Acknowledgments: The authors acknowledge Marisa Silva, Manfred Kaufmann, Manuela Maranhão, Ana Neto, Isadora Moniz and Afonso Prestes for their help during sampling in costal continental Portugal, Madeira and Azores islands, and Vitor Ramos for helping in cyanobacteria identification.

Conflicts of Interest: The authors declare no conflict of interest.

References

1. Adams, D.G.; Duggan, P.S. Tansley review no. 107. Heterocyst and akinete differentiation in cyanobacteria. *New Phytol.* **1999**, *144*, 3–33. [CrossRef]
2. Frazão, B.; Martins, R.; Vasconcelos, V. Are known cyanotoxins involved in the toxicity of picoplanktonic and filamentous north atlantic marine cyanobacteria? *Mar. Drugs* **2010**, *8*, 1908–1919. [CrossRef] [PubMed]
3. Mi, Y.; Zhang, J.; He, S.; Yan, X. New peptides isolated from marine cyanobacteria, an overview over the past decade. *Mar. Drugs* **2017**, *15*, 132. [CrossRef] [PubMed]
4. Buratti, F.M.; Manganelli, M.; Vichi, S.; Stefanelli, M.; Scardala, S.; Testai, E.; Funari, E. Cyanotoxins: Producing organisms, occurrence, toxicity, mechanism of action and human health toxicological risk evaluation. *Arch. Toxicol.* **2017**, *91*, 1049–1130. [CrossRef] [PubMed]

5. Brito, Â.; Ramos, V.; Mota, R.; Lima, S.; Santos, A.; Vieira, J.; Vieira, C.P.; Kaštovský, J.; Vasconcelos, V.M.; Tamagnini, P. Description of new genera and species of marine cyanobacteria from the portuguese atlantic coast. *Mol. Phylogenet. Evol.* **2017**, *111*, 18–34. [CrossRef] [PubMed]
6. Brito, Â.; Ramos, V.; Seabra, R.; Santos, A.; Santos, C.L.; Lopo, M.; Ferreira, S.; Martins, A.; Mota, R.; Frazao, B.; et al. Culture-dependent characterization of cyanobacterial diversity in the intertidal zones of the portuguese coast: A polyphasic study. *Syst. Appl. Microbiol.* **2012**, *35*, 110–119. [CrossRef] [PubMed]
7. Afonso, T.B.; Costa, M.S.; Rezende de Castro, R.; Freitas, S.; Silva, A.; Schneider, M.P.C.; Martins, R.; Leão, P.N. Bartolosides e–k from a marine coccoid cyanobacterium. *J. Nat. Prod.* **2016**, *79*, 2504–2513. [CrossRef] [PubMed]
8. Costa, M.; Garcia, M.; Costa-Rodrigues, J.; Costa, M.S.; Ribeiro, M.J.; Fernandes, M.H.; Barros, P.; Barreiro, A.; Vasconcelos, V.; Martins, R. Exploring bioactive properties of marine cyanobacteria isolated from the portuguese coast: High potential as a source of anticancer compounds. *Mar. Drugs* **2014**, *12*, 98–114. [CrossRef] [PubMed]
9. Costa, M.S.; Costa, M.; Ramos, V.; Leao, P.N.; Barreiro, A.; Vasconcelos, V.; Martins, R. Picocyanobacteria from a clade of marine *Cyanobium* revealed bioactive potential against microalgae, bacteria, and marine invertebrates. *J. Toxicol. Environ. Health Part A* **2015**, *78*, 432–442. [CrossRef] [PubMed]
10. Leão, P.N.; Ramos, V.; Gonçalves, P.B.; Viana, F.; Lage, O.M.; Gerwick, W.H.; Vasconcelos, V.M. Chemoecological screening reveals high bioactivity in diverse culturable portuguese marine cyanobacteria. *Mar. Drugs* **2013**, *11*, 1316–1335. [CrossRef] [PubMed]
11. Martins, R.; Fernandez, N.; Beiras, R.; Vasconcelos, V. Toxicity assessment of crude and partially purified extracts of marine *Synechocystis* and *Synechococcus* cyanobacterial strains in marine invertebrates. *Toxicon* **2007**, *50*, 791–799. [CrossRef] [PubMed]
12. Martins, R.; Pereira, P.; Welker, M.; Fastner, J.; Vasconcelos, V.M. Toxicity of culturable cyanobacteria strains isolated from the portuguese coast. *Toxicon* **2005**, *46*, 454–464. [CrossRef] [PubMed]
13. Martins, R.F.; Ramos, M.F.; Herfindal, L.; Sousa, J.A.; Skaerven, K.; Vasconcelos, V.M. Antimicrobial and cytotoxic assessment of marine cyanobacteria—*Synechocystis* and *Synechococcus*. *Mar. Drugs* **2008**, *6*, 1–11. [CrossRef] [PubMed]
14. Brito, Â.; Gaifem, J.; Ramos, V.; Glukhov, E.; Dorrestein, P.C.; Gerwick, W.H.; Vasconcelos, V.M.; Mendes, M.V.; Tamagnini, P. Bioprospecting portuguese atlantic coast cyanobacteria for bioactive secondary metabolites reveals untapped chemodiversity. *Algal Res.* **2015**, *9*, 218–226. [CrossRef]
15. Hentschel, U.; Usher, K.M.; Taylor, M.W. Marine sponges as microbial fermenters. *FEMS Microbiol. Ecol.* **2006**, *55*, 167–177. [CrossRef] [PubMed]
16. Lemloh, M.L.; Fromont, J.; Brümmer, F.; Usher, K.M. Diversity and abundance of photosynthetic sponges in temperate western australia. *BMC Ecol.* **2009**, *9*, 4. [CrossRef] [PubMed]
17. Rützler, K. Associations between caribbean sponges and photosynthetic organisms. In *New Perspectives in Sponge Biology*; Rutzler, K., Ed.; Smithsonian Institution Press: Washington, DC, USA, 1990; pp. 455–466.
18. Regueiras, A.; Alex, A.; Pereira, S.; Costa, M.S.; Antunes, A.; Vasconcelos, V. Cyanobacterial diversity in the marine sponge *Hymeniacidon perlevis* from a temperate region (portuguese coast, northeast atlantic). *Aquat. Microb. Ecol.* **2017**, *79*, 259–272. [CrossRef]
19. Blunt, J.W.; Copp, B.R.; Munro, M.H.; Northcote, P.T.; Prinsep, M.R. Marine natural products. *Nat. Prod. Rep.* **2010**, *27*, 165–237. [CrossRef] [PubMed]
20. Thomas, T.; Rusch, D.; DeMaere, M.Z.; Yung, P.Y.; Lewis, M.; Halpern, A.; Heidelberg, K.B.; Egan, S.; Steinberg, P.D.; Kjelleberg, S. Functional genomic signatures of sponge bacteria reveal unique and shared features of symbiosis. *ISME J.* **2010**, *4*, 1557–1567. [CrossRef] [PubMed]
21. Konstantinou, D.; Gerovasileiou, V.; Voultsiadou, E.; Gkelis, S. Sponges-cyanobacteria associations: Global diversity overview and new data from the eastern mediterranean. *PLoS ONE* **2018**, *13*, e0195001. [CrossRef] [PubMed]
22. Alex, A.; Antunes, A. Pyrosequencing characterization of the microbiota from atlantic intertidal marine sponges reveals high microbial diversity and the lack of co-occurrence patterns. *PLoS ONE* **2015**, *10*, e0127455. [CrossRef] [PubMed]
23. Alex, A.; Vasconcelos, V.; Tamagnini, P.; Santos, A.; Antunes, A. Unusual symbiotic cyanobacteria association in the genetically diverse intertidal marine sponge *Hymeniacidon perlevis* (demospongiae, halichondrida). *PLoS ONE* **2012**, *7*, e51834. [CrossRef] [PubMed]

24. Usher, K.M.; Kuo, J.; Fromont, J.; Sutton, D.C. Vertical transmission of cyanobacterial symbionts in the marine sponge *Chondrilla australiensis* (demospongiae). *Hydrobiologia* **2001**, *461*, 9–13. [CrossRef]
25. Burgsdorf, I.; Slaby, B.M.; Handley, K.M.; Haber, M.; Blom, J.; Marshall, C.W.; Gilbert, J.A.; Hentschel, U.; Steindler, L. Lifestyle evolution in cyanobacterial symbionts of sponges. *mBio* **2015**, *6*, e00391–e00415. [CrossRef] [PubMed]
26. Gao, Z.M.; Wang, Y.; Tian, R.M.; Wong, Y.H.; Batang, Z.B.; Al-Suwailem, A.M.; Bajic, V.B.; Qian, P.Y. Symbiotic adaptation drives genome streamlining of the cyanobacterial sponge symbiont "*Candidatus* synechococcus spongiarum". *mBio* **2014**, *5*, e00079–e00114. [CrossRef] [PubMed]
27. Han, B.; Gross, H.; Goeger, D.E.; Mooberry, S.L.; Gerwick, W.H. Aurilides b and c, cancer cell toxins from a papua new guinea collection of the marine cyanobacterium lyngbya majuscula. *J. Nat. Prod.* **2006**, *69*, 572–575. [CrossRef] [PubMed]
28. Luesch, H.; Yoshida, W.Y.; Moore, R.E.; Paul, V.J. Lyngbyastatin 2 and norlyngbyastatin 2, analogues of dolastatin g and nordolastatin g from the marine cyanobacterium lyngbya majuscula. *J. Nat. Prod.* **1999**, *62*, 1702–1706. [CrossRef] [PubMed]
29. Luesch, H.; Yoshida, W.Y.; Moore, R.E.; Paul, V.J.; Mooberry, S.L. Isolation, structure determination, and biological activity of lyngbyabellin a from the marine cyanobacterium lyngbya majuscula. *J. Nat. Prod.* **2000**, *63*, 611–615. [CrossRef] [PubMed]
30. Mundt, S.; Kreitlow, S.; Nowotny, A.; Effmert, U. Biochemical and pharmacological investigations of selected cyanobacteria. *Int. J. Hyg. Environ. Health* **2001**, *203*, 327–334. [CrossRef] [PubMed]
31. Papendorf, O.; König, G.M.; Wright, A.D. Hierridin b and 2,4-dimethoxy-6-heptadecyl-phenol, secondary metabolites from the cyanobacterium phormidium ectocarpi with antiplasmodial activity. *Phytochemistry* **1998**, *49*, 2383–2386. [CrossRef]
32. Solis, P.N.; Wright, C.W.; Anderson, M.M.; Gupta, M.P.; Phillipson, J.D. A microwell cytotoxicity assay using *Artemia salina*. *Planta Med.* **1993**, *59*, 250–252. [CrossRef] [PubMed]
33. Lopes, V.R.; Fernández, N.; Martins, R.F.; Vasconcelos, V. Primary screening of the bioactivity of brackishwater cyanobacteria: Toxicity of crude extracts to *Artemia salina* larvae and *Paracentrotus lividus* embryos. *Mar. Drugs* **2010**, *8*, 471–482. [CrossRef] [PubMed]
34. Engström-Öst, J.; Lehtiniemi, M.; Green, S.; Kozlowsky-Suzuki, B.; Viitasalo, M. Does cyanobacterial toxin accumulate in mysid shrimps and fish via copepods? *J. Exp. Mar. Biol. Ecol.* **2002**, *276*, 95–107. [CrossRef]
35. Ferrão-Filho, A.d.S.; Kozlowsky-Suzuki, B.; Azevedo, S.M.F.O. Accumulation of microcystins by a tropical zooplankton community. *Aquat. Toxicol.* **2002**, *59*, 201–208. [CrossRef]
36. Ridley, C.P.; Bergquist, P.R.; Harper, M.K.; Faulkner, D.J.; Hooper, J.N.A.; Haygood, M.G. Speciation and biosynthetic variation in four dictyoceratid sponges and their cyanobacterial symbiont, *Oscillatoria spongeliae*. *Chem. Biol.* **2005**, *12*, 397–406. [CrossRef] [PubMed]
37. Nunes, B.S.; Carvalho, F.D.; Guilhermino, L.M.; Van Stappen, G. Use of the genus *Artemia* in ecotoxicity testing. *Environ. Pollut.* **2006**, *144*, 453–462. [CrossRef] [PubMed]
38. Sakiyama, T.; Ueno, H.; Homma, H.; Numata, O.; Kuwabara, T. Purification and characterization of a hemolysin-like protein, sll1951, a nontoxic member of the rtx protein family from the cyanobacterium *Synechocystis* sp. Strain pcc 6803. *J. Bacteriol.* **2006**, *188*, 3535–3542. [CrossRef] [PubMed]
39. Wang, P.-J.; Chien, M.-S.; Wu, F.-J.; Chou, H.-N.; Lee, S.-J. Inhibition of embryonic development by microcystin-lr in zebrafish, *Danio rerio*. *Toxicon* **2005**, *45*, 303–308. [CrossRef] [PubMed]
40. Pagliara, P.; Caroppo, C. Cytotoxic and antimitotic activities in aqueous extracts of eight cyanobacterial strains isolated from the marine sponge *Petrosia ficiformis*. *Toxicon* **2011**, *57*, 889–896. [CrossRef] [PubMed]
41. Kótai, J. *Instructions for Preparation of Modified Nutrient Solution z8 for Algae*; Norwegian Institute for Water Research b-11769: Oslo, Norway, 1972; p. 5.
42. Rippka, R. Isolation and purification of cyanobacteria. *Meth. Enzymol.* **1988**, *167*, 3–27. [PubMed]
43. Komárek, J. *Süßwasserflora von Mitteleuropa, bd. 19/1: Cyanoprokaryota: Chroococcales*; Springer Spektrum: Berlin/Heidelberg, Germany, 2008; p. 548.
44. Komárek, J.; Anagnostidis, K. *Süßwasserflora von Mitteleuropa, bd. 19/2: Cyanoprokaryota: Oscillatoriales*; Elsevier/Spektrum: Berlin/Heidelberg, Germany, 2005; p. 759.
45. Komárek, J. *Süßwasserflora von Mitteleuropa, bd. 19/3: Cyanoprokaryota: Heterocytous Genera*; Springer Spektrum: Berlin/Heidelberg, Germany, 2013; p. 1131.

46. Castenholz, R.W.; Wilmotte, A.; Herdman, M.; Rippka, R.; Waterbury, J.B.; Iteman, I.; Hoffmann, L. Phylum bx. Cyanobacteria. In *Bergey's Manual® of Systematic Bacteriology: Volume One: The Archaea and the Deeply Branching and Phototrophic Bacteria*; Boone, D.R., Castenholz, R.W., Garrity, G.M., Eds.; Springer: New York, NY, USA, 2001; pp. 473–599.
47. Komárek, J.; Kastovský, J.; Mares, J.; Johansen, J.R. Taxonomic classification of cyanoprokaryotes (cyanobacterial genera) 2014, using a polyphasic approach. *Preslia* **2014**, *86*, 295–335.
48. Fernández, N.; Beiras, R. Combined toxicity of dissolved mercury with copper, lead and cadmium on embryogenesis and early larval growth of the *Paracentrotus lividus* sea-urchin. *Ecotoxicology* **2001**, *10*, 263–271. [CrossRef] [PubMed]
49. Rangel, M.; Malpezzi, E.L.A.; Susini, S.M.M.; De Freitas, J. Hemolytic activity in extracts of the diatom *Nitzschia*. *Toxicon* **1997**, *35*, 305–309. [CrossRef]
50. Slowing, I.I.; Wu, C.W.; Vivero-Escoto, J.L.; Lin, V.S.Y. Mesoporous silica nanoparticles for reducing hemolytic activity towards mammalian red blood cells. *Small* **2009**, *5*, 57–62. [CrossRef] [PubMed]

toxins

MDPI

Article

Multi-Toxin Occurrences in Ten French Water Resource Reservoirs

Frederic Pitois [1,*], Jutta Fastner [2], Christelle Pagotto [3] and Magali Dechesne [4]

1 Limnologie sarl, 16 rue Paul Langevin, 35200 Rennes, France
2 German Federal Environment Agency (UBA), Corrensplatz 1, 14195 Berlin, Germany; jutta.fastner@uba.de
3 Veolia Water, 30 rue Madeleine Vionnet, 93300 Aubervilliers, France; christelle.pagotto@veolia.com
4 Veolia Recherche & Innovation, Chemin de la Digue, 78603 Maisons-Laffitte, France; magali.dechesne@veolia.com
* Correspondence: fred.pitois@limnosphere.com; Tel.: +33-02-9932-1794

Received: 31 May 2018; Accepted: 2 July 2018; Published: 9 July 2018

Abstract: Cyanobacteria are known to produce a wide array of metabolites, including various classes of toxins. Among these, hepatotoxins (Microcystins), neurotoxins (Anatoxin-A and PSP toxins) or cytotoxins (Cylindrospermopsins) have been subjected to numerous, individual studies during the past twenty years. Reports of toxins co-occurrences, however, remain scarce in the literature. The present work is an inventory of cyanobacteria with a particular focus on Nostocales and their associated toxin classes from 2007 to 2010 in ten lakes used for drinking water production in France. The results show that potential multiple toxin producing species are commonly encountered in cyanobacteria populations. Individual toxin classes were detected in 75% of all samples. Toxin co-occurrences appeared in 40% of samples as two- or three-toxin combinations (with 35% for the microcystins–anatoxin combination), whereas four-toxin class combinations only appeared in 1% of samples. Toxin co-occurrences could be partially correlated to species composition and water temperature. Peak concentrations however could never be observed simultaneously and followed distinct, asymmetrical distribution patterns. As observations are the key for preventive management and risk assessment, these results indicate that water monitoring should search for all four toxin classes simultaneously instead of focusing on the most frequent toxins, i.e., microcystins.

Keywords: cylindrospermopsin; anatoxin-a; PSP toxins; microcystins; cyanobacteria; Nostocales; drinking water

Key Contribution: Toxin co-occurrences were common but toxin peak concentrations were anti-correlated, indicating that water quality monitoring should search for all toxin classes simultaneously.

1. Introduction

Cyanobacteria proliferations are a worldwide consequence of lake eutrophication, with potential public health issues for water management, water production and recreational use such as bathing. Despite constant research efforts for the last 30 years, toxin production and occurrence, i.e., why, when and which species will produce any toxin, alone or in any combination with other toxins, is still insufficiently understood. Cyanobacterial toxins include a wide variety of molecules, such as hepatotoxins (microcystins and nodularins), cytotoxins (cylindrospermopsins), neurotoxins such as anatoxin-a, PSP toxins (Paralytic Shellfish Poisoning) or dermatotoxins such as Lyngbyatoxin, a potent dermatitis agent [1].

According to the World Health Organization handbook [2], microcystins (MCs) are the most studied and monitored toxins. MCs are cyclic heptapeptids comprising more than 200 variants [3], have already been reported from most countries, and are known to be produced by many common taxa in continental waters such as *Microcystis*, *Planktothrix*, *Anabaena/Dolichospermum*, etc.

Cylindrospermopsin (CYN) is a hepatotoxic alkaloid first identified in an Australian *Cylindrospermopsis raciborskii* [4], then in tropical or subtropical waters [5–9], and recently also in European lakes in Germany, Italy or France [10–12]. CYN and congeners have been reported to be produced by species such as *Aphanizomenon flos-aquae*, *A. ovalisporum* [13,14], *Raphidiopsis curvata* and *R. mediterranea* [15].

Anatoxin-A (ATX) is a neurotoxic alkaloid observed worldwide and associated with many common taxa: *Dolichospermum* (from *Anabaena*) *flos-aquae*, *Aphanizomenon flos-aquae* [16], *Dolichospermum planctonicum* [17], *Cuspidothrix* (from *Aphanizomenon*) *issatchenkoï* [18], *Raphidiopsis mediterranea* [19], and *Microcystis aeruginosa* [20].

PSP toxins are neurotoxic alkaloids produced by freshwater cyanobacteria and marine dinoflagellates with more than 30 identified variants such as saxitoxin, neosaxitoxin, decarbamoylsaxitoxin, gonyautoxins, etc. [21]. Known potential PSP producing species include *Cuspidothrix issatchenkoï* [22], *Cylindrospermopsis raciborskii* and *Raphidiopsis brookii* [23,24], *Aphanizomenon flos-aquae* and *A. gracile* [25–27], *Dolichospermum circinalis* [1], *Dolichospermum lemmermannii* [28], or *Microcystis aeruginosa* [29].

The dermatotoxins lyngbyatoxin and aplysiatoxin are contact dermatitis and tumor promoting agents produced by benthic species such as *Lyngbya wollei* or *Lyngbya majuscula* [30,31]. These alkaloids have mainly been reported from marine lagoons and subtropical lakes and are under-documented in other contexts, as benthic species are seldom observed in planktic flora surveys.

Besides microcystins, the other toxin classes are less commonly studied and monitored, and data about their simultaneous occurrence are scarce in the literature, with the exception of studies from Italy [32], Germany [33] or the USA [34,35]. ATX, CYN or PSP toxins can thus be considered as "emerging toxins" either because of a recent spread in resource waters, or because of a recent interest for water managers. These toxin classes are mainly produced by taxa from the order Nostocales, i.e., *Dolichospermum*, *Aphanizomenon*, *Cuspidothrix*, *Cylindrospermopsis*, *Raphidiopsis*, etc.

In continental Europe, Nostocales species composition associates autochthonous taxa (*Dolichospermum flos-aquae*, *Aphanizomenon flos-aquae*, and *Aphanizomenon gracile*) with new, invasive species such as *Cylindrospermopsis raciborskii*, *Anabaena bergii* or *Sphaerospermopsis aphanizomenoides* [36]. Various factors have been proposed to explain these invasive species extension, such as transport by migrating birds [37], conjugated with climate change, namely the increase of spring temperature and solar radiation fluxes [38–41]. Although these invasive species do not appear to be the main CYN producers, dedicated studies have shown CYN to become as frequently detected as MCs in German lakes in the recent years [10,36].

In the French regulatory context, MCs are routinely monitored in resource or bathing waters since 2003, whereas CYN, ATX and PSP toxins are only optionally analyzed since 2013. All toxin classes have, however, already been detected individually: ATX [42,43], CYN [12], and PSP [27] are known to occur in French lakes and rivers but large-scale exploration has never been performed. In this context, this work is an inventory of cyanobacteria with a particular focus on Nostocales and associated toxin classes (MCs, ATX, CYN and PSP), conducted from 2007 to 2010 in 10 freshwater lakes in France. These lakes are used as resources for drinking water production.

2. Results

The results presented below were obtained from 10 reservoirs and their associated pre-dams sampled monthly between June and October in 2007, 2008 and 2010. A total of 192 samples were collected, of which 98% contained cyanobacteria, and 70% at least one toxin class.

2.1. Cyanobacteria

Cyanobacteria were observed with relatively low cell densities (Figure 1): 12% of samples were below 1000 cell/mL, whereas 41% of samples were higher than 20,000 cell/mL, i.e., WHO alert level 2, and 24% were higher than 100,000 cell/mL, i.e., WHO alert level 3. The highest peak cell density reached 3,320,500 cell/mL in 2007.

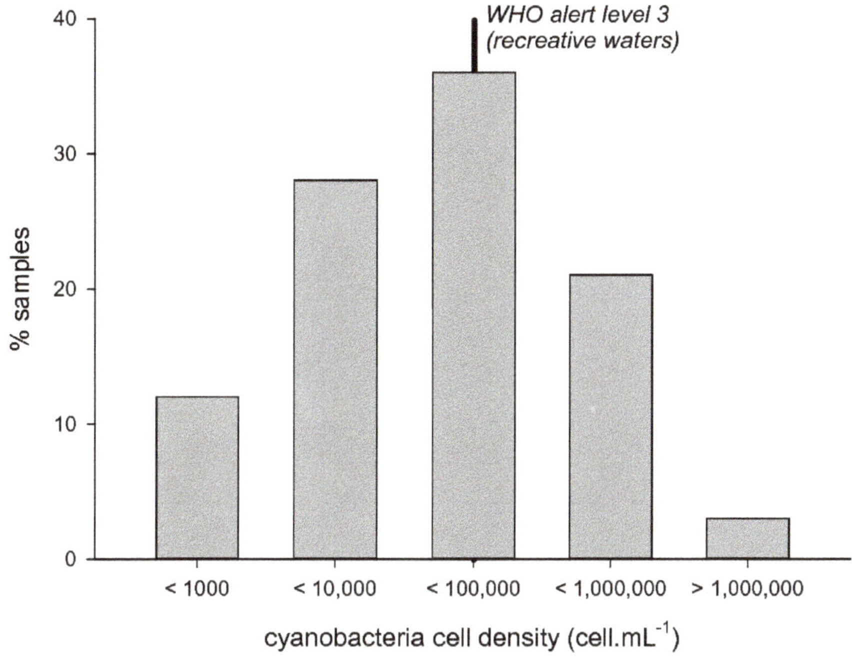

Figure 1. Cyanobacteria cell densities distribution in the samples collected from 2007 to 2010 (n = 192).

The order Chroococcales was the most common order every year, and appeared in 88% of all samples with 18 different taxa. Species from the genera *Aphanothece*, *Snowella*, *Microcystis* (*M. aeruginosa*) or *Worochininia* (*W. compacta*) could be observed in 33–45% of samples. The highest Chroococcales biomass, i.e., 234 mm^3/L, was attributed to a sample dominated by *Microcystis viridis* reaching 1,308,200 cell/mL in 2010 in lake No. 7.

Oscillatoriales, observed in 85% of samples, were the second most frequent order. Species composition was dominated by *Planktothrix agardhii*, in 62% of samples, and *Phormidium splendidum*, in 37% of samples; 10 other taxa were recorded with low cell densities in less than 14% of samples. The maximal Oscillatoriales biomass, 63 mm^3/L, was observed in a sample dominated by *P. agardhii* with 1,550,500 cell/mL in 2010 in lake No. 8.

The order Nostocales was observed in 65% of samples with 17 taxa. The most common species were *Cuspidothrix issatschenkoï* and *Aphanizomenon flos-aquae* in 45–47% of samples. Immature Dolichospermum, i.e., without heterocysts and akinetes, were present in 32% of samples, and all other taxa appeared in less than 12% occasions. The peak Nostocales biomass, 504 mm^3/L, was recorded in a sample dominated by *Dolichospermum flos-aquae* with 3,320,500 cell/mL in 2007 in lake No. 9.

Cyanobacteria distribution is summarized in Figure 2. Most taxa associated with the highest frequencies or cell densities were common species in the French context, such as *Planktothrix agardhii* and *P. rubescens*, *Aphanizomenon flos-aquae*, *Cuspidothrix issatchenkoï*, *Microcystis aeruginosa*, *M. flos-aquae*, *M. viridis*, etc. Some less common species could however be observed, i.e., Nostocales such as *Dolichospermum compactum* and *D. viguieri*, *Sphaerospermopsis eucompacta*, *Anabaenopsis arnoldii*, *Cuspidothrix elenkinii* and *Aphanizomenon schindleri*, in less than 5% of samples and with cell densities lower than 1000 cell/mL. Only one uncommon species, *Raphidiopsis brookii*, could be observed with a significant biomass of 1,824,000 cell/mL and 156 mm^3/L, in August and September 2010 in lake No. 6.

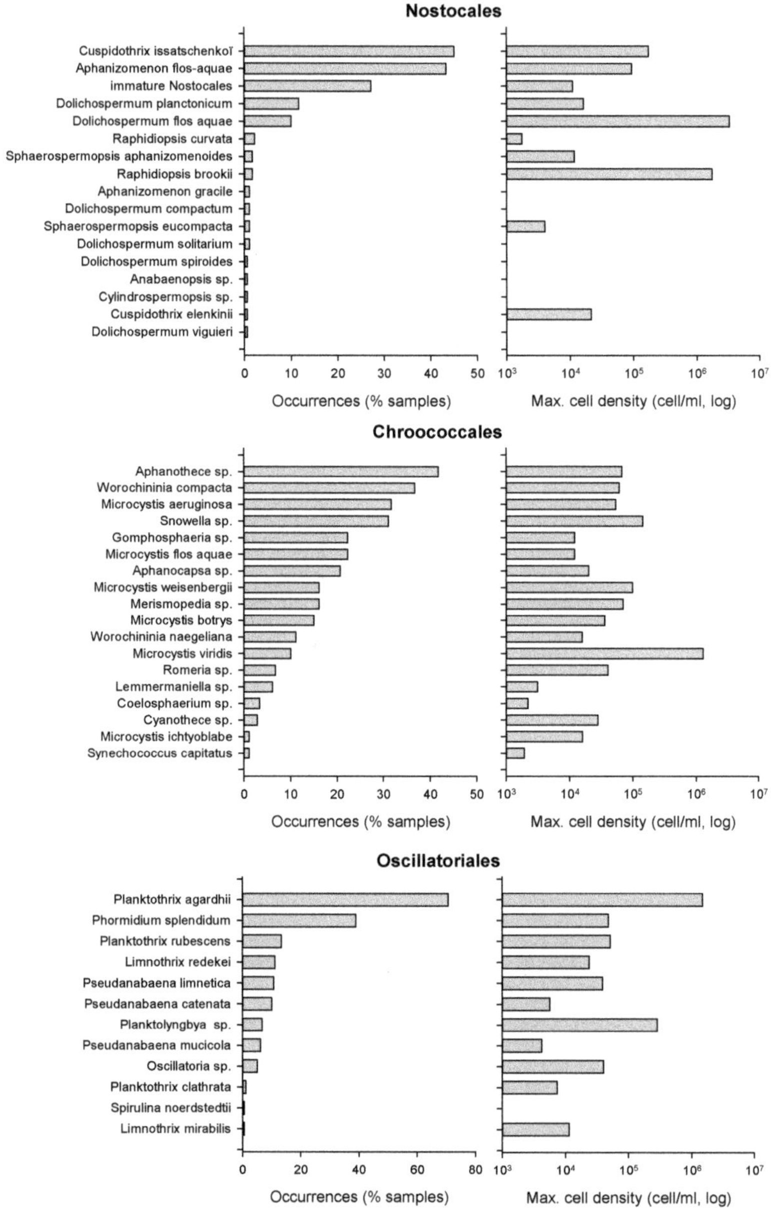

Figure 2. Cyanobacteria taxonomic distribution expressed as occurrence frequency and maximum cell density (n = 185).

From 2007 to 2010, 25 known toxin producing species were recorded, with 18 known MCs producers, 9 known ATX producers, 6 known PSP producers and 4 known CYN producers. Some commons species appear to be potential producers for 2–3 toxin classes. The complete known toxin producing species listing is provided in Table S1.

Compared to analyzed species composition, potential MCs and ATX producers occurred in 92% and 68% of all samples, vs. 65% and 42%, respectively, for known PSP and CYN producers. In the same time, samples containing no cyanobacteria or no known toxin-producing species accounted for, respectively, 3% and 7% of samples.

Samples hosting one toxin-producing species only accounted for 24% of samples, whereas 60% of samples contained species potentially associated with all four toxin classes. This is explained by the widespread distribution of *Aphanizomenon* sp. and *Microcystis aeruginosa*, suspected to be multiple toxin classes producers. These two species were identified in 37% and 29% of samples while occurring concomitantly with other toxin-producing species.

2.2. Toxin Classes

First it should be noted that no extracellular toxin could be detected in any analyzed sample, and all exposed results relate to intracellular toxin concentrations.

2.2.1. Microcystins

Microcystins (MCs) were detected in 64% of all analyzed samples, with high interannual variability: 45% of samples were positive in 2007 vs. 20% in 2008 and 100% in 2010. This is corroborated by another study dedicated to 26 lakes in western France, including two lakes and three monitoring years in common with the present study [44], showing that 2007 and 2008, similar to 2004 and 2011, had distinctly lower MCs detection frequencies compared to 2006, 2009 or 2010 where detection frequencies were the highest. Maximal detection frequencies were recorded from 20,000 to 100,000 cell/mL (i.e., between WHO alert thresholds 1 and 2) with 83% MCs detections, whereas maximal concentrations were observed above 100,000 cell/mL with 12.5 μg/L (Table 1).

Table 1. Toxin detection frequencies (Det: percent of samples) and maximal concentrations (Max.: μg/L) vs. cyanobacteria cell density classes expressed as WHO alert thresholds. **MCs:** total Microcystins, ATX: Anatoxin-a, CYN: Cylindrospermopsin, STX: Saxitoxins

Cell Density Classes	MCs		ATX		CYN		STX	
	Det.	Max.	Det.	Max.	Det.	Max.	Det.	Max.
< 20,000 cell/mL	42%	6.7	18%	0.03	8%	0.03	6%	0.05
20 to 100,000 cell/mL	83%	9.0	47%	0.46	17%	0.01	11%	0.05
> 100,000 cell/mL	79%	12.5	60%	0.34	9%	0.02	35%	0.05

Despite this variability, these frequencies are in broad agreement with already published data, i.e., MCs positive detections in 62–91% of samples analyzed in Italy [32], Germany [33] or the USA [34,35]. MCs concentrations, although substantially higher than the other toxin classes, were mostly lower than 1 μg/L for 60% of samples, and higher than 5 μg/L in 6% of samples (Figure 3). The median (0.55 μg/L) and maximal (12.5 μg/L) values appeared distinctively lower than in other studies, especially compared to Germany [33] and the USA [34,35].

The 2007 and 2008 samples were analyzed for eight MC variants: [Asp3]MC-RR and -LR, and MC-RR, YR, LR, LA, LW and LF. MC-RR and [Asp3]MC-RR, or MC-LR and [Asp3]MC-LR were detected in 94% of samples with mean concentrations of 0.4 ± 1.1 and 0.9 ± 1.3 μg/L respectively. MC-LF was detected in 77% of samples (mean: 0.01 ± 0.02 μg/L), and MC-YR in 71% of samples (mean: 0.05 ± 0.09 μg/L). MC-LW and MC-LA were detected in 63% and 45% of samples respectively, with concentrations lower than 0.03 μg/L.

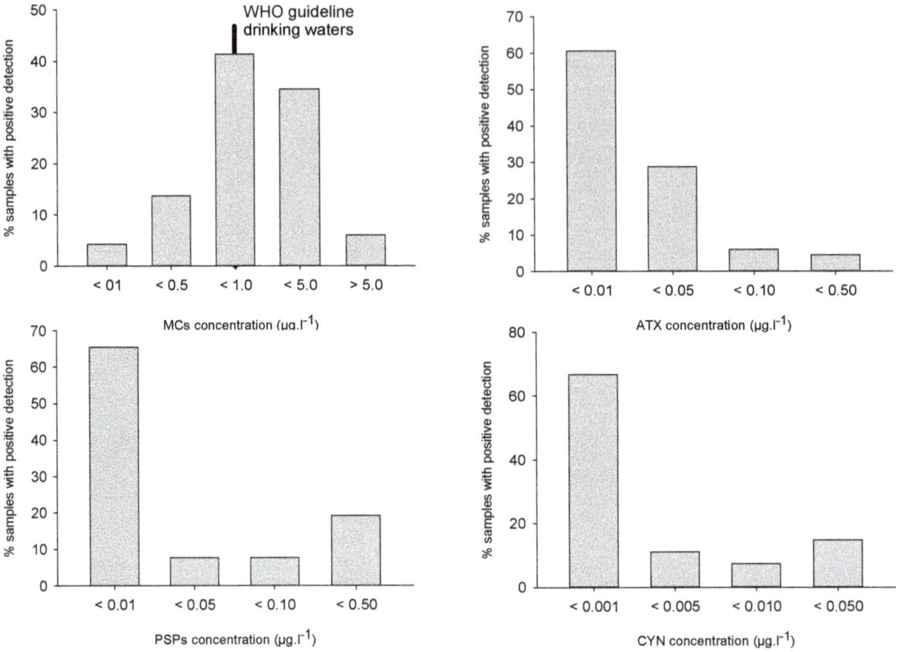

Figure 3. Toxin classes distribution in all analyzed samples (all toxin congeners summed up).

2.2.2. Anatoxin-A

Anatoxin-A (ATX) was detected in 35% of all samples which is similar to or higher than in USA studies [33,35] (7%) but lower than in German studies (57%) [33]. Once again, interannual variations were observed: 22% of samples were positive in 2007 and 2008, and 50% in 2010. Most samples (88%) showed ATX concentrations lower than 0.05 µg/L whereas concentrations higher than 0.1 µg/L could be observed in 1% of samples (Figure 3). Detection frequency tended to increase with cell density, and reached 60% above 100,000 cell/mL (Table 1). Maximal concentration (0.46 µg/L) appeared distinctly lower than median values reported from Germany [33] and the USA [34]

2.2.3. Cylindrospermopsins

Cylindrospermopsins (CYN) were detected in 15% of samples, i.e., 9–11% of samples in 2007–2008 and 19% of samples in 2010, similar to occurrences in the USA [34] and significantly lower than reported from German lakes (83%) [33]. CYN concentrations were always lower than 0.03 µg/L, with 78% of samples lower than 0.01 µg/L (Figure 3). Deoxy-CYN was also investigated but could not be detected in our samples. Maximal detection frequency (17% of samples) was recorded from 20,000 to 100,000 cell/mL, whereas maximal concentrations were observed in samples with cell densities lower than 20,000 cell/mL (Table 1).

It must be noted that dissolved CYN could not be observed in any sample. This is distinctly different from already reported observations where high extra-cellular concentrations tend to be the main fraction of total CYN (see [45] and references therein for example).

2.2.4. PSP Toxins

PSP were detected in 14% of samples, with 23% of samples in 2007, 11% of samples in 2008 and 10% of samples in 2010 (Figure 3). Once again, this is similar to reported frequencies in the

USA [34] and lower than observations from German lakes (69%) [33]. PSP concentrations were mostly (in 73% of samples) lower than 0.005 µg/L, whereas 19% of samples ranged from 0.01 to 0.05 µg/L. Detection frequencies were maximal (i.e., 35% of samples) for cell densities higher than 100,000 cell/mL, whereas similar maximal concentrations could be observed in every cell density class (Table 1). The observed PSP congeners were Saxitoxin (STX) in eight samples, ranging from 0.04 to 0.15 µg/L, and Gonyautoxin-5 (GTX-5) in four samples, ranging from 0.03 to 0.08 µg/L. Other congeners were not detected in any sample.

2.3. Toxin Classes Distribution

Multiple toxin combinations appeared in 40% of all samples, mostly as a two-toxin class combination (27% of samples), whereas three toxins could be detected in 12% of samples and four toxin classes in 1% of samples. At the same time, 35% of samples hosted only one toxin class, mainly MCs. This appears in close agreement with observations in Midwestern USA lakes [34] where multiple toxin combinations were observed in 48% of samples, with two toxin classes in 30% of samples and three toxin classes in 18% of samples.

Toxin combinations tended to increase with cyanobacterial biomass, either expressed as cell density or cell biovolume (Figure 4). Combinations of 3–4 toxin classes could, however, be encountered with a cell density as low as 7605 cell/mL, or a cell volume of 0.59 mm^3/mL.

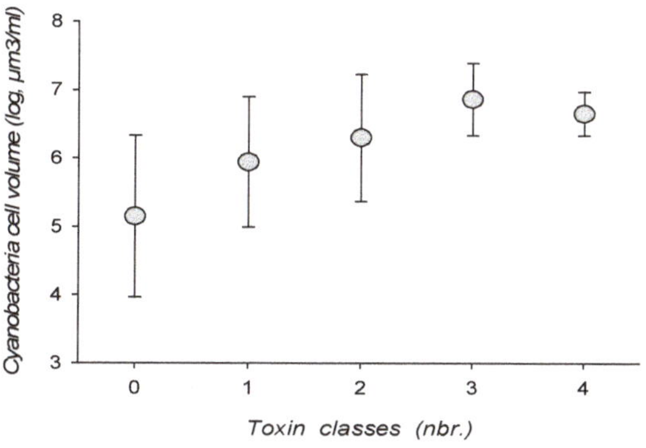

Figure 4. Toxin class associations and cyanobacteria total cell biovolume.

Distinct distribution patterns could be observed for the various toxin classes and toxin variants. Microcystin variants, when expressed as percent of total MCs (Figure 5), could be separated into two groups, with either -RR/-[Asp3]RR or -LR/[Asp3]LR dominated samples. In the -LR-dominated group, MC-YR, LW and LF tended to first increase simultaneously to MC-LR, and then decrease when -LR reached 60% of total MCs. This indicates that MC-positive samples were composed either of (mostly) -RR and -[Asp3]RR variants, or of (mostly) -LR and [Asp3]LR variants associated with low concentrations of -YR, LW and LF microcystins. Only one of all samples had a nearly equal composition with 54% MC-RR vs. 46% MC-LR.

Similar MCs variant distribution patterns can be observed in the results from the USA [34]. In our case, this distribution can be partly attributed to cyanobacteria species composition: MC-RR was correlated with *Planktothrix* and *Aphanizomenon* biomass (r^2 = 0.47 and 0.43, respectively, $p < 0.01$), MC-YR and MC-LF correlated with *Microcystis* biomass (r^2 = 0.38 and 0.34, respectively, $p < 0.01$), and [Asp3]MC-LR with *Dolichospermum* biomass (r^2 = 0.36, $p < 0.01$). MC-LR, on the other hand, could not be correlated with any species group, which is consistent with a possible production by nearly all potentially toxic species observed in the samples. It can thus be hypothesized that MCs variant distribution is a direct consequence of species successional patterns.

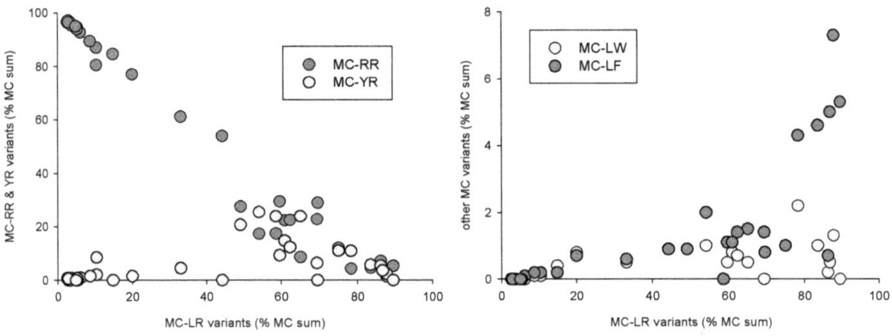

Figure 5. Microcystin variant combinations expressed as percent of total measured MCs.

Other toxin classes could not be analyzed for enough variants or observed in enough samples to show species-controlled distributions. Paired toxin distribution could, however, be compared for MCs vs. ATX (123 samples), MCs vs. CYN (n = 122) and ATX vs. CYN (n = 62), whereas quantified PSP were insufficiently numerous to allow for a comparison (Figure 6).

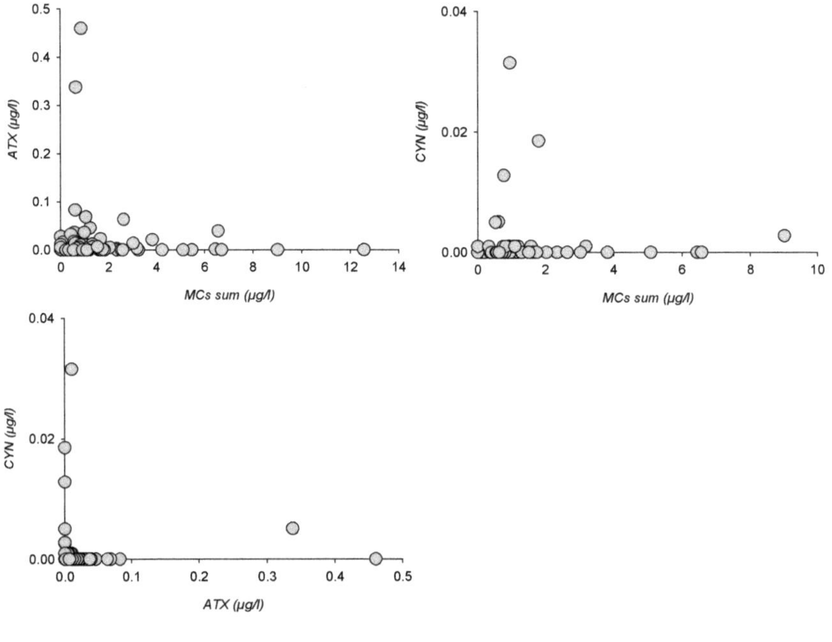

Figure 6. Paired toxin classes distribution for MCs, ATX and CYN.

ATX and CYN peaks appeared asymmetrically distributed when compared to total MCs, as peak concentrations were recorded for lower MCs, whereas MCs peaks corresponded with lower ATX or CYN concentrations. Similarly, ATX and CYN concentrations were conversely distributed relatively to each other. Whichever class is considered, no sample hosted simultaneous peaks with two or more toxins. When compared with MCs distribution patterns, this means that any peak sample was composed of either [Asp3]MC-RR and MC-RR or [Asp3]MC-LR and MC-LR dominated MCs, or ATX, or CYN.

2.4. Measured Environmental Parameters

Statistical analyses were unable to show direct relations among environmental parameters, cyanobacteria and toxins. However, some indirect, non-significant relations appeared: Cyanobacteria cell densities or biomasses were higher in unstratified (vs. stratified) or shallower lakes with higher turbidity (i.e., lower Secchi depths), and tended to increase with water temperatures. Microcystin concentrations increased with higher cyanobacterial biomass. Toxin classes number increased with mean water temperatures (in unstratified lakes) or with mean temperatures in the euphotic layer (in stratified lakes), along with cyanobacteria cell densities or biomass.

These results could be expected as previous studies have shown that individual toxin classes production could be related to abiotic parameters such as temperature or light, whereas shallow, unstratified lakes are known to show higher primary production and higher planktonic biomass than deeper, stratified lakes. However, if we assume that toxin concentration in a lake depends on: (1) the biomass of all potentially toxin-producing species; (2) their genotype composition; and (3) the type and amount of toxin each genotype produces, environmental factors can act on all levels. It is unclear then whether the observed toxin distribution is a direct consequence of water temperature, i.e., higher temperatures leading to higher toxin production, or an indirect consequence of site characteristics on cyanobacteria populations, i.e., temperature and planktonic diversity leading to higher probability for any lake to host one or more species producing one or more toxin classes.

3. Conclusions

Regarding cyanobacteria, expected species such as invasive, allochthonous Nostocales remained rarely encountered during the three sampling years. Some uncommon taxa from the genera *Cylindrospermopsis*, *Anabaenopsis*, *Anabaena* or *Aphanizomenon* were observed but remained lower than 1000 cell/mL. With the exception of a *Raphidiopsis brookii* bloom reaching nearly 2,000,000 cell/mL in 2010, all proliferation episodes were related to common taxa in the French context. The relations between water temperatures, lake depth and water stratification indicate that exposure to high cyanobacterial biomass and multiple toxin classes occurring simultaneously are more likely in shallower, unstratified lakes such as smaller artificial lakes devoted to bathing and other recreative activities. This can also be the case in pre-dam lakes, where waters are often eutrophic because of the nutrient load provided by the lake tributaries, whereas temperatures were often 1–2 °C higher than in the main reservoir. In this sense, pre-dams can act as incubators contaminating the main lakes with dense cyanobacteria inoculum.

Regarding toxin classes, microcystins, as expected, were the most common with 64% positive samples and concentrations similar to already observed values in France and ranging from 1 to 10 µg/L. Anatoxin-A was the second most frequent toxin class with 34% positive samples whereas maximum concentration was inferior to 0.5 µg/L. PSP toxins and cylindrospermopsin, on the other hand, appeared fairly uncommon with 14–15% positive samples and maximum concentrations lower than 0.15 and 0.05 µg/L respectively. Toxin concentrations and toxin class frequencies appeared positively related to cyanobacterial biomass and water temperatures. In this sense, the fairly low multiple toxin occurrence frequencies observed in 2007 and 2008 compared to 2010 could be explained by the relatively unstable meteorological conditions encountered during these summers, as for most sites 2007 and 2008 were the coldest summers since 1993.

The results also show that analyzing MC-LR as an indicator of MCs occurrence or total MCs concentration is inadequate, as MCs variants do not appear evenly distributed in samples, with [Asp3]MC-LR/MC-LR, YR, LA, and LF anticorrelated with [Asp3]MC-RR/MC-RR. Similarly, considering total MCs concentration as indicative of other toxin occurrences is questionable, as peak concentrations for the analyzed toxins were also anticorrelated and simultaneous peaks were never observed.

All our analyses were conducted on concentrated samples and allowed to detect low toxin concentrations. Most present-day toxin monitoring however relies on higher quantification limits, typically 0.2 µg/L for any toxin class in France. If these quantification limits were applied to our results,

MCs would have been detected in 59% of samples vs. 64%, ATX in 1% of samples vs. 34%, and PSP or CYN would not have been observed. Regarding toxin detections, in the French context, toxin analysis tends to be concentrated on samples with cell densities higher than WHO alert level 2 (i.e., 100,000 cell/mL). Our results however show that MCs and ATX may appear with significant frequencies and concentrations even in samples with low cell densities, indicating that toxin monitoring should be extended to WHO alert level 1 (i.e., 20,000 cell/mL).

Finally, although invasive cyanobacteria did not appear, all investigated toxin classes could be observed with significant frequencies and low concentrations which then did not represent an acute risk for drinking water production. However, this indicates that survey efforts should not only be directed toward acute toxin concentrations, but should also encompass the consequences of chronic exposure to low cyanobacterial biomass or to cyanobacterial aerosols, such as allergenicity [46–48], or to subacute toxin concentrations, such as cytotoxicity [49–51].

4. Materials and Methods

4.1. Sites

Sampling campaigns were conducted during the summers 2007, 2008 and 2010 in 10 lakes used as freshwater resources for drinking water production: eight in western France under oceanic climate, one in center-east under semi-continental climate, and one in the south under Mediterranean climate. Lake localization and volumes are summarized in Figure 7 and Table 2.

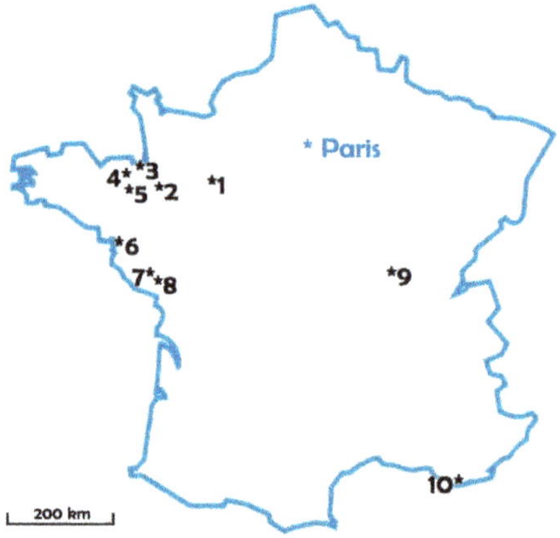

Figure 7. Sampled lakes location.

Table 2. Lake volumes. Pre-dams were included in the monitoring program.

Lake Number	Lake Volume (m^3)	Pre-Dam
1	5,700,000	+
2	204,000	+
3	1,335,000	
4	5,000,000	
5	14,700,000	
6	1,100,000	
7	4,400,000	+
8	5,800,000	+
9	10,000,000	+
10	8,000,000	

4.2. Sampling and Sample Processing

During these three years, 192 integrated water samples were collected. All samples were collected monthly from June to October, as French institutional monitoring data show that cyanobacterial proliferations have the highest probability to occur during these months. Secchi depth was measured with a Secchi disk, and samples were collected in the deepest part of the lakes and of their pre-dams with a Van Dorn sampling bottle and integrated within the euphotic depth. On every occasion, vertical profiles for pH, dissolved oxygen, conductivity and temperature were recorded every 50 cm with an YSI 556 MPS multiparameter probe.

On the sampling day, all samples were separated into two batches: a batch for plankton analysis was fixed with alkaline Lugol solution, whereas a batch for toxin analysis was filtered on Sartorius regenerated cellulose (0.45 μm). Filters and filtered water were then frozen separately until further analysis.

4.2.1. Species Composition Analysis

All samples were analyzed for species distribution by Limnologie sarl on the day following sampling under a Leica DMLS light microscope with phase contrast using common reference floras [52–60]. Results were expressed for all taxa as cell density (cell/mL) and cell biovolume (mm^3/L) calculated from cell density and mean cell dimensions.

4.2.2. Toxin Analysis

Filters were extracted three times with acetonitrile–water–formic acid (80:19.9:0.1), as previously described by Dell'Aversano et al. [61], and the combined supernatants were dried. Frozen filtrates (2 mL) were thawed and acidified with formic acid to a final volume of 0.1% formic acid. Filtrates were then dried by vacuum centrifugation and stored frozen at −20 °C. Prior to analysis extracts were re-dissolved in 500 μL 75% aqueous acetonitrile for analyses of PSPs, 50% aqueous methanol for microcystin analysis, while for analyses of CYN, D-CYN and ATX aliquots were re-dissolved in 0.1% formic acid. In some cases, samples were further concentrated for unequivocal toxin identification.

Analysis by LC-MS/MS: MCs

The extracts were separated using a Purospher STAR RP-18 end-capped column (30 × 4 mm, 3 μm particle size, Merck, Darmstadt, Germany) at 30 °C as described by Spoof et al. [62]. The mobile phase consisted of 0.5% formic acid (A) and acetonitrile with 0.5% formic acid (B) at a flow rate of 0.5 mL/min with the following gradient program: 0 min 25% B, 10 min 70% B, 11 min 70% B. The injection volume was 10 μL. Identification and quantification of the MCs ([Asp3]-MC-RR, MC-RR, MC-YR, [Asp3]-MC-LR, MC-LR, MC-LW, MC-LF, MC-LA, standards purchased at Enzo Life Sciences, Lörrach, Germany) was performed in the MRM (Multiple Reaction Monitoring) mode with the transitions given by Fastner et al. [63].

Analysis by LC-MS/MS: CYN, deoxyCYN, ATX

Analyses for CYN, deoxyCYN and ATX were carried out on an Agilent 2900 series HPLC system (Agilent Technologies, Waldbronn, Germany) coupled to a API 5500 QTrap mass spectrometer (AB Sciex, Framingham, MA, USA) equipped with a turbo-ionspray interface. The extracts were separated using a 5 μm Waters Atlantis C18 column (2.1–150 mm) at 30 °C. The HPLC was set to deliver a linear gradient from 1% to 25% MeOH in water, both containing 0.1% formic acid, within 5 min at a flow rate of 0.25 mL min−1. The mass spectrometer was operated in the multiple reaction-monitoring mode (MRM). For the determination of CYN deoxyCYN, and ATX the transitions given in [63] were used. Quantitation of CYN, deoxyCYN and ATX was performed with the most intensive transition. Standard curves were established for all toxins (CYN was obtained from National Research Council,

Canada, deoxyCYN from Novakits, Nantes, France, and ATX-a from Tocris, Bristol, UK) and analyzed in a line with the unknowns (one calibration curve after 30 unknowns).

Analysis by LC-MS/MS: PSPs

Analyses for paralytic shellfish poisons (PSP) were carried out on an Agilent 1100 series HPLC system (Agilent Technologies, Waldbronn, Germany) coupled to a API 4000 triple quadrupole mass spectrometer (AB Sciex, USA) equipped with a turbo-ionspray interface. The extracts were separated using a 5 μm TSK gel Amide-80, 2 × 250 mm column (Tosoh, Stuttgart, Germany) at 30 °C. The mobile phase consisted of water (A) and acetonitrile-water (95:5, B) both containing 2.0 mM ammonium formate and 3.6 mM formic acid (pH 3.5) at a flow rate of 0.2 mL min^{-1}. For the analysis of multiple toxins (cylindrospermopsin, anatoxin-a, paralytic shellfish poisons) the following gradient program was applied: 75% B for 5 min, 75–65% B over 1 min, hold for 13 min, 65–45% over 4 min, hold for 10 min. The mass spectrometer was operated in the multiple reaction monitoring mode for the detection and quantification of the following toxins as described by Dell'Aversano et al. [61]: saxitoxin (STX); neosaxitoxin (NEO); decarbamoylsaxitoxin (dcSTX) and decarbamoylneosaxitoxin (dcNEO); gonyautoxin-1, -2, -3, -4, and -5 (GTX-1, -2, -3, -4, and -5); decarbamoylgonyautoxin (dcGTX-3, -3); and *N*-sulfogonyautoxins-1 and -2 (C1 and C2). Standard curves were established for all the toxins (PSP standards were obtained from National Research Council, Canada) and analyzed in a line with the unknowns (one calibration curve after 20 unknowns).

Limits of quantification (LOQ) for the different toxins at injection of 10 μL sample are as follows: microcystins 0.04–0.5 μg/L depending on congener, CYN 0.01 μg/L, deoxyCYN 0.02 μg/L, ATX 0.02 μg/L and Saxitoxins 0.1–2 μg/L depending on variants. As particulate toxins have been concentrated by filtration between 25 and 300 mL of lake water, LOQs for the particulate toxins were lower than those given above and have been verified for each sample individually based on a signal to noise ratio of 10.

All toxin values ranging from method detection limit (MDL) and method quantification limit (MQL) were considered as unquantified positive detections and accounted for in toxin frequency calculations.

4.2.3. Data analysis

All results were statistically analyzed in search of relations among field data, species composition, biomass and toxin concentrations.

Tested variables included: lake maximum depth, Secchi depth, euphotic depth/max depth, surface water temperature, mean euphotic zone temperature, thermal gradients, thermal stratification, air temperature, cumulated rain and solar radiation, cyanobacteria cell densities, and toxin classes concentrations. All data were separated into 3–5 classes and subjected to Kruskal–Wallis ANOVA with XLSTAT v. 2011.1 software (Addinsoft sarl, Paris, France).

Supplementary Materials: The following are available online at http://www.mdpi.com/2072-6651/10/7/283/s1, Table S1: Observed cyanobacterial taxa with known or suspected (in parenthesis) potential toxin production.

Author Contributions: J.F., C.P., M.D. and F.P. designed the experiments; F.P. performed the field sampling, J.F. performed the toxin analysis; J.F., C.P., M.D. and F.P. analyzed the data; F.P. and J.F. wrote the paper.

Funding: This study was entirely self-funded by Veolia Water.

Acknowledgments: The authors want to acknowledge J. Komarek (Institute of Botany, Třeboň, Czech Republic) for his valuable comments and Claudia Wiedner (Institute for Aquatic Ecology and Inland Fisheries, Germany) for her scientific support. Observations and analyses in this paper were initiated and organized by Veolia Research & Innovation and Veolia Water.

Conflicts of Interest: The authors declare no conflict of interest.

References

1. Sivonen, K.; Jones, G. Chapter 3: Cyanobacterial toxins. In *Toxic Cyanobacteria in Water: A Guide to Their Public Health Consequences, Monitoring and Management*; WHO Series in Environmental Management; Chorus, I., Bartram, J., Eds.; Routledge: London, UK, 1999; pp. 41–111.
2. Chorus, I.; Bartram, J. Toxic Cyanobacteria. In *Water. A Guide to Their Public Health Consequences, Monitoring and Management*; WHO Series in Environmental Management; Routledge: London, UK, 1999; p. 320.
3. Spoof, L.; Catherine, A. Appendix 3: Tables of Microcystins and Nodularins. In *Handbook of Cyanobacterial Monitoring and Cyanotoxins Analysis*; Meriluoto, J., Spoof, L., Codd, G.A., Eds.; Wiley: Hoboken, NJ, USA, 2017; pp. 526–537.
4. Ohtani, I.; Moore, R.E.; Runnegar, M.T.C. Cylindrospermopsin, a potent hepatotoxin from the blue-green alga *Cylindrospermopsis raciborskii*. *J. Am. Chem. Soc.* **1992**, *114*, 7941–7942. [CrossRef]
5. Burns, J.; Chapman, A.; Williams, C.; Flewelling, L.; Carmichael, W.; Pawlowicz, M. Cyanotoxic Blooms in Florida's (USA) Lakes, Rivers and Tidal River Estuaries: The Recent Invasion of Toxigenic *Cylindrospermopsis raciborskii* and Consequences for Florida's Drinking Water Supplies. In Proceedings of the IX Conference on Harmful Algal Blooms, Hobart, Tasmania, Australia, 7–11 February 2000.
6. Li, R.; Carmichael, W.; Brittain, S.; Eaglesham, G.; Shaw, G.; Mahakhant, A.; Noparatnaraporn, N.; Yongmanitchai, W.; Kaya, K.; Watanabe, M. Isolation and identification of the cyanotoxin cylindrospermopsin and deoxy-cylindrospermopsin from a Thailand strain of *Cylindrospermopsis raciborskii* (Cyanobacteria). *Toxicon* **2001**, *39*, 973–980. [CrossRef]
7. Stirling, D.; Quilliam, M.A. First report of the cyanobacteria toxin cylindrospermopsin in New Zealand. *Toxicon* **2001**, *39*, 1219–1222. [CrossRef]
8. Chonudomkul, D.; Yongmanitchai, W.; Theeragool, G.; Kawachi, M.; Kasai, F.; Kaya, K.; Watanabe, M. Morphology, genetic diversity, temperature tolerance and toxicity of *Cylindrospermopsis raciborskii* (Nostocales, Cyanobacteria) strains from Thailand and Japan. *FEMS Microbiol. Ecol.* **2004**, *48*, 345–355. [CrossRef] [PubMed]
9. Berry, J.P.; Lind, O. First evidence of "paralytic shellfish toxins" and cylindrospermopsin in a Mexican freshwater system, Lago Catemaco, and apparent bioaccumulation of the toxins in "tegogolo" snails (*Pomacea patula catemacensis*). *Toxicon* **2010**, *55*, 930–938. [CrossRef] [PubMed]
10. Fastner, J.; Heinze, R.; Humpage, A.R.; Mischke, U.; Eaglesham, G.K.; Chorus, I. Cylindrospermopsin occurrence in two German lakes and preliminary assessment of toxicity and toxin production of *Cylindrospermopsis raciborskii* (Cyanobacteria) isolates. *Toxicon* **2003**, *42*, 313–321. [CrossRef]
11. Manti, G.; Mattei, D.; Messineo, V.; Melchiorre, S.; Bogialli, S.; Sechi, N.; Casiddu, P.; Luglie, A.; Di Brizio, M.; Bruno, M. First report of *Cylindrospermopsis raciborskii* in Italy. *Harmful Algae News* **2005**, *28*, 8–9.
12. Brient, L.; Lengronne, M.; Bormans, M.; Fastner, J. First occurrence of cylindrospermopsin in freshwater in France. *Environ. Toxicol.* **2009**, *24*, 415–420. [CrossRef] [PubMed]
13. Banker, R.; Teltsch, B.; Sukenik, A.; Carmeli, S. 7-Epicylindrospermopsin, a toxic minor metabolite of the cyanobacterium *Aphanizomenon ovalisporum* from Lake Kinneret, Israel. *J. Nat. Prod.* **2000**, *63*, 387–389. [CrossRef] [PubMed]
14. Preussel, K.; Stüken, A.; Wiedner, C.; Chorus, I.; Fastner, J. First report on cylindrospermopsin producing *Aphanizomenon flos-aquae* (Cyanobacteria) isolated from two German lakes. *Toxicon* **2006**, *47*, 156–162. [CrossRef] [PubMed]
15. McGregor, G.; Sendall, B.; Hunt, L.; Eaglesham, G. Report of the cyanotoxins cylindrospermopsin and deoxy-cylindrospermopsin from *Raphidiopsis mediterranea* Skuja (Cyanobacteria/Nostocales). *Harmful Algae* **2011**, *10*, 402–410. [CrossRef]
16. Rapala, J.; Sivonen, K.; Luukkainen, R.; Niemela, S. Anatoxin-a concentration in Anabaena and Aphanizomenon under different environmental conditions and comparison of growth by toxic and non-toxic Anabaena strains laboratory study. *J. Appl. Phycol.* **1993**, *5*, 581–591. [CrossRef]
17. Bruno, M.; Barbini, D.; Pierdominici, E.; Serse, A.P.; Ioppolo, A. Anatoxin-A and a previously unknown toxin in *Anabaena planctonica* from blooms found in lake Mulargia (Italy). *Toxicon* **1994**, *32*, 369–373. [CrossRef]
18. Wood, S.A.; Selwood, A.I.; Rueckert, A.; Holland, P.; Milne, J.R.; Smith, K.F.; Smits, B.; Watts, L.F.; Cary, C.S. First report of homoanatoxin-a and associated dog neurotoxicosis in New Zealand. *Toxicon* **2007**, *50*, 292–301. [CrossRef] [PubMed]

19. Namikoshi, M.; Murakami, T.; Watanabe, M.F.; Oda, T.; Yamada, J.; Tsujimura, S. Simultaneous production of homoanatoxin-a, anatoxin-a, and a new nontoxic 4-hydroxyhomoanatoxin-a by the cyanobacterium *Raphidiopsis mediterranea* Skuja. *Toxicon* **2003**, *42*, 533–538. [CrossRef]
20. Park, H.D.; Watanabe, M.F.; Harada, K.I.; Nagai, H.; Suzuki, M.; Watanabe, M.; Hayashi, H. Hepatotoxin (microcystin) and neurotoxin (anatoxin-a) contained in natural blooms and strains of cyanobacteria from Japanese freshwaters. *Nat. Toxins* **1993**, *1*, 353–360. [CrossRef]
21. Aráoz, R.; Molgó, J.; Tandeau de Marsac, N. Neurotoxic cyanobacterial toxins. *Toxicon* **2010**, *56*, 813–828. [CrossRef] [PubMed]
22. Nogueira, I.C.; Pereira, P.; Dias, E.; Pflugmacher, S.; Wiegand, C.; Franca, S.; Vasconcelos, V.M. Accumulation of paralytic shellfish toxins (PST) from the cyanobacterium *Aphanizomenon issatschenkoi* by the cladoceran *Daphnia magna*. *Toxicon* **2004**, *44*, 773–780. [CrossRef] [PubMed]
23. Lagos, N.; Onodera, H.; Zagatto, P.A.; Andrinolo, D.; Azevedo, S.M.F.Q.; Oshima, Y. The first evidence of paralytic shellfish toxins in the fresh water cyanobacterium *Cylindrospermopsis raciborskii* isolated from Brazil. *Toxicon* **1999**, *37*, 1359–1373. [CrossRef]
24. Yunes, J.S.; De La Rocha, S.; Giroldo, D.; Da Silveira, S.B.; Comin, R.; Bicho, M.D.S.; Melcher, S.S.; Santanna, C.L.; Vieira, A.A.H. Release of carbohydrates and proteins by a subtropical strain of *Raphidiopsis brookii* (Cyanobacteria) able to produce saxitoxin at three nitrate concentrations. *J. Phycol.* **2009**, *45*, 585–591. [CrossRef] [PubMed]
25. Pereira, P.; Li, R.; Carmichael, W.; Dias, E.; Franca, S. Taxonomy and production of paralytic shellfish toxins by the freshwater cyanobacterium *Aphanizomenon gracile*. *Eur. J. Phycol.* **2004**, *39*, 361–368. [CrossRef]
26. Ballot, A.; Fastner, J.; Wiedner, C. Paralytic shellfish poisoning toxin-producing cyanobacterium *Aphanizomenon gracile* in Northeast Germany. *Appl. Environ. Microbiol.* **2010**, *76*, 1173–1180. [CrossRef] [PubMed]
27. Ledreux, A.; Thomazeau, S.; Catherine, A.; Duval, C.; Yéprémian, C.; Marie, A.; Bernard, C. Evidence for saxitoxins production by the cyanobacterium *Aphanizomenon gracile* in a French recreational water body. *Harmful Algae* **2010**, *10*, 88–97. [CrossRef]
28. Rapala, J.; Robertson, A.; Negri, A.P.; Berg, K.A.; Tuomi, P.; Lyra, C.; Erkomaa, K.; Lahti, K.; Hoppu, K.; Lepistö, L. First Report of Saxitoxin in Finnish Lakes and Possible Associated Effects on Human Health. *Environ. Toxicol.* **2005**, *20*, 331–340. [CrossRef] [PubMed]
29. Santanna, C.; de Carvalho, L.R.; Fiore, M.F.; Silva-Stenico, M.E.; Lorenzi, A.S.; Rios, F.R.; Konno, K.; Lagos, N. Highly Toxic *Microcystis aeruginosa* Strain, Isolated from Sao Paulo—Brazil, Produce Hepatotoxins and Paralytic Shellfish Poison Neurotoxins. *Neurotox Res.* **2011**, *19*, 389–402. [CrossRef] [PubMed]
30. Quiblier, C.; Wood, S.; Echenique-Subiabre, I.; Heath, M.; Villeneuve, A.; Humbert, J.-F. A review of current knowledge on toxic benthic freshwater cyanobacteria—Ecology, toxin production and risk management. *Water Res.* **2013**, *47*, 5464–5479.
31. Pearson, L.A.; Dittmann, E.; Mazmouza, R.; Ongley, S.A.; D'Agostino, P.M.; Neilan, B.A. The genetics, biosynthesis and regulation of toxic specialized metabolites of cyanobacteria. *Harmful Algae* **2016**, *54*, 98–111. [CrossRef] [PubMed]
32. Messineo, V.; Bogialli, S.; Melchiorre, S.; Sechi, N.; Lugliè, A.; Casiddu, P.; Mariani, M.A.; Padedda, B.M.; Di Corcia, A.; Mazza, R.; et al. Cyanobacterial toxins in Italian freshwaters. *Limnologica* **2009**, *39*, 95–106. [CrossRef]
33. Dolman, A.M.; Rücker, J.; Pick, F.; Fastner, J.; Rohrlack, T.; Mischke, U.; Wiedner, C. Cyanobacteria and cyanotoxins: The influence of nitrogen versus phosphorus. *PLoS ONE* **2012**, *7*, e38757. [CrossRef] [PubMed]
34. Graham, J.; Loftin, K.A.; Meyer, M.T.; Ziegler, A.C. Cyanotoxin mixtures and taste-and-odor compounds in cyanobacterial blooms from the Midwestern United States. *Environ. Sci. Technol.* **2010**, *44*, 7361–7368. [CrossRef] [PubMed]
35. Backer, L.C.; Manassaram-Baptiste, D.; LePrell, R.; Bolton, B. Cyanobacteria and algae blooms: Review of health and environmental data from the Harmful Algal Bloom-Related Illness Surveillance System (HABISS) 2007–2011. *Toxins* **2015**, *7*, 1048–1064. [CrossRef] [PubMed]
36. Stüken, A.; Rücker, J.; Endrulat, T.; Preußel, K.; Hemm, M.; Nixdorf, B.; Karsten, U.; Wiedner, C. Distribution of three alien cyanobacterial species (Nostocales) in northeast Germany: *Cylindrospermopsis raciborskii*, *Anabaena bergii* and *Aphanizomenon aphanizomenoides*. *Phycologia* **2006**, *45*, 696–703. [CrossRef]

37. Cellamare, M.; Leitão, M.; Coste, M.; Dutartre, A.; Haury, J. Tropical phytoplankton taxa in Aquitaine lakes (France). *Hydrobiologia* **2010**, *639*, 129–145. [CrossRef]
38. Gugger, M.; Molica, R.; Le Berre, B.; Dufour, P.; Bernard, C.; Humbert, J.F. Genetic diversity of Cylindrospermopsis strains (cyanobacteria) isolated from four continents. *Appl. Environ. Microbiol.* **2005**, *71*, 1097–1100. [CrossRef] [PubMed]
39. Wiedner, C.; Rucker, J.; Bruggemann, R.; Nixdorf, B. Climate change affects timing and size of populations of an invasive cyanobacterium in temperate regions. *Oecologia* **2007**, *152*, 473–484. [CrossRef] [PubMed]
40. Rücker, J.; Tingwey, E.I.; Wiedner, C.; Anu, C.M.; Nixdorf, B. Impact of the inoculum size on the population of Nostocales cyanobacteria in a temperate lake. *J. Plankton Res.* **2009**, *31*, 1151–1159. [CrossRef]
41. O'Neil, J.M.; Davis, T.W.; Burford, M.A.; Gobler, C.J. The rise of harmful cyanobacteria blooms: The potential roles of eutrophication and climate change. *Harmful Algae* **2012**, *14*, 313–334. [CrossRef]
42. Gugger, M.; Lenoir, S.; Berger, C.; Ledreux, A.; Druart, J.C.; Humbert, J.F.; Guette, C.; Bernard, C. First report in a river in France of the benthic cyanobacterium *Phormidium favosum* producing anatoxin-a associated with dog neurotoxicosis. *Toxicon* **2005**, *45*, 919–928. [CrossRef] [PubMed]
43. Cadel-Six, S.; Peyraud-Thomas, C.; Brient, L.; Tandeau de Marsac, N.; Rippka, R.; Méjean, A. Different genotypes of anatoxin-producing cyanobacteria coexist in the Tarn River, France. *Appl. Environ. Microbiol.* **2007**, *73*, 7605–7614. [CrossRef] [PubMed]
44. Pitois, F.; Vezie, C.; Thoraval, I.; Baurès, E. Improving Microcystin monitoring relevance in recreative waters: A regional case-study (Brittany, Western France, Europe). *Int. J. Hygiene Environ. Health* **2016**, *219*, 288–293. [CrossRef] [PubMed]
45. Rücker, J.; Stüken, A.; Nixdorf, B.; Fastner, J.; Chorus, I.; Wiedner, C. Concentrations of particulate and dissolved cylindrospermopsin in 21 Aphanizomenon-dominated temperate lakes. *Toxicon* **2007**, *50*, 800–809. [CrossRef] [PubMed]
46. Bernstein, J.A.; Ghosh, D.; Levin, L.S.; Zheng, S.; Carmichael, W.W.; Lummus, Z.; Bernstein, L. Cyanobacteria: An unrecognized ubiquitous sensitizing allergen? *Allergy Asthma Proc.* **2011**, *32*, 106–110. [CrossRef] [PubMed]
47. Genitsaris, S.; Kormas, K.; Moustaka-Gouni, M. Airborne Algae and Cyanobacteria: Occurrence and Related Health Effects. *Front. Biosci.* **2011**, *E3*, 772–787.
48. Ohkouchi, Y.; Tajima, S.; Nomura, M.; Itoh, S. Inflammatory responses and potencies of various lipopolysaccharides fromb acteria and cyanobacteria in aquatic environments and water supply systems. *Toxicon* **2015**, *97*, 23–31. [CrossRef] [PubMed]
49. Gacsi, M.; Antal, O.; Vasas, G.; Mathe, C.; Borbely, G.; Saker, M.L.; Gyori, J.; Farkas, A.; Vehovszky, A.; Banfalvi, G. Comparative study of cyanotoxins affecting cytoskeletal and chromatin structures in CHO-K1 cells. *Toxicol. In Vitro* **2009**, *23*, 710–718. [CrossRef] [PubMed]
50. Nováková, K.; Bláhaa, L.; Babica, P. Tumor promoting effects of cyanobacterial extracts are potentiated by anthropogenic contaminants—Evidence from in vitro study. *Chemosphere* **2012**, *89*, 30–37. [CrossRef] [PubMed]
51. Kozdeba, M.; Borowczyk, J.; Zimolag, E.; Wasylewski, M.; Dziga, D.; Madeja, Z.; Drukala, J. Microcystin-LR affects properties of human epidermal skin cells crucial for regenerative processes. *Toxicon* **2014**, *80*, 38–46. [CrossRef] [PubMed]
52. Geitler, L. *Cyanophyceae von Europa, Kryptogamen Flora von Deutschland, Osterreich und der Schweiz*; Koeltz Scientific Books (reprint 1985) Koeningstein: Oberreifenberg, Germany, 1932; p. 1196.
53. Hill, H. A new Raphidiopsis species (Cyanophyta, Rivulariaceae) from Minnesota lakes. *Phycologia* **1972**, *11*, 73–77. [CrossRef]
54. Komárek, J.; Anagnostidis, K. *Cyanoprokaryota—1 Teil: Chroococcales, Süßwasserflora von Mitteleuropa*; Spektrum Akademischer Verlag: Berlin, Germany, 1998; p. 548.
55. Komárek, J.; Anagnostidis, K. *Cyanoprokaryota—2 Teil: Oscillatoriales, Süßwasserflora von Mitteleuropa*; Spektrum Akademischer Verlag: Berlin, Germany, 2005; p. 759.
56. Watanabe, M. Studies on the Planktonic Blue-Green Algae 3. Some Aphanizomenon Species in Hokkaido, Northen Japan. *Bull. Nat. Sci. Mus. Tokyo Ser. B* **1991**, *17*, 141–150.
57. Watanabe, M. Studies on Planktonic Blue-green Algae 4. Some Anabaena species with straight Trichomes in Japan. *Bull. Nat. Sci. Mus. Tokyo Ser. B* **1992**, *18*, 123–137.

58. Watanabe, M. Studies on Planktonic Blue-green Algae 8. Anabaena species with twisted Trichomes in Japan. *Bull. Nat. Sci. Mus. Tokyo Ser. B* **1998**, *24*, 1–13.
59. Watanabe, M.; Niiyama, Y.; Tuji, A. Studies on Planktonic Blue-green Algae 10. Classification of Planktonic Anabaena with coiled Trichomes maintained In the National Science Museum. *Tokyo. Bull. Nat. Sci. Mus. Tokyo Ser. B* **2004**, *30*, 135–149.
60. Komárek, J.; Komárkova, J. Diversity of Aphanizomenon-like cyanobacteria. *Czech Phycol. Olomouc* **2006**, *6*, 1–32.
61. Dell'Aversano, C.; Eaglesham, G.K.; Quilliam, M.A. Analysis of cyanobacterial toxins by hydrophilic interaction liquid chromatography-mass spectrometry. *J. Chromatogr. A* **2004**, *1028*, 155–164. [CrossRef] [PubMed]
62. Spoof, L.; Vesterkvist, P.; Lindholm, T.; Meriluoto, J. Screening for cyanobacterial hepatotoxins, microcystins and nodularin in environmental water samples by reversed-phase liquid chromatography–electrospray ionisation mass spectrometry et al. *J. Chromatogr. A* **2003**, *1020*, 105–119. [CrossRef]
63. Fastner, J.; Beulker, C.; Geiser, B.; Hoffmann, A.; Kröger, R.; Teske, K.; Hoppe, J.; Mundhenk, L.; Neurath, H.; Sagebiel, D.; et al. Fatal Neurotoxicosis in Dogs Associated with Tychoplanktic, Anatoxin-a Producing Tychonema sp. in Mesotrophic Lake Tegel, Berlin. *Toxins* **2018**, *10*, 60. [CrossRef] [PubMed]

toxins

MDPI

Article

Effects of Dietary Astaxanthin Supplementation on Energy Budget and Bioaccumulation in *Procambarus clarkii* (Girard, 1852) Crayfish under Microcystin-LR Stress

Zhenhua An *, Yingying Zhang and Longshen Sun

College of Animal Science and Technology, Yangzhou University, Yangzhou 225009, China; zhangyingying@yzu.edu.cn (Y.Z.); lssun@yzu.edu.cn (L.S.)

* Correspondence: anzhenhua@yzu.edu.cn; Tel.: +86-514-8797-2208; Fax: +86-514-8735-0440

Received: 7 May 2018; Accepted: 21 June 2018; Published: 4 July 2018

Abstract: This research aimed to study the effects of astaxanthin on energy budget and bioaccumulation of microcystin-leucine-arginine (microcystin-LR) in the crayfish *Procambarus clarkii* (Girard, 1852). The crayfish (21.13 $\pm$ 4.6 g) were cultured under microcystin-LR stress (0.025 mg/L) and were fed with fodders containing astaxanthin (0, 3, 6, 9, and 12 mg/g) for 8 weeks in glass tanks (350 mm $\times$ 450 mm $\times$ 150 mm). Accumulations of microcystin-LR were measured in different organs of *P. clarkii*. The results suggested that astaxanthin can significantly improve the survival rate and specific growth rate (SGR) of *P. clarkii* ($p < 0.05$). The dietary astaxanthin supplement seems to block the bioaccumulation of microcystin-LR in the hepatopancreas and ovaries of *P. clarkii* to some extent ($p < 0.05$). Astaxanthin content of 9–12 mg/g in fodder can be a practical and economic choice.

Keywords: microcystin-LR; *Procambarus clarkii*; energy budget; astaxanthin

Key Contribution: Astaxanthin in the diet was used to degrade the oxidative stress caused by microcystin-LR, and bioaccumulation was significantly affected by the astaxanthin content. The appropriate content of astaxanthin in feed may be around 9–12 mg/g and it may make the crayfish *Procambarus clarkii* a more secure food for people.

1. Introduction

In China, cyanobacteria blooms have often been observed in some large lakes around Jiangsu province. These lakes and nearby ponds are also characterized by the presence of cultured red swamp *Procambarus clarkii* (Girard, 1852). As the most invasive freshwater crayfish in the world, *P. clarkii* is supposed to be tolerant to extreme conditions and is easily cultured, even in water where some cyanobacteria blooms occur [1]. Microcystins, as a group of cyclic polypeptides, are the most common cyanotoxins produced by several genera of cyanobacteria. Several studies have shown that exposure to microcystins can either directly kill organisms or decrease their resistance to bacterial or viral infections [2] and to some extent postpone the creatures' growth speed. Perhaps that is why *P. clarkii* in nutrient-enriched waters with cyanobacteria blooms always grow slowly and weakly. However, there are also studies suggesting that the *P. clarkii* individuals can feed on microalgae and accumulate toxins in their tissues, without showing any apparent changes in their behavior or vitality [3]. Considering the enormous crayfish consumption in China, this makes the presence of microcystins a serious problem for *P. clarkii* culture and safe consumption in this area.

Astaxanthin, as a useful antioxidant [4] is supposed to be a key composition that can make a great contribution to growth performance, maturation, and carapace color of crayfish [5]. It is also reported that astaxanthin can relieve the negative circumstance stress on juvenile kuruma shrimp

Marsupenaeus japonicus [6] and astaxanthin contents increased significantly within the interval between the juvenile stages I and II in the embryonic development of crayfish *Astacus leptodactylus* [7]. In this study, the effects of astaxanthin on energy budget and microcystin-leucine-arginine (microcystin-LR) bioaccumulation of the crayfish *P. clarkii* were measured. Our aims were to develop protocols to accelerate the microcystin-LR depuration of crayfish *P. clarkii* and assess the effects of astaxanthin. The results may provide a reference for the astaxanthin promotion in *P. clarkii* cultures.

2. Results

2.1. Growth

During the experiment, some crayfish stopped wiggling antennae and crouched. The experiment showed that after nearly two months of poisoning, the reactivity and movement of some tested crayfish became slower and they subsequently died. The survival ratios of the experiment are shown in Table 1.

Table 1. The survival ratios of *Procambarus clarkii* (Girard, 1852) fed on a diet with different astaxanthin (Ax) contents (0, 3, 6, 9, 12 mg/g) under microcystin-leucine-arginine (microcystin-LR) stress (25 ug/L) (mean ± S.E).[1]

Ax Content mg/g	0	3	6	9	12
Survival ratio %	77.78 ± 4.81^{a}	83.33 ± 8.33^{ab}	$86.11 \pm 12.73^{a,b}$	94.44 ± 4.81^{b}	97.22 ± 4.81^{b}

[1] Values (expressed as mean ± S.E., n = 3) with different letters superscript are significantly different from each other ($p < 0.05$).

The maximum and minimum specific growth rate (SGR) (SGR_w and SGR_e, respectively) occurred in treatment E (12 mg/L astaxanthin concentration ration). Stepwise regression analysis showed that SGR_w and SGR_e increased with increasing astaxanthin concentration (Figure 1). The relationship among SGR_w and SGR_e and astaxanthin concentration (Ac %) can be described by the regression equations:

$$SGR_w = 0.191 + 25.867 \text{ Ac } (r^2 = 0.899, n = 5) \quad (1)$$

$$SGR_e = 0.153 + 25.113 \text{ Ac } (r^2 = 0.863, n = 5) \quad (2)$$

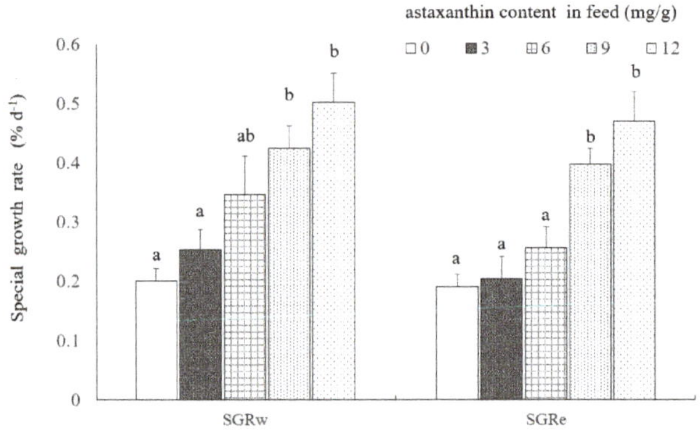

Figure 1. The effects of astaxanthin content (0, 3, 6, 9, 12 mg/g) in feed on the specific growth rate (SGR_w and SGR_e) (expressed as mean ± S.E., n = 3) of the crayfish *Procambarus clarkii* (Girard, 1852) under microcystin-LR stress (25 ug/L). Histograms sharing a common letter on top are not significantly different ($p > 0.05$).

The results of one-way ANOVA analysis showed astaxanthin concentrations in the diet had significant effects on SGR_W and SGR_e values of *P. clarkii* ($p < 0.05$). The SGR_W and SGR_e values in treatments D and E were significantly higher than the treatments A and B ($p < 0.05$) (Figure 2).

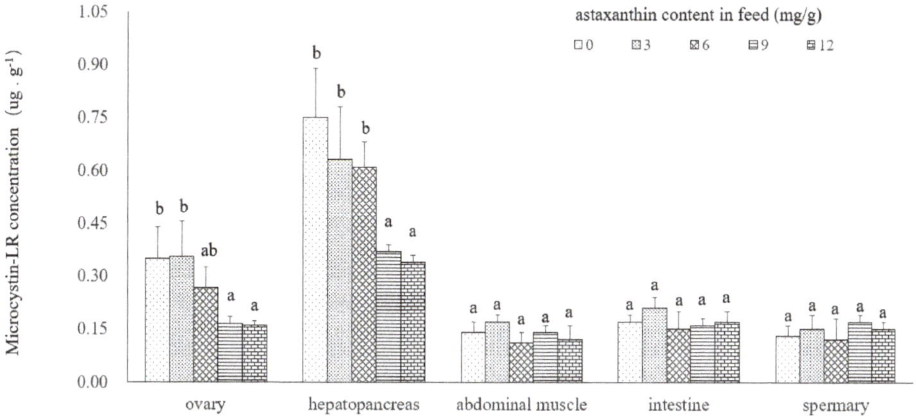

Figure 2. Biological enrichment of microcystin-LR (expressed as mean ± S.E., $n = 3$) in the hepatopancreas, muscle, intestine, ovary and spermary of *Procambarus clarkii* (Girard, 1852). Histograms sharing a common letter on top in the same tissue are not significantly different ($p > 0.05$).

2.2. Microcystin Depuration

In the experiment, microcystin-LR was observed in all organs/tissues analyzed, but mostly in the hepatopancreas and ovary. The microcystin contents were observed to decrease in the following order: hepatopancreas > ovary > intestine > abdominal muscle (Figure 2). The microcystin-LR contents in hepatopancreas and ovary were significantly higher than others and they can be considered as indicators of microcystin-LR stress. The results in abdominal muscle were slightly different to those of the spermary. The results of one-way ANOVA analysis showed that the astaxanthin concentration in diet had a significant effect on bioaccumulation of *P. clarkii* in the ovary and hepatopancreas ($p < 0.05$).

2.3. Energy Allocation

The energy budget in the tested crayfish changed with the different astaxanthin concentrations in diet (Table 2). At treatments E and D the ratio of G/C in treatment were significantly higher than in the other groups. The ratio of F/C in treatments suggested no significant differences between each other ($p > 0.05$). The highest ratio of R/C occurred in treatment A and it was significantly higher than in treatments D and E ($p < 0.05$).

Table 2. Daily energy budgets of *Procambarus clarkii* fed on diets with different astaxanthin contents (0, 3, 6, 9, 12 mg/g) (mean ± S.E) [1].

Astaxanthin Content mg/g	C [2]	G [3]	F [4]	E [5]	U [6]	R [7]
0	4304.24 ± 177.48 ^{a}	612.47 ± 95.0 ^{a}	1027.15 ± 105.80 ^{a}	120.56 ± 14.02 ^{a}	182.17 ± 14.02 ^{a}	2361.89 ± 146.96 ^{b}
3	4413.37 ± 151.21 ^{a}	678.24 ± 107.11 ^{a}	1079.89 ± 77.55 ^{a}	141.24 ± 23.59 ^{a}	186.47 ± 13.59 ^{a}	2327.53 ± 139.19 ^{b}
6	4315.56 ± 147.14 ^{a}	845.24 ± 125.24 ^{a,b}	1024.54 ± 86.78 ^{a}	155.73 ± 16.10 ^{a}	196.59 ± 16.10 ^{a}	2093.46 ± 170.37 ^{a,b}
9	4532.36 ± 284.82 ^{a}	1048.12 ± 76.37 ^{b}	1263.78 ± 97.73 ^{a}	221.47 ± 31.68 ^{b}	201.21 ± 21.68 ^{a}	1797.78 ± 114.54 ^{a}
12	4557.24 ± 147.97 ^{a}	1175.11 ± 92.45 ^{b}	1243.45 ± 107.73 ^{a}	233.45 ± 21.68 ^{b}	214.53 ± 21.68 ^{a}	1690.70 ± 114.54 ^{a}

[1] Values (expressed as mean ± S.E., $n = 3$) with different letters superscript in the same column are significantly different from each other ($p < 0.05$); [2] C ($J \cdot g^{-1} \cdot d^{-1}$) = energy consumed in food; [3] G ($J \cdot g^{-1} \cdot d^{-1}$) = energy deposited as growth; [4] F ($J \cdot g^{-1} \cdot d^{-1}$) = energy lost from feces; [5] E ($J \cdot g^{-1} \cdot d^{-1}$) = energy lost from molt; [6] U ($J \cdot g^{-1} \cdot d^{-1}$) = energy lost from excretion; [7] R ($J \cdot g^{-1} \cdot d^{-1}$) = energy used in respiration.

The results suggested that ratios of G/C and R/C were significantly affected by the astaxanthin concentration in the diet ($p < 0.05$). Due to the different numbers of molts, the ratios of E/C in treatment A, B, and C were significantly lower than D and E ($p < 0.05$).

3. Discussion

As an organic toxin, microcystin-LR is difficult to degrade and accumulates easily in some aquatic products [8,9], causing public concern about the pollution problems caused by microcystin. Although microcystins are produced by several genera of cyanobacteria, such as *Microcystis*, *Anabaena*, and *Oscillatoria*, the most commonly reported species is *Microcystis aeruginosa*. This species is inclined to live in relatively quiet waters and forms surface water blooms between summer and autumn [10]. In China, around the Changjiang plain, most *P. clarkii* crayfish are cultured in ponds without fast water pump facilities, and that makes the microcystin problem of *P. clarkii* more serious in this area.

There are many studies suggesting that the sensitivity of aquatic animals to microcystins changes depending on the organism [11], the variant, and the mode of exposure [12]. The content of microcystins in different organs can be an indicator of microcystins stress and the organism can also perform depuration of microcystins to some extent [13]. There are several studies on depuration of microcystins in fish and other aquatic organisms, showing a decrease in microcystin content in several organs (liver and muscle) in a time-dependent manner [14,15]. It has been suggested that the most affected organ with regard to the lipid peroxidation caused by microcystin-LR is the liver [16]. In this study, we used the purified microcystin-LR and found the hepatopancreas and ovary to accumulate with toxins than other organs; and this may be the reason why the occurrence of cyanobacteria blooms in the crayfish culture ponds often results in a fall in production and size of captures. The spermary seems to differ from the ovary and its toxin concentration was almost the same as in the muscles. Considering the abundant lipid content of the crayfish ovary during breeding season [17], this finding may be due to the difference of lipid content and the metabolic regulation mechanisms of *P. clarkii*.

Microcystin exposure has been reported to cause oxidative stress to animals [18,19]. In our research the energy allocated for respiration (R) was significantly decreased with the increasing dietary astaxanthin supplementation ($p < 0.05$). This could implicate that astaxanthin could degrade the oxidative stress caused by microcystin-LR to some extent. There was some research suggesting that astaxanthin could also enhance the specific growth rate and relieve the fresh water-osmotic stress in juvenile kuruma shrimp *Marsupenaeus japonicus* [7]. Some research also reported that an astaxanthin-supplemented diet could not only shorten the molting cycle of the juvenile crustacean but also the postlarval stages of some shrimps such as *Penaeus japonicus* [20]. This research showed the similar results that the energy allocated for molt (E) was significantly higher in the treatments where the crayfish were fed with high dietary astaxanthin supplementation ($p < 0.05$). In sum, astaxanthin could apparently reduce the energy used for respiration and increase the growth and molting ratios in total energy consumed in food.

With the rapidly development of the astaxanthin compositing industry and the falling costs of astaxanthin, it is more and more practical to use the astaxanthin in crustacean culture. Based on the growth and energy allocation results, the appropriate supplement of astaxanthin in feed may be around 10 mg/g. In this way, astaxanthin, which is usually considered responsible for crayfish coloration, will also make the crayfish *P. clarkii* a safer food for human consumption.

4. Material and Methods

4.1. Source of Animals and Acclimation

The experiments were conducted from 1 September to 27 October 2015, at the Aquaculture Research Laboratory, Yangzhou University. The crayfish were captured in the suburb of Baoying lake, Yangzhou city and cultivated at a temperature 24 ± 2 °C in several fiberglass tanks with tap water aerated for 48 h, and the water pH adjusted to 7.5 ± 0.5. During the acclimation period, the animals were fed twice a day (at 08.00 h and 18.00 h) with commercial feed provided by Fuyuda corporation (41.70% crude protein, 7.67% crude lipid, 7.89% ash, moisture < 2.70%; energy 21.55 kJ/g dry mass). Aeration was provided continuously and one-third of the water volume in all the experiment tanks was exchanged every day. Dissolved oxygen was maintained above 4.0 mg/L.

4.2. Experimental Design and Procedure

The original pure lyophilized microcystin-LR was bought from Express Technology Co., Ltd., (Beijing, China) and was prepared into 10 mg/L of mother solution with double-distilled water. Microcystin analysis was conducted by ELISA test. The ELISA test kits were bought from J & Q Environment Corporation (Beijing, China). This method had a sensitivity of 0.1 ng/mL. The astaxanthin capsules were brought from Fujian Corona Technology Corporation (Fuzhou, Fujian Province, China). Each capsule contained astaxanthin 60 mg (alga extraction), and soybean oil 440 mg. In making the feed, we squeezed the capsules and mixed the contents well with the same batch commodity feed which was also used in acclimation. The feed slowly absorbed the oil through its capillaries and within 1 h there was no discernible solution in water. We also added a little soybean oil in different feeds to compensate for the energy differences caused by astaxanthin adjunction.

Finally, five astaxanthin feed concentrations (0, 3, 6, 9 and 12 mg/g; treatments A, B, C, D, and E) were made, and each treatment had three parallel groups. Each parallel group had 12 crayfish. Totally 180 crayfish (with average weight 21.13 ± 4.6 g and male: female = 1:1) were used and there were no significant differences between groups. The acclimation was followed by starvation for 24 h. The initial body weight of each experimental animal was measured. The experiments were carried out in glass aquaria (35 cm width × 45 cm length × 15 cm depth), the solution in each experimental group was over 15 L with 0.025 mg/L microcystin-LR concentration and one-third of the water volume in all the experiment tanks was exchanged every day with siphoned feces and uneaten feed to stabilize the microcystin-LR concentration. Aeration was provided continuously, and dissolved oxygen was maintained above 4.0 mg/L.

During the experiment the animals were fed ad libitum twice a day (08:00 h and 18:00 h) with the feed. The experimental temperature was 24 ± 2 °C. The number of dead crayfish was recorded per 24 h and at finally 22 crayfish (12.22%) were dead in the experiment. At the end of the culture, six living crayfish (half male and half female) were picked up in each parallel group. In total, 18 crayfish were prepared for bioaccumulation test in each treatment. The concentrations of microcystin-LR in the hepatopancreas, intestine, gonads (ovary or spermary), and abdominal muscle of the crayfish were measured by the means of ELISA. The analysis of microcystin content was conducted on samples of similar fresh weight (0.5 g for intestine and spermary; 1.0 g for hepatopancreas, abdomen, and ovaries) obtained by pooling the organs/tissues of three crayfish. Each tissue was tested with three samples in each treatment. The ovaries or spermaries were prepared separately from nine females and nine males. Other tissues were randomly sampled from nine of the former crayfish from which gonads had been obtained. Half the crayfish provided both gonads and tissues (hepatopancreas, abdomen muscle, and intestine) for analysis.

4.3. Energy Determination and Estimation of Energy Budget

During the 56-day course of the experiment, the weight of each ration was recorded. Uneaten feed, feces, and molt (exuvia) were separated and removed by siphon to avoid decomposition, dried at 65 °C, shattered, weighed, and kept for analysis of energy and nitrogen content. At the end of the experiment, the animals were starved for 24 h, and then weighed and dried at 65 °C for 48 h and shattered for measurement. The energy contents of the crayfish bodies, feed, and feces were measured with a Parr 6300 Oxygen Bomb Calorimeter (Parr, Moline, IL, USA). The energy budget was calculated by the following equation [21]:

$$C = G + F + U + R + E \quad (3)$$

where C is the energy consumed in food; G is the energy deposited as growth; F is the energy lost in feces; E is the energy lost in molt; U is the energy lost in excretion; and R is the energy used for respiration. The value of C and F can be calculated by the weight of the samples of food intake, feces weight, and energy content per gram. G can be calculated by the following equation:

$$G = (Fw \times Fe) - (Iw \times Ie) \quad (4)$$

where Fw and Iw are final body weight and initial body weight of the crayfish, respectively; Fe and Ie are the energy content per germ of final body and initial body of the crayfish, respectively.

F and E were calculated by the following equation:

$$F = Pw \times Pe; \quad (5)$$

$$E = Ew \times Ee \quad (6)$$

where Pw and Ew are final collected feces weight and crustaceous membrane weight of the crayfish, respectively; Pe and Ee are the energy content per germ of feces and crustaceous membrane of the crayfish, respectively.

The nitrogen contents of the crayfish bodies, food, and feces were measured with a Vario Elcube elemental analyzer (Elementar, Germany) at the test center of Yangzhou University. The estimation of U was based on the nitrogen budget equation [22]:

$$U = (CN - GN - FN - EN) \times 24{,}830 \quad (7)$$

where CN is the nitrogen consumed from food; FN is the nitrogen lost in feces; GN is the nitrogen deposited in the body; EN is the nitrogen deposited in the molt; and 24,830 is the energy content ($J \cdot g^{-1}$) of excreted nitrogen.

The value of R was calculated by the energy budget equation:

$$R = C - G - F - U - E \quad (8)$$

4.4. Calculation and Data Analysis

Specific growth rate in terms of weight (SGRw) and energy (SGRe) were calculated as:

$$SGRw\ (\%\ day^{-1}) = 100 \times (\ln W_2 - \ln W_1)/D \quad (9)$$

$$SGRe\ (\%\ day^{-1}) = 100 \times (\ln E_2 - \ln E_1)/D \quad (10)$$

where W_2 and W_1 are the final and initial wet body weight of the crayfish, respectively; E_2 and E_1 are the final and initial body energy of the crayfish, respectively; and D is the duration of the experiment. The data were analyzed by SPSS for Windows (Version 19.0) statistical package (SPSS Inc., Chicago, IL, USA). Inter-treatment differences of survival ratios, SGRw, SGRe, concentration of microcystin-LR, and energy allocation were analyzed with one-way ANOVA followed by post-hoc Tukey multiple range tests. Differences were considered significant if $p < 0.05$.

Author Contributions: Z.H.A. designed the study. L.S.S. conceived and designed the experiments and collected the samples. Z.H.A. and Y.Y.Z. analyzed the data and contributed reagents, materials, and analysis tools, and provided valuable input into the study design and the manuscript.

Acknowledgments: Jiangsu Fisheries Reasearch System (Red Swamp Crayfish) JFRS-03 and Jiangsu Agricultural Science and Technology Innovation Fund (JASTIF) supported this study.

Conflicts of Interest: The authors declare no conflict of interest.

References

1. Gherardi, F.; Lazzara, L. Effects of the density of an invasive crayfish (*P. clarkii*) on a pelagic and surface microalgae in a Mediterranean wetland. *Arch. Hydrobiol.* **2006**, *165*, 401–414. [CrossRef]
2. Carmichael, W.W. *Toxic Microcystis and the Environment*; Watanabe, M.F., Harada, K., Carmichael, W.W., Fujiki, H., Eds.; CRC Press: Boca Raton, FL, USA, 1996; p. 111.
3. Vasconcelos, V.; Oliveira, S.; Teles, F.O. Impact of a toxic and non-toxic strain of *Microcystis aeruginosa* on the crayfish *Procambarus clarkii*. *Toxicon* **2001**, *39*, 1461–1470. [CrossRef]
4. Yuan, J.P.; Peng, J.; Yin, K.; Wang, J.H. Potential health-promoting effects of astaxanthin: A high-value carotenoid mostly from microalgae. *Mol. Nutr. Food Res.* **2011**, *55*, 150–165. [CrossRef] [PubMed]
5. Kaldre, K.; Haugjärv, K.; Liiva, M.; Gross, R. The effect of two different feeds on growth, carapace colour, maturation and mortality in marbled crayfish (*Procambarus fallax* f. *virginalis*). *Aquac. Int.* **2015**, *23*, 185–194. [CrossRef]
6. Wang, W.; Ishikawa, M.; Koshio, S.; Yokoyama, S.; Hossain, M.S.; Moss, A.S. Effects of dietary astaxanthin supplementation on juvenile kuruma shrimp, *Marsupenaeus japonicus*. *Aquaculture* **2018**, *491*, 197–204. [CrossRef]
7. Berticat, O.; Nègre-Sadargues, G.; Castillo, R. The metabolism of astaxanthin during the embryonic development of the crayfish *Astacus leptodactylus* Eschscholtz (Crustacea, Astacidea). *Comp. Biochem. Physiol. Part B Biochem.* **2000**, *127*, 309–318. [CrossRef]
8. Xu, H.B.; Sui, H.X.; Gao, S.R. Primary experimental study on bioaccumulation of microcystin in *Cyprinus carpio* L. *Chin. J. Food Hyg.* **2003**, *15*, 202–204.
9. Li, X.G.; Zhou, G.; Zhou, J.L. Preliminary study on the Accumulation and Biodepuration of Microcystins in Tilapia. *J. Hydroecol.* **2010**, *3*, 67–70.
10. Krishnamurthy, T.; Carmichael, W.W.; Sarver, E.W. Toxic peptides from freshwater cyanobacteria (blue-green-algae). I. Isolation, purification and characterization of peptides from *Microcystis aeruginosa* and *Anabaena flos-aquae*. *Toxicon* **1986**, *24*, 865–873. [CrossRef]
11. Hansson, L.A.; Gustafsson, S.; Rengefors, K.; Bomark, L. Cyanobacterial chemical warfare affects zooplankton community composition. *Fresh Water Biol.* **2007**, *52*, 1290–1301. [CrossRef]
12. Codd, G.A.; Poon, G.K. Cyanobacterial toxins. In *Biochemistry of the Algae and Cyanobacteria*; Rogers, L.J., Gallon, J.R., Eds.; Oxford Science Publishers, Clarendon Press: Oxford, UK, 1998; pp. 283–296.
13. Chen, J.; Xie, P. Tissue distributions and seasonal dynamics of the hepatotoxic microcystins-LR and-RR in two freshwater shrimps, *Palaemon modestus* and *Macrobrachium nipponensis*, from a large shallow, eutrophic lake of the subtropical China. *Toxicon* **2005**, *45*, 615–625. [CrossRef] [PubMed]
14. Smith, J.L.; Haney, J.F. Foodweb transfer, accumulation, and depuration of microcystins, a cyanobacterial toxin, in pumpkinseed sunfish (*Lepomis gibbosus*). *Toxicon* **2006**, *48*, 580–589. [CrossRef] [PubMed]
15. Elena, T.; Silvia, B.; Sara, B. Depuration of microcystin-LR from the red swamp crayfish *Procambarus clarkii* with assessment of its food quality. *Aquaculture* **2008**, *285*, 90–95.
16. Huang, C.H.; Chang, R.J.; Huang, S.L.; Chen, W.L. Dietary vitamin E supplementation affects tissue lipid peroxidation of hybrid tilapia. *Oreochromis niloticus*×*O. aureus*. *Comp. Biochem. Physiol. Part B* **2003**, *134*, 265–270. [CrossRef]
17. Garciaguerrero, M.; Racotta, I.S.; Villarreal, H. Variation in lipid, protein, and carbohydrate content during the embryonic development of the crayfish *Cherax Quadricarinatus* (Decapoda: Parastacidae). *J. Crustacean Biol.* **2003**, *23*, 1–6. [CrossRef]
18. Cazenave, J.; Bistoni, M.A.; Pesce, S.F.; Wunderlin, D.A. Differential detoxification and antioxidant response in diverse organs of *Corydoras paleatus* experimentally exposed to microcystin-RR. *Aquatic Toxicol.* **2006**, *76*, 1–12. [CrossRef] [PubMed]
19. Prieto, A.I.; Pichardo, S.; Jos, A.; Moreno, I.; Camean, A.M. Time-dependent oxidative stress responses after acute exposure to toxic cyanobacterial cells containing microcystins in tilapia fish (*Oreochromis niloticus*) under laboratory conditions. *Aquatic Toxicol.* **2007**, *84*, 337–345. [CrossRef] [PubMed]
20. Petit, H.; Negre-Sadargues, G.; Castillo, R.; Trilles, J.P. The effects of dietary astaxanthin on growth and moulting cycle of postlarval stages of the prawn, *Penaeus Japonicus* (Crustacea, Decapoda). *Comp. Biochem. Physiol. Part A Physiol.* **1997**, *117*, 539–544. [CrossRef]

21. Carfoot, T.H. *Animal Energetics*; Academic Press: New York, NY, USA, 1987; pp. 407–515.
22. Levine, D.M.; Sulkin, S.D. Partitioning and utilization of energy during the larval development of the xanthid crab, *Rithropanopeus harrisii* (Gould). *J. Exp. Mar. Biol. Ecol.* **1979**, *40*, 247–257. [CrossRef]

toxins

MDPI

Article

The Presence of Toxic and Non-Toxic Cyanobacteria in the Sediments of the Limpopo River Basin: Implications for Human Health

Murendeni Magonono [1], Paul Johan Oberholster [2], Addmore Shonhai [3], Stanley Makumire [3] and Jabulani Ray Gumbo [1,*]

1 Department of Hydrology and Water Resources, School of Environmental Sciences, University of Venda, Thohoyandou 0950, South Africa; murendy22@gmail.com

2 Council for Scientific and Industrial Research, Natural Resources and the Environment, Stellenbosch 7600, South Africa; poberholster@csir.co.za

3 Department of Biochemistry, School of Mathematical and Natural Sciences, University of Venda, Thohoyandou 0950, South Africa; addmore.shonhai@univen.ac.za (A.S.); stanmakster@gmail.com (S.M.)

* Correspondence: jabulani.gumbo@univen.ac.za; Tel.: +27-15-962-8563

Received: 10 May 2018; Accepted: 21 June 2018; Published: 3 July 2018

Abstract: The presence of harmful algal blooms (HABs) and cyanotoxins in drinking water sources poses a great threat to human health. The current study employed molecular techniques to determine the occurrence of non-toxic and toxic cyanobacteria species in the Limpopo River basin based on the phylogenetic analysis of the 16S rRNA gene. Bottom sediment samples were collected from selected rivers: Limpopo, Crocodile, Mokolo, Mogalakwena, Nzhelele, Lephalale, Sand Rivers (South Africa); Notwane (Botswana); and Shashe River and Mzingwane River (Zimbabwe). A physical-chemical analysis of the bottom sediments showed the availability of nutrients, nitrates and phosphates, in excess of 0.5 mg/L, in most of the river sediments, while alkalinity, pH and salinity were in excess of 500 mg/L. The FlowCam showed the dominant cyanobacteria species that were identified from the sediment samples, and these were the *Microcystis* species, followed by *Raphidiopsis raciborskii*, *Phormidium* and *Planktothrix* species. The latter species were also confirmed by molecular techniques. Nevertheless, two samples showed an amplification of the cylindrospermopsin polyketide synthetase gene (S3 and S9), while the other two samples showed an amplification for the microcystin/nodularin synthetase genes (S8 and S13). Thus, these findings may imply the presence of toxic cyanobacteria species in the studied river sediments. The presence of cyanobacteria may be hazardous to humans because rural communities and farmers abstract water from the Limpopo river catchment for human consumption, livestock and wildlife watering and irrigation.

Keywords: cyanobacteria; cyanotoxins; nutrient enrichment; akinetes; harmful algal blooms; PCR; phylogenetic analyses

Key Contribution: Presence of viable cyanobacteria akinetes and cysts in river sediments; a source of inoculum of cyanobacteria growth in the Limpopo river basin. Some of the cyanobacteria species are toxic.

1. Introduction

Toxic and non-toxic cyanobacteria species are on the increase in most parts of the world, including in South Africa. The emergence and resurgence of harmful algal blooms (HABS) is due to eutrophication. The toxic cyanobacteria are known to carry genes that produce cyanotoxins which are lethal to humans. However, the toxic and non-toxic cyanobacteria species merely differ in the *mcy* gene content, which is the peptide synthetase producing microcystin [1]. This may explain the observation of non-detectable microsystin toxin despite the presence of *mcy* gene [2]. A study by Frazao et al. [3] used the PCR method to determine molecular analysis of genes involved in the production of known cyanotoxins, microcystins, nodularins and cylindrospermopsin. The toxic strains of the cyanobacteria genera, *Leptolyngbya*, *Oscillatoria*, *Microcystis*, *Planktothrix* and *Anabaena* are known to have in common the *mcy* (*A–E*, *G*, *J*) genes that are involved in the biosynthesis of microcystin [1,3]. The nodularin cyanotoxin is linked to the *nda* synthetase gene, a polyketide synthase (PKS) and nonribosomal peptide synthetase (NRPS) and biosynthesized by *Nodularia spumigena* NSOR10 cyanobacteria [4]. The review studies carried out by Pearson et al. [4] and Sinha [5] showed that cyanotoxin cylindrospermopsin is linked to the genes *aoa* or *cyr* (*A–O*) and is now known to be biosynthesized by a number of cyanobacteria genera, such as *Cylindrospermopsis* and *Umezakia natans* in Japan; *Aphanizomenon ovalisporum* in Israel, Australia, USA and Spain; *Anabaena bergii* in Australia; *Raphidiopsis raciborskii* in Thailand, China and Australia; *Raphidiopsis curvata* in China; *Aphanizomenon flos-aquae* in Germany; *Anabaena lapponica* in Finland; *Lyngbya wollei* in Australia; *Aphanizomenon gracile* in Germany; *Oscillatoria* sp. in the USA *Aphanizomenon* sp. in Germany; and *Raphidiopsis mediterranea*, *Dolichospermum mendotae* and *Chrysosporum ovalisporum* in Turkey.

The emergence of toxic cyanobacteria species during a bloom period is linked to environmental factors such as light, nutrient enrichment or nutrient depletion, and the presence or non-presence of predators [4]. Eutrophication, a build-up of organic matter produced by phototrophs, such as cyanobacteria [6,7], is often seen as algal blooms and driven by inputs of nitrogen and phosphorus. Cyanobacteria blooms are a major concern worldwide due to the production of cyanotoxins which are harmful to humans [8]. Cyanobacteria tend to proliferate during the summer when concentrations of total phosphorus fall to 100–1000 µg/L [9]. A variety of hypotheses explain why cyanobacteria blooms are becoming increasingly prevalent [10–12]. The most common hypotheses focus on nutrient conditions [10,11,13–17] and nutrient cycling [18] within a water body, as well as aspects of cyanobacteria cell physiology, such as their ability to migrate vertically within the water column, fix atmospheric nitrogen and to produce cyanotoxins [19–22].

Cyanobacterial blooms are often associated with eutrophic conditions [23–25]. Various studies have documented the relationship between nitrogen and phosphorus concentrations, speciation and stoichiometry, and cyanobacteria occurrence [10,13]. A recent study reported that *Microcystis* growth response increases in relation to nitrogen over phosphorus [26]. The same study [26] also reported that the growth response of toxic *Microcystis* to nitrogen was greater than non-toxic strains. Some species of cyanobacteria are known for their ability to fix nitrogen and thus giving them high chances of producing cyanotoxins [27]. Other studies have shown that microcystin toxicity is also influenced by changes in pH, temperature and light intensity [28–30]. A study conducted by Beversdorf et al., [27] indicated that some of the non-nitrogen fixing cyanobacteria may produce toxins because of nitrogen stress events.

However, a review of the available literature shows that there is limited information on the occurrence of toxic and non-toxic cyanobacteria species in relationship with river basin sediments on the African continent [31]. Thus, the main objectives of the study were to: (1) assess the physical-chemical characteristics of river sediments and how these contribute to the resurgence and growth of cyanobacteria species should ideal river flow conditions return; (2) use the FlowCam and molecular techniques to identify toxic and non-toxic cyanobacteria genes in the river sediments; and (3) to use the 16S rRNA in identifying the cyanobacteria species and explore relationships among the cyanobacteria species in the river sediments.

2. Results

2.1. The Physical-Chemical Characteristics of the River Sediments

The physical characteristics of the river sediments drawn from the different tributaries of the Limpopo River basin and the Limpopo River itself showed considerable variation (Figure 1).

The electrical conductivity (EC) and total dissolved solids (TDS) values in the river sediments ranged between 21.2 and 1269 µS cm^{-1} throughout the sampling sites. The pH values were between 6.4 and 8.5, while the total phosphorus concentration values in the river sediments ranged from 0.5 mg/L to 6.3 mg/L (Figure 2). The highest total phosphorus value was recorded for the sediments from the Nzhelele River (S12) near the Mphephu Resort and downstream of the Siloam oxidation ponds. The second highest phosphorus concentrations were measured at the Shashe River (S13), with the total phosphorus measuring 1.2 ± 0.5 mg/L at sample point S18.

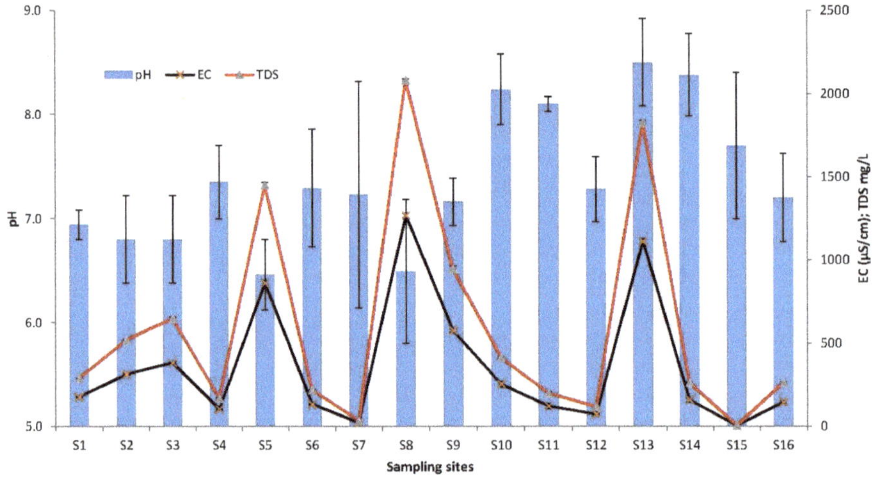

Figure 1. Average values of the physical characteristics of the river sediments of the 18 sampling sites. Whiskers reflect standard error. EC: electrical conductivity; TDS: total dissolved solids.

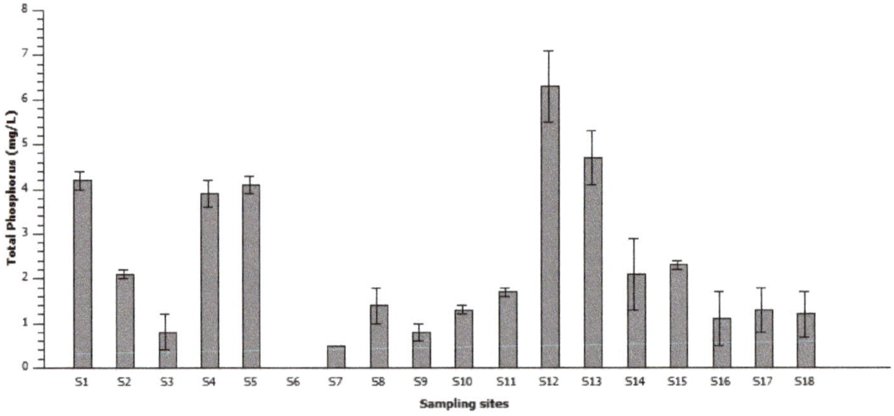

Figure 2. Average values of total phosphorus in the river sediments of the 18 sampling sites. Whiskers reflect standard error.

Finally, the nitrogen concentration values in the river sediments ranged from 1.5 mg/L to 6.5 mg/L (Figure 3). The highest concentrations were recorded for the sample from the Nzhelele River (S12) near the Mphephu Resort and downstream of the Siloam Hospital oxidation ponds, while the total nitrogen at site S18 was 6.25 mg/L.

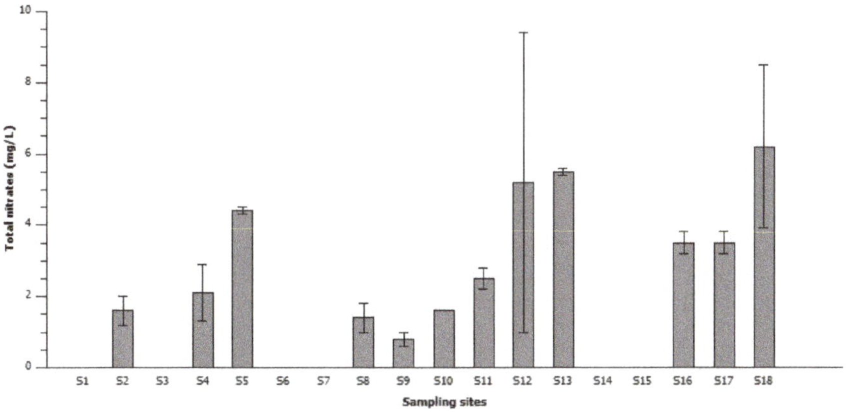

Figure 3. Average values of total nitrogen in the river sediments. Whiskers reflect standard error.

2.2. The Presence of Cyanobacteria in the River Sediments

Table 1 shows the presence of toxic and non-toxic cyanobacteria species that were detected in the Limpopo River basin. The dominant cyanobacteria observed is the filamentous *Leptolyngbya* species, followed by the Synechocystis species (non-toxic and toxic strains), toxigenic *Microcystis* species, and toxigenic Raphidiopsis raciborskii species. The FlowCam showed the presence of the different cyanobacteria species in the Limpopo River basin, as shown in Table 1 and Figure 4. The dominant cyanobacteria species identified from the samples were the *Microcystis* species, followed by the Raphidiopsis raciborskii, Calothrix, Phormidium and Planktothrix species. Finally, no cyanobacteria species were detected in the Mokolo River (S7).

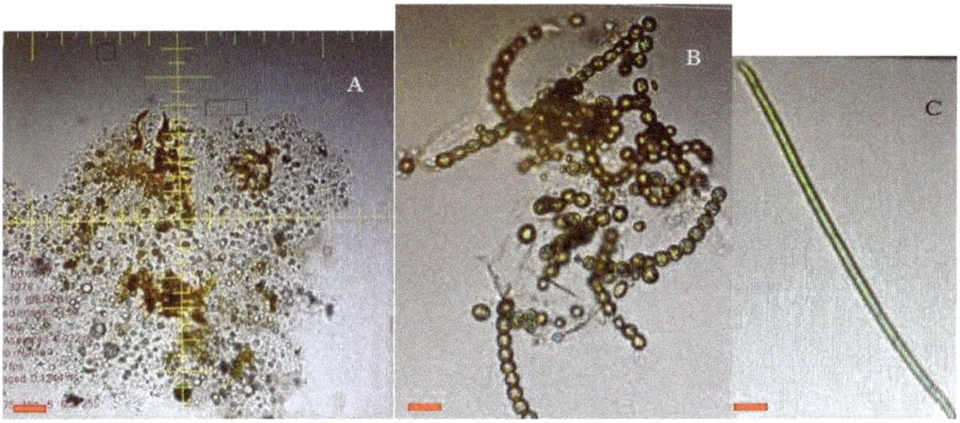

Figure 4. The (**A**) *Microcystis*, (**B**) *Anabaena* and (**C**) *Oscillatoria* species in the river sediments. Red scale bar = 20 μm.

Table 1. Summary of toxic and non-toxic cyanobacteria species in the Limpopo river basin.

Cyanobacteria Species/Sample Sites	S1	S2	S3	S4	S5	S6	S7	S8	S9	S10	S11	S12	S13	S14	S15	S16	S17	S18
Aphanizomenon sp.			+ *															
Raphidiopsis curvata			+ *															
Microcystis aeruginosa	+		+	+	+			++ *		+	+	+	++ *	+	+		+	
Microcystis panniformis								+ *										
Synechocystis PCC 6803			+ *				+		+									
Cylindrospermopsis sp.			+ *						+ *									
Lyngbya sp.	+																	
Leptolyngbya sp.		+														+		
Leptolyngbya boryana								+						+				
Calothrix sp.		+						+	++ *				+	+				
Oscillatoria sp.		+	+						++ *									
Phormidium sp.		+			+					+					+			
Phormidium uncinatum			+															
Nostoc sp.			+ *		+								+ *					
Anabaena circinalis					+													
Anabaena oscillarioides					+													
Chroococcus						+												
Anabaechopsis circularis						+												
Spirulina laxissima SAG 256.80										+								
Planktothrix rubescens								+ *					+ *					+
Alkalinema pantanalense																		+
Gloeocapsa sp.									+ *		+							
Arthrospira sp. str *PCC8005*			+ *															

Notes: + FlowCam analysis and + Molecular techniques with toxic genes * expression.

2.3. PCR Analysis of the 16S rRNA Gene

Multiple fragments were obtained for each sample by sequencing with both forward and reverse primers, and these samples were edited and assembled using the Staden package [32]. All assembled sequences were aligned in BioEdit v7.0.9 [33]. However, the sample collected from the Limpopo River (S1) was not shown in Figure 5, since it was used as the test sample. It was also noted that the amplified fragment from the test sample only produced 100 bp, while around 650 bp was expected. Other samples, such as the samples Limpopo River (S15), Limpopo River (S17) and Musina borehole abstraction point (S16), did amplify, but failed to assemble in the Staden package [32]. The assembled sequences were run on the BLAST algorithm [34] to identify closely similar sequences already deposited in GenBank via NCBI, and the outcomes are shown in Table 2.

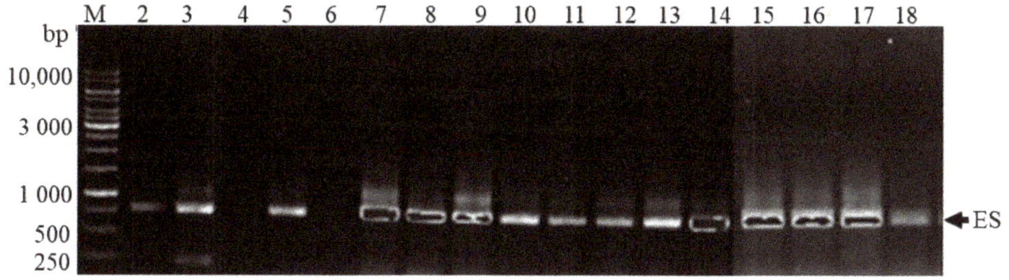

Figure 5. PCR amplification using 27F and 740R primer pair for 16S rRNA gene. ES (estimated fragments); M (Standard Marker), 2–18 Sample numbers. Lane 2 = Notwane River; 3 = Sand River upstream; 4 = Mogalakwena River; 5 = Mawoni River; 6 = Lephalale River; 7 = Mokolo River; 8 = Crocodile River downstream of Hartbeespoort Dam; 9 = Nzhelele River downstream; 10 = Sand River downstream; 11 = Crocodile River downstream (near the bridge on road D1235); 12 = Nzhelele River upstream; 13 = Mzingwane River; 14 = Shashe River; 15 = Limpopo River (next to Thuli Coalmine); 16 = Limpopo River (abstraction point at 0.0 m); 17 = Limpopo River (abstraction point at 1.0 m); 18 = Limpopo River (abstraction point at 1.68 m).

Table 2. Results from the BLAST search showing the similarity between the GenBank sequences and the sample sequences from this study. The families of each species are shown in a separate column.

Samples	Similarity %	Species Similar to	Family	Accession No
S2	93	Uncultured *Leptolyngbya* sp. Clone	Leptolyngbyaceae	KM108695.1
S3	94	*Synechocystis PCC 6803*	Oscillatoriophycideae	CP012832.1
S5	97	*Anabaena oscillarioides*	Nostocaceae	AJ630428.1
S7	99	*Synechocystis* sp. *PCC 6803*	Oscillatoriophycideae	CP012832.1
S8	99	*Leptolyngbya boryana*	Leptolyngbyaceae	AP014642.1
S9	97	*Synechocystis PCC 6803*	Oscillatoriophycideae	CP012832.1
S9	100	*Cylindrospermopsis raciborskii CHAB3438*	Oscillatoriophycideae	KJ139743.1
S9	100	*Aphanizomenon* sp.	Nostocaceae	GQ385961.1
S9	100	*Raphidiopsis curvata*	Nostocaceae	KJ139745.1
S10	96	*Spirulina laxissima SAG 256.80*	Spirulinaceae	DQ393278.1
S11	87	Uncultured Cyanobacterium clone	-	AM159315.1
S12	83	Uncultured Cyanobacterium clone	-	HQ189039.1
S13	90	Uncultured Cyanobacterium clone	-	JX041703.1
S14	98	*Leptolyngbya boryana*	Leptolyngbyaceae	AP014642.1
S16	83	*Leptolyngbya*	Leptolyngbyaceae	KJ654311.1
S18	96	*Alkalinema pantanalense*	Pseudanabaenaceae	KF246497.2

It must be understood from the BLAST algorithm [34] that more than 98% similarity obtained matches the sample to the correct species, more than 90% similarity obtained matches the sample to the correct genus, while more than 80% similarity obtained matches the sample at the Family level.

The PCR products that were separated by gel electrophoresis are shown in Figure 5. The presence of the different bands indicated a positive amplification, whereas a blank sample indicated a negative amplification. The blank samples where repeated several times and failed to amplify. Almost all of the samples showed positive amplification, which confirmed the presence of cyanobacterial DNA in the samples. The two samples which showed no amplification were drawn from the Mogalakwena (S4) and Lephalale Rivers (S6). The BLAST algorithm [34] showed that more than 98% similarity obtained matches the sample to the correct species, more than 90% similarity obtained matches the sample to the correct genus, while more than 80% similarity obtained matches the sample to the correct family.

BLAST data from the samples drawn from the Mokolo River (S7), Crocodile River downstream of Hartbeespoort Dam (S8) and the Shashe River (S14) did not identify the cyanobacteria up to the species level. However, the cyanobacteria from the samples drawn from the Notwane River (S2), Sand River upstream (S3), Mawoni River (S5), Nzhelele River downstream (S9), Sand river downstream and the Limpopo River (S18) (abstraction point at 1.68 m) were identified up to genus level. Lastly, the cyanobacteria from the sampled Limpopo River (S16) (abstraction point at 0.0 m) was identified up to the family level. Furthermore, the samples from Crocodile River downstream (S11) (near bridge on road D1235), Nzhelele River upstream (S12) and Mzingwane River (S13) showed similarities and there were no families that could be detected for these samples.

2.4. Detection of Genes Involved in Toxin Production

The detection of cyanotoxins was done through detecting genes for the proteins that make toxins. This was achieved with PCR by amplification of microcystin/nodularin synthetase using the HEP primer pairs and cylindrospermopsin polyketide synthetase genes using a PKS primer pair. It must be noted that the detection of the genes involved in the biosynthesis of toxins does not confirm the production of the toxins in the field. The *mcyA-C* primer pair and M13 and M14 primer pair were also used to determine the presence of the genes that contain the proteins for toxins production. However, the genes were not detected, as there was no amplification in most of the samples of any of the genes associated with the proteins that produce toxins. Nevertheless, a few samples, such as the Sand River (S3) upstream and Nzhelele River (S9) downstream (Figure 6), showed the amplification of the cylindrospermopsin polyketide synthetase gene. This confirmed the presences of cyanotoxin, cylindrospermopsin in the sediment samples and was attributed to the cyanobacteria species, *Raphidiopsis raciborskii* (Table 3).

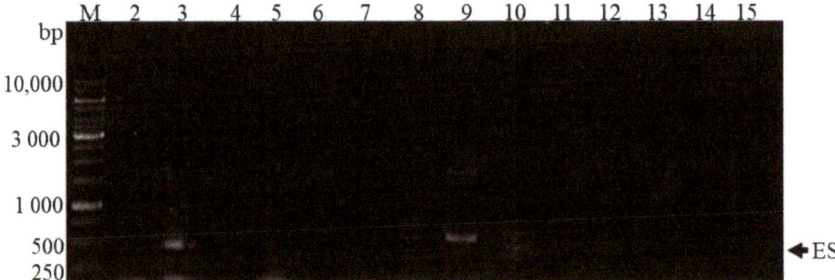

Figure 6. PCR products using PKS primers for cylindrospermopsin polyketide synthetase gene. ES (estimated fragment); M (Standard Marker), 2–18 Samples number. Lane 2 = Notwane River; 3 = Sand River upstream; 4 = Mogalakwena River; 5 = Mawoni River; 6 = Lephalale River; 7 = Mokolo River; 8 = Crocodile River downstream of Hartbeespoort Dam; 9 = Nzhelele River downstream; 10 = Sand River downstream; 11 = Crocodile River downstream (near the bridge on road D1235); 12 = Nzhelele River upstream; 13 = Mzingwane River; 14 = Shashe River; 15 = Limpopo River (next to Thuli Coal Mine).

Table 3. Results from the BLAST search showing the similarity between the GenBank sequences and sample sequenced using PKS and HEP primers for toxin gene identification.

Primers	Sample No	Similarity %	Species Similar to	Accession No
PKS	S3	100	*Aphanizomenon* sp. *10E6*	GQ385961.1
	S3	100	*Raphidiopsis curvata*	KJ139745.1
	S3	100	*Cylindrospermopsis raciborskii*	AF160254.1
	S3	100	*Arthrospira* sp. *str. PCC 8005*	FO818640.1
	S3	100	*Nostoc* sp. *NIES-4103*	AP018288.1
	S9	93	*Calothrix* sp. *336/3*	CP011382.1
	S9	89	*Oscillatoria nigro-viridis PCC 7112*	CP003614.1
	S9	100	*Gloeocapsa* sp. *PCC 7428,*	CP003646.1
	S9	100	*Cylindrospermum* sp. *NIES-4074*	AP018269.1
HEP	S8	100	Uncultured *Microcystis* sp. *clone msp microcystin synthetase E (mcyE) gene, partial cds*	KF687998
	S8	100	*Microcystis panniformis FACHB-1757*	CP011339.1
	S8	100	*Microcystis aeruginosa PCC 7806*	AF183408.1
	S8	100	*Nostoc* sp. *152*	KC699835.1
	S8	100	*Planktothrix rubescens NIVA-CYA 98*	AM990462.1
	S13	100	*Nostoc* sp. *152*	KC699835.1
	S13	100	*Planktothrix rubescens NIVA-CYA 98*	AM990462.1
	S13	100	Uncultured *Microcystis* sp. *from Uganda*	FJ429839.2
	S13	100	*Microcystis aeruginosa PCC 7806SL*	CP020771.1
	S13	100	Uncultured *Microcystis* sp. *clone mw microcystin synthetase E (mcyE) gene, partial cds*	KF687997.1

The HEP primer pair produced two positive results for samples from Crocodile River (S8) and Mzingwane River (S13). Both positive results were attributed to the presence of toxigenic *Microcystis* sp. (Table 3). The latter is an important finding, since water supplies from the Limpopo River basin are used by water utilities for drinking water supplies, and by commercial and subsistence irrigation farmers for the production of food crops and livestock watering (Figure 7).

Figure 7. A scenario involving boreholes drilled inside the Limpopo river channel and contamination with cyanobacteria (green dots) cysts and akinetes for (**A**) irrigation farmers & (**B**) water utility raw water supply for human consumption.

2.5. Phylogenetic Relationship

The relationship between the samples and their most similar species, as noted from the BLAST search, was confirmed by the phylogenetic tree, and the relationships between some cyanobacteria species from different samples (Figure 8). The first confirmation was on the similarity of samples from the Crocodile River (S8) downstream Hartbeespoort Dam and the Shashe River (S14) to *Leptolyngbya* boryana with 99% bootstrap confidence. Secondly the similarity of the Musina borehole extraction (S16) sample to Alkalinema pantanalense, with 98% bootstrap, and the similarity of the samples from Sand River (S3) upstream. Another similarity was evident in the Nzhelele River (S9) downstream near Tshipise and Mokolo River to Synechocystis sp. PCC 6803. The other similarity was Mawoni River (S5) downstream of Makhado oxidation pond to *Leptolyngbya* sp. with 97% bootstrap confidence, with the Notwane River (S2) to uncultured *Leptolyngbya* sp. with 99% bootstrap confidence, and lastly the Sand River (S10) downstream to the filamentous cyanobacterial species Spirulina laxissima with 100% bootstrap confidence.

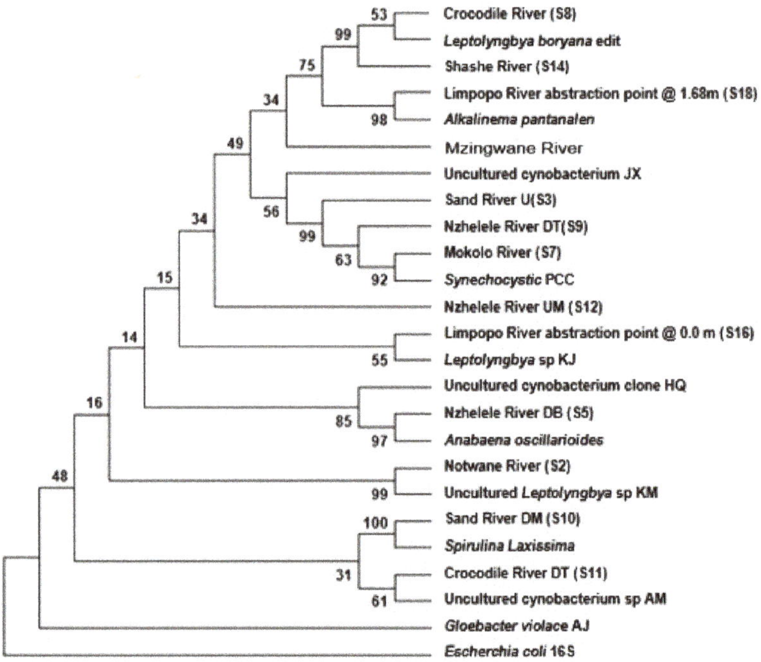

Figure 8. The evolutionary history was inferred using the Neighbor-Joining method. U: upstream; UM: Upstream; DT: Downstream; DB: Downstream; DM: downstream; PCC: Pasteur Culture Collection of Cyanobacteria.

However, the detection of cyanobacteria at the Musina borehole extraction (S16) and the Sand River (S3) may suggest that there were aquatic animals such as fish, dispersal or transportation of cyanobacterial cells from the entrance (mouth) of Sand River towards the Musina abstraction point (Figure 9). In simple terms, there was an upstream transport of cyanobacteria species that was facilitated by aquatic animals, but this requires further investigation. The other matches from the BLAST search include the Musina Borehole extraction point (S16) to *Leptolyngbya* sp., as well as the Crocodile River (S11) near bridge on road D1235 and upstream of Thabazimbi town to uncultured Cyanobacteria clone. However, the latter bootstrap confidence levels were between 55 and 61%, respectively. It was evident from the data that a divergence matrix can be used to verify the truth of both the BLAST search and phylogenetic tree. The divergence matrix confirmed that cyanobacteria from the Crocodile River (S8) downstream of Hartbeespoort Dam and from Shashe River (S14) were the same *Leptolyngbya* boryana species, since they both showed at least 98% similarity to this species.

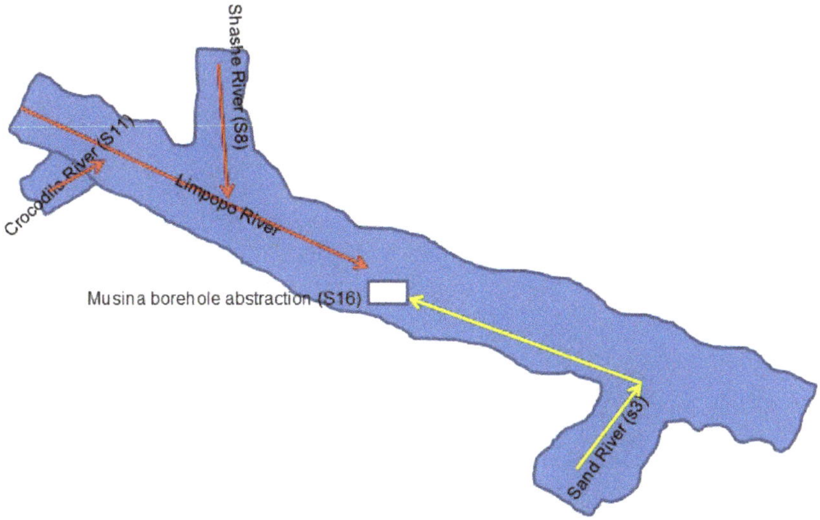

Figure 9. A scenario involving the movement of cyanobacteria species during water flows in the Limpopo River (red arrow) towards the Musina abstraction borehole (White Square). The possible upstream movement (yellow arrow) from the Sand River (S3) to the Musina borehole (S16) may involve cyanobacteria 'hitching a ride' on aquatic animals such as fish and crocodiles.

Thus, the current study indicates that there is DNA evidence to suggest a similarity between the cyanobacteria at the Musina abstract point and that from the Crocodile River system. The possibility thereof lies in the fact that the Musina abstraction point was downstream from the Crocodile River, which flows into the Limpopo River (Figure 7). However, samples from the Nzhelele River upstream near the Mphephu Resort and Mzingwane River (Zimbabwe) did not match. Nevertheless, the relationship between the cyanobacteria species from specific locations were identified by a Divergence Matrix (Table 4). In addition, the same species were detected by the difference co-efficient of 0.00, whereas completely unrelated species were detected by the co-efficient of 1.00.

Table 4. Divergence matrix for reflection of similarity.

	S2	S3	S5	S7	S8	S9	S10	S11	S12	S13	S14	S16	S18
Notwane River (S2)	-												
Sand River (S3)	0.216	-											
Nzhelele River (S5)	0.191	0.187	-										
Mokolo River (S7)	0.167	0.064	0.130	-									
Crocodile River (S8)	0.166	0.160	0.149	0.119	-								
Nzhelele River (S9)	0.184	0.095	0.152	0.028	0.140	-							
Sand River (S10)	0.155	0.216	0.153	0.156	0.169	0.169	-						
Crocodile River (S11)	0.257	0.295	0.280	0.244	0.278	0.254	0.236	-					
Nzhelele River (S12)	0.391	0.394	0.365	0.351	0.350	0.361	0.364	0.492	-				
Mzingwane River (S13)	0.190	0.180	0.184	0.130	0.134	0.139	0.168	0.267	0.377	-			
Shashe River (S14)	0.173	0.163	0.156	0.119	0.006	0.140	0.173	0.278	0.355	0.134	-		
Musina borehole (S16)	0.376	0.359	0.312	0.314	0.342	0.321	0.343	0.414	0.555	0.371	0.347	-	
Musina borehole (S18)	0.183	0.184	0.179	0.136	0.128	0.150	0.186	0.285	0.366	0.173	0.131	0.348	-

The cyanobacteria species from the Crocodile River (S8) were the same species as the cyanobacteria from the sampled Shashe River (S14), since they had less than 1% difference (0.006). This may have been expected, since the Shashe River is downstream of the Crocodile River (Figure 9). The cyanobacteria species from the Mokolo River (S7) and the Nzhelele River (S9) share undetectable differences. However, a comparison shows that there was a difference between the cyanobacteria species from the Notwane (S2) and Mawoni Rivers (S5). The cyanobacteria which differed the most from the other species were the cyanobacteria species from the Notwane River (S2) and Limpopo River (S16). The latter sampling sites' species also differed from each other, with 28%, while their comparison co-efficient range from 0.17 to 0.28. Furthermore, the Nzhelele River upstream (S12) and Limpopo River (S16) species differed from each other with 28%, while their comparison co-efficient ranged between 0.312 to 0.492, which was the highest for all species.

The first observation was that Uncultured Cyanobacterium clone HQ189039.1 could not be used for the phylogenetic tree because of its length (about 480 bp). This is because the t complete deletion option of gaps and missing information in MEGA 7 [35] was used. The second observation was that two outgroup sequences had been used in phylogenetic alignment.

3. Discussion

The physical chemical data generated in the current study shows that there were large variations in sediment EC between the different sampling sites while the sediment temperature was $\geq$22 °C during all the sampling trips. High temperatures arising from climate change have been reported as an important factor in the global expansion of harmful algal bloom worldwide [36]. Rising temperature exceeding 20 °C can promote the growth rate of cyanobacteria, whereas the growth rate other freshwater eukaryotic phytoplankton decreases, which is regarded as a competitive advantage for cyanobacteria [37]. A study by O'Neil et al. [9] reported that higher temperatures promote the dominance of cyanobacteria and favor the production of microcystins, as well as resulting in an increase in their concentration.

The high pH value measured during the current study may have a competitive advantage for many cyanobacteria, because of their strong carbon-concentrating abilities compared to eukaryotic phytoplankton species [38]. A laboratory experiment carried out by Jahnichen et al. [39] on *Microcystis aeruginosa* showed that microcystin production started when pH exceeded 8.4, thus indicating a lack of free carbon dioxide (CO_2).

The increased input of nutrients into the surface water is the main factor responsible for massive proliferations of cyanobacteria in fresh water, brackish and coastal marine ecosystems. However, phosphorus and nitrogen nutrients in high levels lead to accelerated growth of cyanobacteria [40,41].

Thus, the higher concentration of phosphorus measured in the current study downstream of the Siloam oxidation ponds may be due to the discharge of sewage effluent [42]. The low concentration of phosphorus measured in the Lephalale River (S6) is possibly related to less anthropogenic land use activities upstream of this sample site [43]. Phosphorus has been implicated more widely than nitrogen as a limiting nutrient of phytoplankton and cyanobacteria in freshwater systems [44]. A minimum amount of phosphorus entering or becoming soluble in a water body can trigger a significant algal bloom [45]. The impact of excess phosphorus on receiving rivers or streams is evident from the green coloration of surface water owing to the presence of phytoplankton or cyanobacteria. The Limpopo River (S1) receives inflows from both the Notwane and Crocodile Rivers, and these contribute significantly to the phosphorus loading of the Limpopo River. Furthermore, the main source of phosphorus in the Notwane River (S2) is the municipal discharge from the Glen Valley sewage plant and agricultural runoff from irrigated farms and livestock ranching in the river's catchment [23,46]. The Crocodile River receives sewage effluent from upstream catchment land use activities such as the discharge of sewage effluent into tributaries of the Crocodile River, discharge into Crocodile River itself and agricultural runoff [47,48]. The Sand River (S3) receives municipal nutrient discharge from the Polokwane sewage plants and rainwater runoff that would contain fertilizer from agricultural activities in the river sub catchment [49]. The sample point on the Mogalakwena River (S4) was downstream of the Mokopane, Modimolle and Mookgophong towns' sewage plants, golf courses, game farming, livestock farming and irrigated farmlands [50]. After the town of Mokopane, the Nyl River is renamed to the Mogalakwena River. The Mzingwane River (S13) receives municipal discharge from the Filabusi, Gwanda and West Nicholson sewage plants and agricultural runoff from irrigated farms and livestock ranching in these areas [51], and this may be attributed to sewage plants upstream in Francistown and agricultural runoff from irrigated farms and livestock ranching [48,52]. These rivers are part of the Limpopo River's tributaries and contribute to the successive loading of phosphorus in the Limpopo River (S15–S16) [53].

The highest value of nitrogen was recorded in the Nzhelele River (S12) near the Mphephu Resort and downstream of Siloam hospital oxidation ponds. The reason for the detection of these high nitrogen values is possibly related to the discharge of sewage effluent from the Siloam hospital [42]. Filamentous cyanobacteria can obtain nitrogen by fixing the atmospheric nitrogen gas and converting it to nitrate for their growth [54]. Nitrogen is a common gas (79%) that is found in the atmosphere. Thus, cyanobacteria genera such as *Anabaena* are able to utilize atmospheric nitrogen in addition to nitrate originating from the river sediments [54,55]. The other sample sites with nitrates in excess of 2 mg/L are Sand River (S4), Mawoni River (S5), Crocodile River (S11), Mzingwane River (S13), and Limpopo River (S16 to S18). All these tributaries have one in common source of pollution upstream, namely, a municipal sewage plant, and are also surrounded by farmland where commercial irrigation farming is practiced, as in the case of the Crocodile, Notwane, Shashe, Mzingwane and Sand rivers. Subsistence agriculture is practiced in the case of the Mawoni and Mzingwane rivers [46,48–52]. The Crocodile River also receives inflows from eutrophic Hartbeespoort Dam [47]. The Limpopo River (S16) is downstream of all the sample points, and this shows the cumulative discharge of nitrates originating from the tributaries, causing an increase in concentration of nitrogen. The Musina local municipality has drilled 8 boreholes in the Limpopo river bed, and most of these boreholes are located close to S16, and thus there is a possibility of cyanotoxin contamination of the borehole water [56]. Thus, further research is required to determine if there is cyanotoxin contamination of borehole water.

Botha and Oberholster [57] performed a survey of South African freshwater bodies between 2004 and 2007, using RT-PCR and PCR technology to distinguish toxic and non-toxic *Microcystis* strains bearing *mcy* genes, which correlate with their ability to synthesize the cyanotoxin microcystin. The study revealed that 99% of South Africa's major impoundments contained toxicogenic strains of *Microcystis*. The study by Su et al. [58] in the Shanzi impoundment, China showed that the sediments were the source of cyanobacteria inoculum. This implies that the cyanobacteria flocculates in the sediments during periods of adverse environmental conditions, such as cessation in river flows. These cyanobacteria cysts or spores then reactivate during periods of river flow. The cyanobacteria cysts and spores are related to the summer environmental conditions in the Limpopo river basin, where the majority of tributaries are perennial and river flows commence during the period of summer rainfall. The river flows disturb the sediments, thus bringing into the water column the cyanobacteria cysts or spores [58]. The source of nutrients in the Limpopo river basin may be attributed directly to sewage discharge of municipal waste water plants such as Glen Valley and Mahalpye on the Botswana side, and on the South Africa side, the western and northern parts of the city of Johannesburg to the town of Musina and indirectly to the agricultural practices of fertilizer application and animal waste [46,48–52].

Cyanobacteria undergo distinct developmental stages [59]. For example, they differentiate into resting cells, spores, akinetes and cysts which represent a survival strategy under unfavorable environmental conditions [55,60]. Under favorable conditions, the cell will germinate again [61]. The ability of cyanobacteria to adapt to adverse dry periods allows them to inhabit the river sediments, as shown by studies by Perez at al. [60], Kim et al. [55] and this study (Figure 10). The study of Kim et al. [55] further illustrated the viable nature of cysts and akinetes in providing the next inoculum of *Microcystis, Anabaena, Aphanizomenon* and *Oscillatoria* is Bukhan, Namhan Rivers and Lake Paldang and Kyeongan stream, in South Korea.

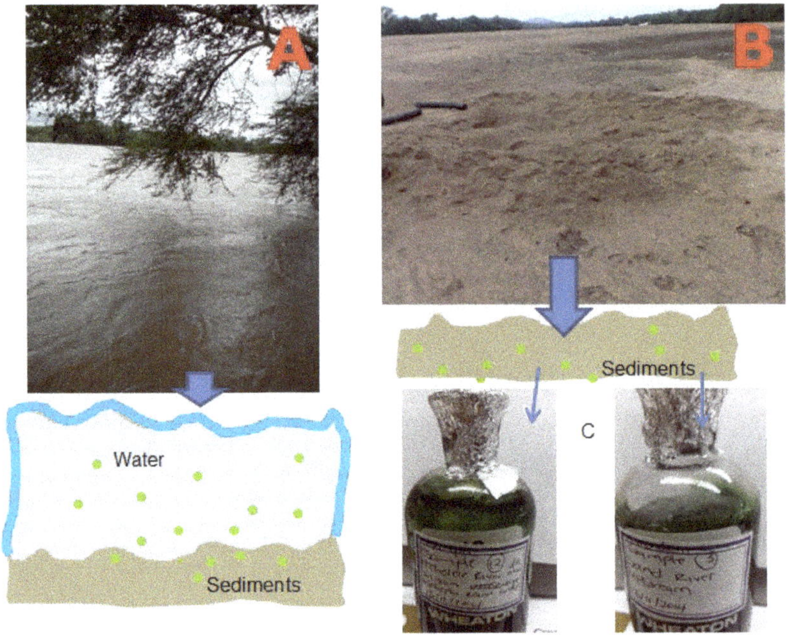

Figure 10. Scenario involving sedimentation of cyanobacteria (green dots) cysts and akinetes (**A**) during flood and flow conditions in Limpopo River and (**B**) during non-flow (DRY) conditions in the Limpopo River and (**C**) growth of cyanobacteria under continuous lighting and provision of BG medium at room temperature.

As expected, the toxigenic *Microcystis* species was found in the Crocodile River, downstream of the Hartbeespoort Dam, a eutrophic water impoundment known for the regular occurrence of *Microcystis* dominated harmful algal blooms [40]. However, our two toxigenic *Microcystis* strains were different from the seventeen toxigenic *Microcystis* strains studied by Mbukwa et al. [24] from the Hartbeespoort Dam. The differences may be explained by the different use of *mcy* primers in identifying the genes expressing toxicity and differences in experimental approach. Our study on the *mcyA-Cd* primer did not amplify, whereas Mbukwa et al. [24] reported the amplification of the *mcyA-Cd* genes, showing seventeen toxigenic *Microcystis* strains. However, during our study, the *mcyE* genes were positively expressed with the HEP primer and did amplify, but the outcomes were also different from the toxigenic *Microcystis* strains studies by Mbukwa et al. [24]. During our study, the total genomic DNA was not extracted directly from the sediments, but from cyanobacteria that was cultured in the laboratory. Laboratory culture conditions have been known to alter the toxicity of *Microcystis* species, as shown by the study of Scherer et al. [62]. In the latter study, the authors mimicked a temperature increase of 10 °C. Under these increased temperature conditions, *Microcystis* was able to express *mcyB* gene related to production of toxicity instead of the *mcyD* gene. This may imply the biodiversity of toxigenic *Microcystis* strains in Hartbeespoort Dam and the Crocodile River and the Limpopo River basin.

Mbukwa et al. [24] used DNA molecular techniques to identify the two species of *Microcystis* as *M. aeruginosa* (origins from Hartbeespoort Dam, South Africa) and *M. novacekii* (origins from Phakalane effluent, Gaborone, Botswana). The molecular techniques showed the presence of the *mcy* genes responsible for microcystin encoding, thus confirming that the two *Microcystis* species did have the potential to produce toxins. The Phakalane pond effluent is discharged into the Notwane River, a tributary of the Limpopo [23]. An earlier study by Basima [51] upstream of the Mzingwane S13 sample point showed the abundance of cyanobacteria genera dominated by *Microcystis* species followed by *Anabaena* and *Nostoc* species in water impoundments situated inside the Mzingwane River. In the lower Limpopo River, situated in Mozambique, at the Chokwe irrigation scheme, which receives irrigation waters from Maccaretane Dam, Pedro et al. [25] reported the presence of *Microcystis* species and microcystin-LR concentrations of 0.68 ppb. The latter concentrations were linked to the presence of the *mcyB* and *mcyA* genes in collected water samples. Mikalsen et al. [63] identified eleven *Microcystis* species containing different variants of the *mcyABC* (toxic species), and seven *Microcystis* species that lacked the *mcyABC* gene (non-toxic species). A study by Davies et al. [64] on four temperate lakes in the northwest of the USA showed that the increase in water temperature contributed to an increase in toxic *Microcystis* species (possessing the *mcyD* gene). Yamamoto [65] and Oberholster et al. [66] have shown that the *Microcystis* species adopt survival strategies to mitigate harsh external environments such as reduced river flow, a major characteristic of the Limpopo River, by sinking into the sediments.

The Limpopo river basin is characterized by extreme weather events such as heatwaves, floods and drought [67]. Could the latter, including extreme heatwaves, possibly have contributed to toxic or non-toxic *Microcystis* species? The presence of microcystins in the rivers may constitute a health risk, especially for the communities that may be in contact or drink the polluted water without any form of treatment or suitable treatment that is able to remove the toxins in the water. The convectional method for water treatment is not convenient for the removal of microcystins in water [68]. Drinking water treatment processes might trigger the release of hepatoxin into drinking water by disrupting the trichomes of cyanobacteria [69]. Thus, the presence of cyanotoxins can also poison the livestock and game animals (wildlife) in transfrontier parks such as Kruger National Park, Gona-re-zhou National Park and Mapungubwe National Park [70]. Microcystins have already been implicated in the death of wildlife in the Kruger National Park [71]. Cyanotoxins have been implicated in the negative growth (stunting) of plants, and this may have serious repercussions for irrigation farmers [72].

The evolutionary tree constructed could not be used for phylogenetic purposes because of two important reasons: (1) the number of samples used for PCR per river site was not enough to make a conclusive argument; and (2) the cyanobacteria were the expected products which needed to be identified. Hence the tree was used to verify the identification as done by BLAST search; however, the phylogenetic relationship was basically done by divergence matrix, and combined discussion followed the divergence matrix.

4. Conclusions

Many countries in Africa have reported cases of intoxication and death of animals that may have been caused by cyanobacterial toxins. Monitoring and or reducing the nutrient loads into the river system will decrease the threat of cyanobacteria blooms to human and animal health. The results obtained in the current study indicated the presence of toxic and non-toxic cyanobacteria species in the bottom sediments of the Limpopo River and its tributaries. Molecular tools were used in the present study to determine non-toxic and toxic cyanobacteria based on genes that produce proteins related to cyanotoxins. The presence of nutrients, phosphates and nitrates in the river sediments did stimulate the growth of the cyanobacteria during summer river flow periods. Furthermore, the expression of genes that have the potential to produce toxins, for example cylindrospermopsin and microcystin/nodularin in the river sediments indicate a potential risk to the environment and human health. The cyanotoxins are harmful to humans who consume the water originating from boreholes located inside the Limpopo River basin or drilled along the Limpopo River basin. Secondly, the water supplies from the Limpopo River basin are used by commercial and subsistence irrigation farmers for growing food crops and livestock watering. Thus, presence of cyanotoxins can also adversely affect the livestock and game animals (wildlife) in transfrontier parks. Furthermore, cyanotoxins have been implicated in the negative growth (stunting) of plants, and this may have serious repercussions for the irrigation farmers in this region.

5. Future Research Work

Further studies are required to determine the level and types of cyanotoxins in the Limpopo river basin, since water resources are used for a variety of purposes, such as human consumption, irrigation, livestock and wildlife watering and impact if any on aquatic biodiversity.

6. Materials and Methods

6.1. The Study Area

The study area is the Limpopo River and its major tributaries (Figure 11). The Limpopo River basin consists of four countries: Botswana, South Africa, Zimbabwe and Mozambique [73].

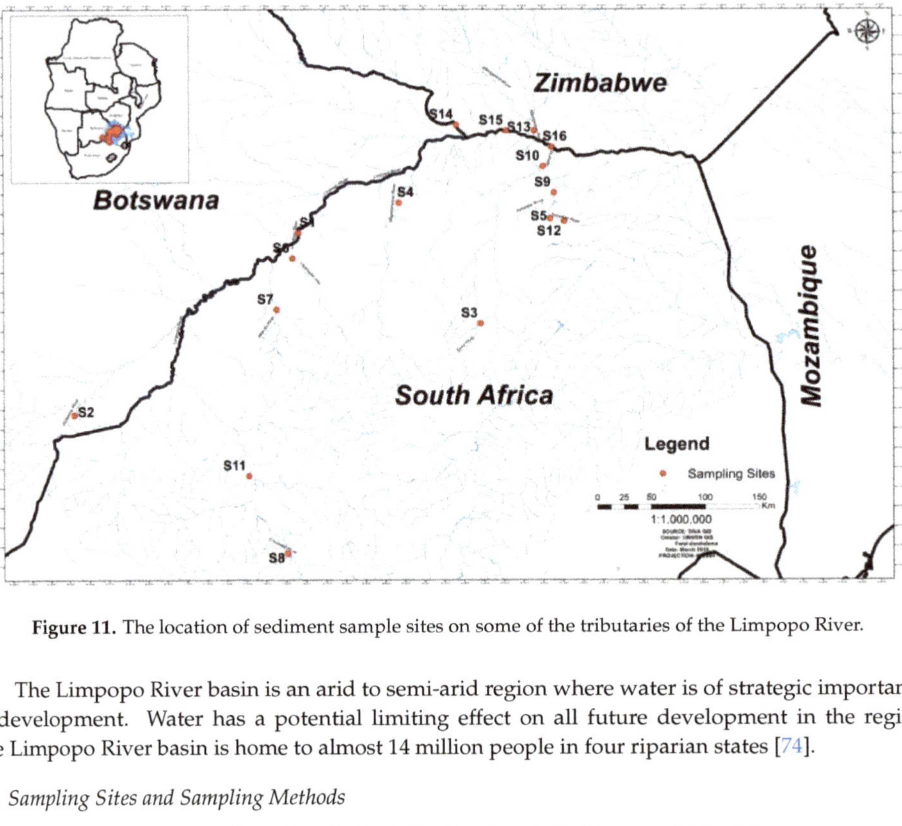

Figure 11. The location of sediment sample sites on some of the tributaries of the Limpopo River.

The Limpopo River basin is an arid to semi-arid region where water is of strategic importance to development. Water has a potential limiting effect on all future development in the region. The Limpopo River basin is home to almost 14 million people in four riparian states [74].

6.2. Sampling Sites and Sampling Methods

Sampling sites were selected with the following in mind: (a) accessibility, biotype, e.g., sandy bottom sediment; (b) canopy cover and depth, and (c) river sites receiving inflows of municipal sewage discharges. The 18 grab river sediment samples were collected in October and November 2014. The river sediment samples (~500 g) were collected in sterile glass containers from rivers and tributaries of the Limpopo River (Table 5). The use of river sediments was chosen because most suspended material, including cyanobacteria spores and cysts, settles at the river bottom, where they become part of the sediments in river systems.

Table 5. The location of sample sites and sample codes.

River Names	Samples Numbers
Limpopo River (Groblers' bridge)	S1
Notwane River (Odi Bridge-Matabeleng)	S2
Sand River upstream	S3
Mogalakwena River next to Tolwe	S4
Mawoni River downstream Makhado oxidation ponds	S5
Lephalale river	S6
Mokolo River	S7
Crocodile River downstream Hartbeespoort dam	S8
Nzhelele River downstream near Tshipise	S9
Sand River downstream (at bridge on N1 road towards Musina)	S10
Crocodile River downstream (near bridge on road D1235) near Thabazimbi	S11
Nzhelele River upstream near Mphephu resort (downstream of Siloam oxidation ponds)	S12
Mzingwane River (Zimbabwe)	S13
Shashe River (near Irrigation scheme, Zimbabwe)	S14
Limpopo River next to Thuli coal mine	S15
Limpopo River abstraction point @ 0.0 m	S16
Limpopo River abstraction point @ 1.0 m	S17
Limpopo River abstraction point @ 1.68 m	S18

6.3. *Physical-Chemical Measurements*

In the laboratory, the physical measurement of pH, Total dissolved solids (TDS) and electric conductivity (EC) was carried out using Portable pH meter Crison MM40 (Crison Instruments SA, Alella, Spain) on the river sediments. It was first calibrated as per the manufacturer's guidelines. The pH, TDS and EC of the sediments were determined by the method of Islam et al. [75], in which 50 g of sediment was mixed with 50 mL of distilled water in a 100 mL beaker to produce a ratio of 1:1. The mixture was stirred with a stirring rod to homogenize the mixture and was then left for 30 min to settle. EC, pH and TDS were then measured by inserting the electrodes in the soil solution and readings were taken.

6.4. *Nutrient Analysis*

The air-dried sediments were subjected to nutrient analysis, and this involved determining Total Phosphate (TP) and Total Nitrogen (TN). The analyses were done in duplicates and the aliquot of all digested samples was analyzed with Merck Spectroquant® Pharo 100 spectrophotometer with a wavelength of 320–1100 nm purchased from Merck (Darmstadt, Germany).

6.4.1. Total Phosphorus Analysis

Total phosphorus was determined by using the perchloric acid digestion method as described by American Public Health Association (APHA) [76]: 2 g of air-dried sediment was acidified to methyl orange with concentrated HNO_3, another 5 mL of concentrated HNO_3 was added and evaporated on a hotplate until dense fumes appeared. 10 mL each of concentrated HNO_3 and $HClO_4$ was added and evaporated gently until dense white fumes of $HClO_4$ appeared. The solution was then neutralized with 6N NaOH and made up to 100 mL with distilled water. Aliquots of the samples were then analyzed with spectrophotometer using phosphate cell test kit (Merck, Darmstadt, Germany).

6.4.2. Total Nitrogen Analysis

Total nitrogen was determined per APHA [76] as ammonia: 1 g of each air-dried sediment sample was treated with 2 mL of sulphuric acid. The sample was heated on a hotplate for 2 h. Aliquots of 50 mL of deionized water were added to each sample. The sample was filtrated through No. 41 Whatman filter paper. The filtrate of each sample was made up to 250 mL with deionized water and 55 mL of 1 M sodium hydroxide solution. Aliquots of the samples were then analyzed with spectrophotometer using a nitrate cell test kit (Merck, Darmstadt, Germany).

6.5. Data Analysis

The physico-chemical and cyanotoxin measurements were conducted in duplicates, and the standard deviation and mean were calculated, using a Microsoft (MS) Excel 2010 spreadsheet for each sampling point. The graphs were plotted using MS Excel.

6.6. The Culture of Cyanobacteria Species in River Sediments

The modified BG11 medium was laboratory-prepared as per Gumbo et al. [77] for cyanobacteria culturing. The 200 mL sterile modified BG11 medium was transferred to sterile 250 mL laboratory jars under sterile conditions and then 200 g of river sediments was added. A total of 18 laboratory jars were incubated for 30 days under continuous light (1100 lux) fluorescent lamps at room temperature. The harvested cyanobacteria cells were subsequently used for identification and molecular characterization.

6.7. The Identification of Cyanobacterial Species Using the FlowCam

The harvested cyanobacteria cells were used to identify cyanobacterial species present in the samples, a bench top FlowCam (Model vs. IV) was used. In the FlowCam system, the sample is drawn into the flow chamber by a pump. Using the laser in trigger mode, the photomultiplier and scatter detector monitor the fluorescence and light scatter of the passing particles. When the particles passing through the laser fan have sufficient fluorescence values and/or scatter, the camera is triggered to take an image of the field of view. The fluorescence values were then saved by the Visual Spreadsheet. The computer, digital signal processor, and trigger circuitry work together to initiate, retrieve and process images of the field of view. Groups of pixels that represented the particles were then segmented out of each raw image and saved as a separate collage image. The image was then captured and compared to the image of cyanobacteria as per the procedure of van Vuuren et al. [78].

6.8. The Identification of Cyanobacterial Species Using Molecular Characterization

The cyanobacteria cells were harvested and used for molecular characterization as per the following procedures, as outlined below:

6.8.1. DNA Extraction and Purification

Samples were freeze-dried and stored at −20 °C for DNA extraction. Total genomic DNA was extracted using the ZR-Duet^{TM} DNA/RNA Miniprep DNA extraction kit from Inqaba Biotech Laboratories South Africa (Pretoria, South Africa). Sample preparation and DNA extraction was carried out following the protocol supplied by the manufacturer.

6.8.2. Detection and Amplification of 16S rRNA by Polymerase Chain Reaction

The PCR method was performed for detection and amplification of 16S rRNA as described briefly by Frazao et al. [3]. The PCR amplification of the cyanobacteria 16S rRNA gene was determined using set of primers 27F/809R (Table 2). Thermal cycling conditions were 1 cycle at 95 °C for 5 min, 35 cycles at 95 °C for 30 s, 55.4 °C for 30 s and 72 °C for 60 s and 1 cycle at 72 °C for 10 min. Reactions were carried out in a 50 µL reaction volume that consisted of 0.5 pmol of each primer (10 pM/µL), 25 µL of Dream Taq master mix (Inqaba Biotech), 19 µL sterile ultra-pure water and 5 µL of DNA sample.

6.8.3. Toxin Gene Detection

The presence of cyanotoxins was determined by PCR using primers that were used for detection of genes involved in the production of nodularins (NOD), microcystins (MC) and cylindrospermopsin (CYN) (Table 6). The NOD gene cluster, *nda*, consists of nine open reading frames (*ndaA-I*) [79]. The MC gene cluster, *mcy*, comprises 10 genes in two transcribed operons, *mcyA-C* and *mcyD-J* [80]. The HEP primer pair was used for detection of genes involved in MC and NOD production.

These primers are responsible for sequencing the aminotransferase (AMT) domain, which is located on the modules *mcyE* and *ndaF* of the MC and NOD synthetase enzyme complexes, respectively [80,81]. Primers *mcyA-C* were used to detect the *mcyA*, *mcyB* and *mcyC* genes [82,83]. For detection of CYN production (*cyr*) genes, the polyketide synthase PKS M4 and M5 primers and the peptide synthetase M13 and M14 primers were used as designed by Schembri et al. [84], who demonstrated a direct link between the presence of the peptide synthetase and polyketide synthase genes and the ability of cyanobacteria to produce CYN.

Table 6. The PCR primers used for amplification of 16S rRNA gene for cyanobacteria identification and for the amplification of genes related to cyanotoxins production. A—Individual annealing temperature, B—Reference annealing temperature, bp = base pairs.

Primers	Target Genes	Sequence (5′-3′)	A	B	Size (bp)	Amplified Gene	Ref.
27F 809R	-	AGAGTTTGATCCTGGCTCAG GCTTCGGCACGGCTCGGGTCGATA	52 64	60	780	16S rRNA	[85,86]
mcyA-Cd F *mcyA-Cd* R	*mcyA*	AAAATTAAAAGCCGTATCAAA AAAAGTGTTTTATTAGCGGCTCAT	51 43	59	297	Microcystin synthetase	[83]
HEPF HEPR	*mcyE/ndaF*	TTTGGGGTTAACTTTTTTGGGCATAGTC AATTCTTGAGGCTGTAAATCGGGTTT	57 55	52	472	Microcystin/nodularin synthetase	[81]
PKS M4 PKS M5	*cyr*	GAAGCTCTGGAATCCGGTAA AATCCTTACGGGATCCGGTGC	52 56	55	650	Cylindrospermopsin polypeptide synthase	[84]
M13 M14	*ps*	GGCAAATTGTGATAGCCACGAGC GATGGAACATCGCTCACTGGTG	57 57	55	597	Cylindrospermopsin peptide synthetase	[84]

The PCR reaction conditions that were used were those described for the amplification of the 16S rRNA gene [81]. Concerning the cycling conditions, for *mcyA-Cd* genes, the thermal cycling conditions were 1 cycle at 95 °C for 2 min, 35 cycles at 95 °C for 90 s, 56 °C for 30 s and 72 °C for 50 s and 1 cycle at 72 °C for 7 min. For HEP and CYN as genes, the thermal cycling conditions were as those for the amplification of the 16S rRNA with an exception for HEP gene annealing temperature of 58.15 °C for 30 s. Positive control was used.

Electrophoresis

PCR products were electrophoresed in 0.8% agarose gel by adding prepared 1.2 g of agarose powder into 150 mL 1X TAE buffer (48.4 g Tris, 11.4 mL Glacial acetic acid, 3.7 g EDTA disodium salt topped up to 1000 mL with deionized water). The mixture was heated until there was complete dissolution. Exactly 10 µL of ethidium bromide was added and mixed thoroughly. The mixture was transferred to the gel-casting tray with the comb already in position and allowed to solidify. The solidified gel was transferred to the running trays. The gel in the tray was covered with 1X TAE buffer. In the first well 3 µL 100 bp of the molecular weight marker was loaded and the samples were loaded from the second well onwards. The gel was run at 100 V and 250 mA for 60 min. The gel was viewed using the Gel doc (Biorad, Hercules, CA, USA) and the picture was taken.

6.8.4. PCR Purification and Sequencing

PCR products were purified using the GeneJet Gel Extraction Kit Thermo Scientific (Pretoria, South Africa) under room temperature as per the protocol provided by the kit manufacturer. The purified DNA was stored at −20 °C. PCR products were sent for sequencing at Inqaba biotech laboratory (Pretoria, South Africa). Sequences were analyzed using the BLAST system (http://www.ncbi.nlm.nih.gov/BLAST/).

Primers

Primers used for PCR amplification were synthesized at Inqaba Biotech (Pretoria, South Africa). Details of primer sequences, their specific targets and amplicon sizes are summarized (Table 6) below.

6.8.5. Phylogenetic Relationship

Additional sequences were downloaded in FASTA format from GenBank through NCBI and combined with assembled sequences. The evolutionary history was inferred using the Neighbor-Joining method [87]. The bootstrap consensus tree inferred from 1000 replicates [88] is taken to represent the evolutionary history of the taxa analyzed [88]. Branches corresponding to partitions reproduced in fewer than 50% bootstrap replicates are collapsed. The percentage of replicate trees in which the associated taxa clustered together in the bootstrap test (1000 replicates) is shown next to the branches [88]. The evolutionary distances were computed using the Kimura 2-parameter method [35] and are in the units of the number of base substitutions per site. The analysis involved 25 nucleotide sequences. Codon positions included were 1st + 2nd + 3rd + Noncoding. All positions containing gaps and missing data were eliminated. There were a total of 640 positions in the final dataset. Evolutionary analyses were conducted in MEGA7 [34].

6.8.6. Divergence Matrix

PCR products for the 16S rRNA gene, identified on agarose gels, were selected for subsequent identification by sequencing (Inqaba Biotech, Pretoria, South Africa). The obtained sequenced data were used to conduct homology searches on GenBank using BLAST (http://blast.ncbi.nlm.nih.gov/blast.cgi) [89], and for further bioinformatic analyses to perform divergence matrix using BioEdit v7.0.9 [33]). Sequences were exported to and analyzed with the MEGA 7 package [34].

Author Contributions: M.M., P.J.O., J.R.G. conceived and designed the study; M.M., J.R.G. collected samples and performed the study; M.M., A.S., S.M., J.R.G. analyzed the data; M.M., P.J.O., J.R.G., A.S., S.M. wrote the paper.

Acknowledgments: We acknowledge financial support from National Research Foundation (NRF) and University of Venda Research and Publication Committee (RPC) and Eskom Tertiary Support Program (TESP) for the research study.

Conflicts of Interest: The authors declare no conflict of interest.

References

1. Christiansen, G.; Molitor, C.; Philmus, B.; Kurmayer, R. Nontoxic strains of cyanobacteria are the result of major gene deletion events induced by a transposable element. *Mol. Biol. Evol.* **2008**, *25*, 1695–1704. [CrossRef] [PubMed]
2. Janse, I.; Kardinaal, W.E.A.; Meima, M.; Fastner, J.; Visser, P.M.; Zwart, G. Toxic and nontoxic *Microcystis* colonies in natural populations can be differentiated on the basis of rRNA gene internal transcribed spacer diversity. *Appl. Environ. Microbiol.* **2004**, *70*, 3979–3987. [CrossRef] [PubMed]
3. Frazao, B.; Martins, R.; Vasconcelos, V. Are Known Cyanotoxins Involved in the Toxicity of Picoplanktonic and Filamentous North Atlantic Marine Cyanobacteria? *Mar. Drugs* **2010**, *8*, 1908–1919. [CrossRef] [PubMed]
4. Pearson, L.A.; Dittmann, E.; Mazmouz, R.; Ongley, S.E.; D'Agostino, P.M.; Neilan, B.A. The genetics, biosynthesis and regulation of toxic specialized metabolites of cyanobacteria. *Harmful Algae* **2016**, *54*, 98–111. [CrossRef] [PubMed]
5. Sinha, R.; Pearson, L.A.; Davis, T.W.; Muenchhoff, J.; Pratama, R.; Jex, A.; Neilan, B.A. Comparative genomics of *Cylindrospermopsis raciborskii* strains with differential toxicities. *BMC Genom.* **2014**, *15*, 83. [CrossRef] [PubMed]
6. Gumbo, R.J.; Ross, G.; Cloete, E.T. Biological control of *Microcystis* dominated harmful algal blooms. *Afr. J. Biotechnol.* **2008**, *7*, 4765–4773.
7. Paerl, H.W.; Fulton, R.S.; Moisander, P.H.; Dyble, J. Harmful freshwater algal blooms with an emphasis on cyanobacteria. *Sci. World* **2001**, *1*, 76–113. [CrossRef] [PubMed]
8. Boyer, G.L. Toxic Cyanobacteria in the Great Lakes: More than just the Western Basin of Lake Erie. *GLRC Great Lakes Res. Rev.* **2006**, *7*, 2–7.
9. O'Neil, J.M.; Davis, T.W.; Burford, M.A.; Gobler, C.J. The rise of harmful cyanobacteria blooms: The potential roles of eutrophication and climate change. *Harmful Algae* **2012**, *14*, 313–334. [CrossRef]

10. Vitousek, P.M.J.; Aber, R.W.; Howarth, G.E.; Likens, P.A.; Matson, D.W.; Schindler, W.H.; Tilman, G.D. Human alteration of the global nitrogen cycle: Causes and consequences. *Issues Ecol.* **1997**, *1*, 1–17.
11. Scheffer, M. The story of some shallow lakes. In *Ecology of Shallow Lakes*; Springer: Dordrecht, The Netherlands, 2004; pp. 1–19.
12. Reynolds, C.S. *Ecology of Phytoplankton*; Cambridge University Press: Cambridge, UK, 2006; 550p.
13. Smith, V.H. Low nitrogen to phosphorus ratios favor dominance by blue-green algae in lake Phytoplankton. *Science* **1983**, *221*, 669–671. [CrossRef] [PubMed]
14. Hyenstrand, P. *Factors Influencing the Success of Pelagic Cyanobacteria*; Uppsala University: Uppsala, Sweden, 1999.
15. Berman, T. The role of DON and the effect of N: P ratios on occurrence of cyanobacterial blooms: Implications from the outgrowth of *Aphanizomenon* in Lake Kinneret. *Limnol. Oceanogr.* **2001**, *46*, 443–447. [CrossRef]
16. Downing, J.A.; Watson, S.B.; McCauley, E. Predicting cyanobacteria dominance in lakes. *Can. J. Fish. Aquat. Sci.* **2001**, *58*, 1905–1908. [CrossRef]
17. Von Ruckert, G.; Giani, A. Effect of nitrate and ammonium on the growth and protein concentration of *Microcystis viridis* Lemmermann (Cyanobacteria). *Rev. Brasiliera Bot.* **2004**, *27*, 325–331.
18. McCarthy, M.J.; Gardner, W.S.; Lavrentyev, P.J.; Moats, K.M.; Joehem, F.J.; Klarer, D.M. Effects of hydrological flow regime on sediment-water interface and water column nitrogen dynamics in a great lakes coastal wetland (Old Woman Creek, Lake Erie). *J. Great Lakes Res.* **2007**, *33*, 219–231. [CrossRef]
19. Andersen, K.; Shanmugam, K. Energetics of biological nitrogen fixation: Determination of the ratio of formation of H2 to NH4+ catalyzed by nitrogenase of *Klebsiella pneumoniae* in vivo. *J. Gen. Microbiol.* **1977**, *103*, 107–122. [CrossRef] [PubMed]
20. Visser, P.M. Growth and Vertical Movement of the Cyanobacterium *Microcystis* in Stable and Artificially Mixed Water Columns. Ph.D. Thesis, University of Amsterdam, Amsterdam, The Netherland, 1995.
21. Thiel, T.; Pratte, B. Effect on heterocyst differentiation of nitrogen fixation in vegetative cells of the cyanobacterium *Anabaena variabilis* ATCC 29413. *J. Bacteriol.* **2001**, *183*, 280–286. [CrossRef] [PubMed]
22. Chan, F.; Pace, M.L.; Howarth, R.W.; Marino, R.M. Bloom formation in heterocystic nitrogen-fixing cyanobacteria: The dependence on colony size and zooplankton grazing. *Limnol. Oceanogr.* **2004**, *49*, 2171–2178. [CrossRef]
23. Mbukwa, E.; Msagati, T.A.; Mamba, B.B.; Boussiba, S.; Wepener, V.; Leu, S.; Kaye, Y. Toxic *Microcystis novacekii* T20-3 from Phakalane Ponds, Botswana: PCR Amplifications of Microcystin Synthetase (*mcy*) Genes, Extraction and LC-ESI-MS Identification of Microcystins. *J. Environ. Anal. Toxicol.* **2015**. [CrossRef]
24. Mbukwa, E.A.; Boussiba, S.; Wepener, V.; Leu, S.; Kaye, Y.; Msagati, T.A.; Mamba, B.B. PCR amplification and DNA sequence of *mcyA* gene: The distribution profile of a toxigenic *Microcystis aeruginosa* in the Hartbeespoort Dam, South Africa. *J. Water Health* **2013**, *11*, 563–572. [CrossRef] [PubMed]
25. Pedro, O.; Rundberget, T.; Lie, E.; Correia, D.; Skaare, J.U.; Berdal, K.G.; Neves, L.; Sandvik, M. Occurrence of microcystins in freshwater bodies in Southern Mozambique. *J. Res. Environ. Sci. Toxicol.* **2012**, *1*, 58–65.
26. Vézie, C.; Rapala, J.; Vaitomaa, J.; Seitsonen, J.; Sivonen, K. Effect of nitrogen and phosphorus on growth of toxic and nontoxic *Microcystis* strains and on intracellular microcystin concentrations. *Microb. Ecol.* **2002**, *43*, 443–454. [CrossRef] [PubMed]
27. Beversdorf, L.J.; Miller, T.R.; McMahon, K.D. The role of nitrogen fixation in cyanobacterial bloom toxicity in a temperate, eutrophic lake. *PLoS ONE* **2013**, *8*, e56103. [CrossRef] [PubMed]
28. Celeste, C.M.M.; Lorena, R.; Oswaldo, A.J.; Sandro, G.; Daniela, S.; Dario, A.; Leda, G. Mathematical modeling of *Microcystis aeruginosa* growth and [D-Leu1] microcystin-LR production in culture media at different temperatures. *Harmful Algae* **2017**, *67*, 13–25. [CrossRef] [PubMed]
29. Van der Westhuizen, A.J.; Eloff, J.N. Effect of temperature and light on the toxicity and growth of the blue-green alga *Microcystis aeruginosa* (UV-006). *Planta* **1985**, *163*, 55–59. [CrossRef] [PubMed]
30. Geada, P.; Pereira, R.N.; Vasconcelos, V.; Vicente, A.A.; Fernandes, B.D. Assessment of synergistic interactions between environmental factors on *Microcystis aeruginosa* growth and microcystin production. *Algal Res.* **2017**, *27*, 235–243. [CrossRef]
31. Ndlela, L.L.; Oberholster, P.J.; Van Wyk, J.H.; Cheng, P.H. An overview of cyanobacterial bloom occurrences and research in Africa over the last decade. *Harmful Algae* **2016**, *60*, 11–26. [CrossRef] [PubMed]
32. Staden, R.; Judge, D.P.; Bonfield, J.K. Analysing sequences using the Staden package and EMBOSS. In *Introduction to Bioinformatics. A Theoretical and Practical Approach*; Human Press Inc.: Totawa, NJ, USA, 2003.

33. Hall, T.A. BioEdit: A user-friendly biological sequence alignment editor and analysis program for Windows 95/98/NT. *Nucleic Acids Symp. Ser.* **1990**, *41*, 95–98.
34. Kumar, S.; Stecher, G.; Tamura, K. MEGA7: Molecular Evolutionary Genetics Analysis version 7.0 for bigger datasets. *Mol. Biol. Evol.* **2016**, *33*, 1870–1874. [CrossRef] [PubMed]
35. Kimura, M. A simple method for estimating evolutionary rate of base substitutions through comparative studies of nucleotide sequences. *J. Mol. Evol.* **1980**, *16*, 111–120. [CrossRef] [PubMed]
36. Oberholster, P.J.; Botha, A.M.; Myburgh, J.G. Linking climate change and progressive eutrophication to incidents of clustered animal mortalities in different geographical regions of South Africa. *Afr. J. Biotechnol.* **2009**, *8*, 5825–5832.
37. Peperzak, L. Climate change and harmful algal bllom in the North Sea. *Acta Oecol.* **2003**, *24*, 139–144. [CrossRef]
38. Rantala, A.; Rajaniemi-Wacklin, P.; Lyra, C.; Lepisto, L.; Rintala, J.; Mankiewiez-Boczek, J.; Sivonen, K. Detection of microcystin-producing cyanobacteria in Finnish Lakes with genus-specific microcystin synthetase Gene E (*mcyE*) PCR and associations with environmental factors. *Appl. Environ. Microbiol.* **2006**, *72*, 6101–6110. [CrossRef] [PubMed]
39. Jahnichen, S.; Petzoldt, T.; Benndorf, J. Evidence for control of microcystin dynamics in BautzenReservoir (Germany) by cyanobacterial population growth rates and dissolved inorganic carbon. *Arch. Hydrobiol.* **2001**, *150*, 177–196. [CrossRef]
40. Bartram, J.; Chorus, I.; Carmichael, W.W.; Jones, G.; Skulberg, O.M. *Toxic Cyanobacteria in Water. A Guide to Their Public Health Consequences, Monitoring and Management*; Chorus, I., Bartram, J., Eds.; World Health Organization: Geneva, Switzerland, 1999; pp. 1–14.
41. Xu, H.; Paerl, H.W.; Qin, B.; Zhu, G.; Gaoa, G. Nitrogen and phosphorus inputs control phytoplankton growth in eutrophic Lake Taihu, China. *Limnol. Oceanogr.* **2010**, *55*, 420–432. [CrossRef]
42. Edokpayi, J.N.; Odiyo, J.O.; Popoola, E.O.; Msagati, T.A. Evaluation of Microbiological and Physicochemical Parameters of Alternative Source of Drinking Water: A Case Study of Nzhelele River, South Africa. *Open Microbiol. J.* **2018**, *12*, 18–27. [CrossRef] [PubMed]
43. Burne, C. Macro-Nutrient and Hydrological Trends in Some Streams of the Waterberg, Limpopo: Investigating the Effects of Land-Use Change on Catchment Water Quality. Master's Dissertation, University of Witwatersrand, Johannesburg, South Africa, 2016.
44. Correll, D.L. Phosphorus: A rate limiting nutrient in surface waters. *Poultry Sci.* **1999**, *78*, 674–682. [CrossRef] [PubMed]
45. Oberholster, P.J.; Dabrowski, J.; Botha, A.M. Using modified multiple phosphorus sensitivity indices for mitigation and management of phosphorus loads on a catchment level. *Fundam. Appl. Limnol. Arch. Hydrobiol.* **2013**, *182*, 1–16. [CrossRef]
46. Mosimanegape, K. Integration of Physicochemical Assessment of Water Quality with Remote Sensing Techniques for the Dikgathong Damin Botswana. Master's Dissertation, University of Zimbabwe, Harare, Zimbabwe, 2016.
47. Matthews, M.W. Eutrophication and cyanobacterial blooms in South African inland waters: 10 years of MERIS observations. *Remote Sens. Environ.* **2014**, *155*, 161–177. [CrossRef]
48. Swanepoel, A.; Du Preez, H.H.; Cloete, N. The occurrence and removal of algae (including cyanobacteria) and their related organic compounds from source water in Vaalkop Dam with conventional and advanced drinking water treatment processes. *Water SA* **2017**, *43*, 67–80. [CrossRef]
49. Seanego, K.G.; Moyo, N.A.G. The effect of sewage effluent on the physico-chemical and biological characteristics of the Sand River, Limpopo, South Africa. *Phys. Chem. Earth Parts A/B/C* **2013**, *66*, 75–82. [CrossRef]
50. Musa, R.; Greenfield, R. Nutrient loads on an important watercourse. Pre-and Post-Acid spill. In Proceedings of the 7th International Toxicology Symposium, Johannesburg, South Africa, 31 August 2015; p. 89.
51. Basima, L.B. An Assessment of Plankton Diversity as a Water Quality Indicator in Small Man-m Ade Reservoirs in the Mzingwane Catchment, Limpopo Basin, Zimbabwe. Master's Dissertation, University of Zimbabwe, Harare, Zimbabwe, 2005.
52. Mupfiga, E.T.; Munkwakwata, R.; Mudereri, B.; Nyatondo, U.N. Assessment of sedimentation in Tuli Makwe Dam using remotely sensed data. *J. Soil Sci. Environ. Manag.* **2016**, *7*, 230–238.

53. Mavhunga, M. The Presence of Cyanobacteria & Diatoms in Limpopo River Sediment Profile: Implications for Human Health. Unpublished Hons mini Thesis, University of Venda, Johannesburg, South Africa, 2015.
54. Oberholster, P.J.; Jappie, S.; Cheng, P.H.; Botha, A.M.; Matthews, M.W. First report of an *Anabaena Bory* strain containing microcystin-LR in a freshwater body in Africa. *Afr. J. Aquat. Sci.* **2015**, *40*, 21–36. [CrossRef]
55. Kim, Y.J.; Baek, J.S.; Youn, S.J.; Kim, H.N.; Lee, B.C.; Kim, G.; Park, S.; You, K.A.; Lee, J.K. Cyanobacteria Community and Growth Potential Test in Sediment of Lake Paldang. *J. Korean Soc. Water Environ.* **2016**, *32*, 261–270. [CrossRef]
56. Dzebu, W. Statement in Relation to Water Supply Challenges in Musina, 14 March 2017. Available online: www.musina.gov.za/index.php/public-notices?download=1028...in...to...musina (accessed on 26 March 2014).
57. Botha, A.M.; Oberholster, P.J. *PCR-Based Markers for Detection and Identification of Toxic Cyanobacteria*; WRC Report No. K5/1502/01/07; Water Research Commission: Pretoria, South Africa, 2007; p. 70.
58. Su, Y.; You, X.; Lin, H.; Zhuang, H.; Weng, Y.; Zhang, D. Recruitment of cyanobacteria from the sediments in the eutrophic Shanzi Reservoir. *Environ. Technol.* **2016**, *37*, 641–651. [CrossRef] [PubMed]
59. Maldener, I.; Summers, M.L.; Sukenik, A. Cellular differentiation in filamentous cyanobacteria. In *The Cell Biology of Cyanobacteria*; Flores, E., Herrero, A., Eds.; Caister Academic Press: Norwich, UK, 2014; pp. 263–291.
60. Perez, R.; Forchhammer, K.; Salerno, G.; Maldener, I. Clear differences in metabolic and morphological adaptations of akinetes of two Nostocales living in different habitats. *Microbiology* **2016**, *162*, 214–223. [CrossRef] [PubMed]
61. Adam, D.G.; Duggan, P.S. Heterocyst and akinete differentiation in cyanobacteria. *New Phytol.* **1999**, *144*, 3–33. [CrossRef]
62. Scherer, P.I.; Raeder, U.; Geist, J.; Zwirglmaier, K. Influence of temperature, mixing, and addition of microcystin-LR on microcystin gene expression in *Microcystis aeruginosa*. *Microbiol. Open* **2017**, *6*. [CrossRef] [PubMed]
63. Mikalsen, B.; Boison, G.; Skulberg, O.M.; Fastner, J.; Davies, W.; Gabrielsen, T.M.; Rudi, K.; Jakobsen, K.S. Natural variation in the microcystin synthetase operon *mcyABC* and impact on microcystin production in *Microcystis* strains. *J. Bacteriol.* **2003**, *185*, 2774–2785. [CrossRef] [PubMed]
64. Davis, T.W.; Berry, D.L.; Boyer, G.L.; Gobler, C.J. The effects of temperature and nutrients on the growth and dynamics of toxic and non-toxic strains of *Microcystis* during cyanobacteria blooms. *Harmful Algae* **2009**, *8*, 715–725. [CrossRef]
65. Yamamoto, Y. Effect of temperature on recruitment of cyanobacteria from the sediment and bloom formation in a shallow pond. *Plankton Benthos Res.* **2009**, *4*, 95–103. [CrossRef]
66. Oberholster, P.J.; Botha, A.-M.; Cloete, T.E. Use of molecular markers as indicators for winter zooplankton grazing on toxic benthic cyanobacteria colonies in an urban Colorado lake. *Harmful Algae* **2006**, *5*, 705–716. [CrossRef]
67. Mosase, E.; Ahiablame, L. Rainfall and Temperature in the Limpopo River Basin, Southern Africa: Means, Variations, and Trends from 1979 to 2013. *Water* **2018**, *10*, 364. [CrossRef]
68. Hoeger, S.J.; Dietrick, D.R.; Hitzfeld, B.C. Effect of ozonation on the removal of cyanobacteria toxins during drinking water treatment. *Environ. Health Perspect.* **2002**, *110*, 1127–1132. [CrossRef] [PubMed]
69. Brittain, S.M.; Wang, J.; Babcock-Jackson, L.; Carmichael, W.W.; Rinehart, K.L.; Culver, D.A. Isolation and characterization of microcystins, cyclic heptapeptide hepatotoxins from Lake Erie strain of *Microcystis aeruginosa*. *J. Great Lakes Res.* **2000**, *26*, 241–249. [CrossRef]
70. Andersson, J.A.; de Garine-Wichatitsky, M.; Cumming, D.H.; Dzingirai, V.; Giller, K.E. People at wildlife frontiers in Southern Africa. In *Transfrontier Conservation Area: People Living on the Edge*; Routledge: Abingdon-on-Thames, UK, 2013; pp. 1–11.
71. Oberholster, P.J.; Myburgh, J.G.; Govender, D.; Bengis, R.; Botha, A.M. Identification of toxigenic *Microcystis* strains after incidents of wild animal mortalities in the Kruger National Park, South Africa. *Ecotoxicol. Environ. Saf.* **2009**, *72*, 1177–1182. [CrossRef] [PubMed]
72. McCollough, B. Toxic Algae and Other Marine Biota: Detection, Mitigation, Prevention and Effects on the Food Industry. Master's Dissertation, Kansas State University, Manhattan, KS, USA, 2016.
73. Zhu, T.; Ringler, C. Climate change impacts on water availability and use in the Limpopo River Basin. *Water* **2012**, *4*, 63–84. [CrossRef]

74. Department of Water Affairs. *Joint Water Quality Baseline Report for Limpopo Basin between Botswana and South Africa*; Department of Water Affairs: Pretoria, South Africa, 2011.
75. Islam, M.S.; Ahmed, M.K.; Raknuzzaman, M.; Habibullah-Al-Mamun, M.; Islam, M.K. Heavy metal pollution in surface water and sediment: A preliminary assessment of an urban river in a developing country. *Ecol. Indic.* **2015**, *48*, 282–291. [CrossRef]
76. APHA; AWWA; WPCF. *Standard Methods for Examination of Water and Wastewater*, 20th ed.; American Publishing Health Association: Washington, DC, USA, 1998.
77. Gumbo, J.R.; Ross, G.; Cloete, T.E. The Isolation and identification of Predatory Bacteria from a *Microcystis* algal Bloom. *Afr. J. Biotechnol.* **2010**, *9*, 663–671.
78. Van Vuuren, S.J.; Taylor, J.; Gerber, A. *A Guide for the Identification of Microscopic Algae in South Africa Freshwaters*; Department of Water Affairs and Forestry, North-West University: Potchefstroom, South Africa, 2006.
79. Moffitt, M.C.; Neilan, B.A. Characterization of the nodularin synthetase gene cluster and proposed theory of the evolution of cyanobacterial hepatotoxins. *Appl. Environ. Microbiol.* **2004**, *70*, 6353–6362. [CrossRef] [PubMed]
80. Pearson, L.A.; Neilan, B.A. The molecular genetics of cyanobacterial toxicity as a basis for monitoring water quality and public health risk. *Curr. Opin. Biotechnol.* **2008**, *19*, 281–288. [CrossRef] [PubMed]
81. Jungblut, A.D.; Neilan, B.A. Molecular identification and evolution of the cyclic peptide hepatotoxins, microcystin and nodularin, synthetase genes in three orders of cyanobacteria. *Arch. Microbiol.* **2006**, *185*, 107–114. [CrossRef] [PubMed]
82. Fergusson, K.M.; Saint, C.P. Multiplex PCR assay for *Cylindrospermopsis raciborskii* and cylindrospermopsin-producing cyanobacteria. *Environ. Toxicol.* **2003**, *18*, 120–125. [CrossRef] [PubMed]
83. Hisbergues, M.; Christiansen, G.; Rouhiainen, L.; Sivonen, K.; Borner, T. PCR-based identification of microcystin-producing genotypes of different cyanobacterial genera. *Arch. Microbiol.* **2003**, *180*, 402–410. [CrossRef] [PubMed]
84. Schembri, M.A.; Neilan, B.A.; Saint, C.P. Identification of genes implicated in toxin production in the cyanobacterium *Cylindrospermopsis raciborskii*. *Environ. Toxicol.* **2001**, *16*, 413–421. [CrossRef] [PubMed]
85. Neilan, B.A.; Jacobs, D.; Del Dot, T.; Blackall, L.L.; Hawkins, P.R.; Cox, P.T.; Goodman, A.E. rRNA sequences and evolutionary relationships among toxic and nontoxic cyanobacteria of the genus *Microcystis*. *Int. J. Syst. Bacteriol.* **1997**, *47*, 693–697. [CrossRef] [PubMed]
86. Jungblut, A.D.; Hawes, I.; Mountfort, D.; Hitzfeld, B.; Dietrich, D.R.; Burns, B.P.; Neilan, B.A. Diversity within cyanobacterial mat communities in variable salinity meltwater ponds of McMurdo Ice Shelf, Antarctica. *Environ. Microbiol.* **2005**, *7*, 519–529. [CrossRef] [PubMed]
87. Saitou, N.; Nei, M. The neighbor-joining method: A new method for reconstructing phylogenetic trees. *Mol. Biol. Evol.* **1987**, *4*, 406–425. [PubMed]
88. Felsenstein, J. Confidence limits on phylogenies: An approach using the bootstrap. *Evolution* **1985**, *39*, 783–791. [CrossRef] [PubMed]
89. Altschul, S.F.; Gish, W.; Miller, W.; Myers, E.W.; Lipman, D.J. "Gapped Blast and PSI-BLAST" A generation of protein database search programs. *Nucleic Acids Res.* **1997**, *25*, 3389–3402. [CrossRef] [PubMed]

toxins

MDPI

Article

Development of Time-Resolved Fluoroimmunoassay for Detection of Cylindrospermopsin Using Its Novel Monoclonal Antibodies

Lamei Lei, Liang Peng, Yang Yang and Bo-ping Han *

Department of Ecology and Institute of Hydrobiology, Jinan University, Guangzhou 510632, China; tleilam@jnu.edu.cn (L.L.); tpengliang@jnu.edu.cn (L.P.); yangy@jnu.edu.cn (Y.Y.)

* Correspondence: tbphan@jnu.edu.cn; Tel.: +86-020-3837-4065; Fax: +86-020-3837-4600

Received: 29 May 2018; Accepted: 15 June 2018; Published: 21 June 2018

Abstract: Cylindrospermopsin (CYN) is a cyanotoxin that is of particular concern for its potential toxicity to human and animal health and ecological consequences due to contamination of drinking water. The increasing emergence of CYN around the world has led to urgent development of rapid and high-throughput methods for its detection in water. In this study, a highly sensitive monoclonal antibody N8 was produced and characterized for CYN detection through the development of a direct competitive time-resolved fluorescence immunoassay (TRFIA). The newly developed TRFIA exhibited a typical sigmoidal response for CYN at concentrations of 0.01–100 ng mL^{-1}, with a wide quantitative range between 0.1 and 50 ng mL^{-1}. The detection limit of the method was calculated to be 0.02 ng mL^{-1}, which is well below the guideline value of 1 μg L^{-1} and is sensitive enough to provide an early warning of the occurrence of CYN-producing cyanobacterial blooms. The newly developed TRFIA also displayed good precision and accuracy, as evidenced by low coefficients of variation (4.1–6.5%). Recoveries ranging from 92.6% to 108.8% were observed upon the analysis of CYN-spiked water samples. Moreover, comparison of the TRIFA with an ELISA kit through testing 76 water samples and 15 *Cylindrospermopsis* cultures yielded a correlation r^2 value of 0.963, implying that the novel immunoassay was reliable for the detection of CYN in water and algal samples.

Keywords: cylindrospermopsin; monoclonal antibody; time-resolved fluoroimmunoassay; method validation; detection

Key Contribution: This is an original work on the production of a highly specific monoclonal antibody to CYN. A novel immunoassay with a low detection limit was developed for CYN measurement.

1. Introduction

Cyanobacterial blooms occur frequently in eutrophic freshwater lakes, reservoirs and rivers throughout the world. Many cyanobacterial species are capable of producing cyanotoxins that pose a significant threat to both water quality and human health [1]. Cyanotoxins are a variety of secondary metabolites that include microcystins (MCs), nodularin, cylindrospermopsin (CYN), anatoxin-a, and saxitoxins [2]. Cylindrospermopsin is becoming one of the most commonly studied cyanotoxins because of its toxicity and increasing presence in different environments [3,4]. Several cyanobacterial species, such as *Cylindrospermopsis raciborskii*, several *Aphanizomenon* species and *Raphidiopsis curvata*, have been reported to be potent CYN-producers [5–7]. Cylindrospermopsin-producing cyanobacteria has been detected in Australia and New Zealand, Asia, South and North America, West Africa, and Europe [4,8].

Cylindrospermopsin is a guanidine alkaloid that has an LD_{50} of 2.1 mg kg^{-1} over 24 h after intraperitoneal administration to mice [5]. Exposure to CYN rapidly increased the production of reactive oxygen species (ROS) and may result in serious cytotoxic and genotoxic effects [3,9,10]. Oxidative stress is one of the key mechanisms involved in CYN toxicity [9,10]. Moreover, CYN was found to suppress lymphocytes proliferation and could be classified as a potential immunotoxicant [9,11,12]. The alkaloid is probably more hazardous to human and animal health than microcystins (MCs) because of its cell transforming potential [13]. When compared to other cyanotoxins, CYN is more stable under a wide range of pH and temperatures, and may present significant consequences for aquatic environments [14]. Thus, qualitative and quantitative analytical tools need to be developed for long-term monitoring of CYN in freshwater to minimize its risks to water quality and human health.

Common approaches to the detection of cyanobacteria and their toxins in the environment are currently chemical-, biochemical-, or molecular-based methods [3,15]. Specifically, analytical methods include high-performance liquid chromatography-photo-diode array (HPLC-PDA), liquid chromatography-mass spectrometry (LC-MS/MS), enzyme-linked immunosorbent assay (ELISA), and conventional or real-time PCR assays [16–19], with LC-MS/MS the most commonly used [16,20,21]. However, LC-MS/MS relies on specialized and expensive equipment. The development of immunological approaches has yielded more sensitive, rapid and high-throughput tools for the detection and quantification of cyanotoxins in all kinds of water and cultured samples [14]. To establish immunoassays, monoclonal or polyclonal antibodies are required to be raised against CYN. Two commercial ELISA kits based on rabbit anti-CYN polyclonal antibodies are currently available with low detection limits of 0.1 ng/mL (Beacon Analytical Systems Inc., Saco, ME, USA) and 0.05 ng/mL (Abraxis LLC, Warminster, PA, USA), respectively [15]. Both kits have been used to determine CYN in raw water and cyanobacterial extracts [19,20,22–27]. Elliott et al. [28] published the first detailed report on the production of polyclonal and monoclonal antibodies to CYN. These antibodies were employed in competitive indirect ELISA, an optical biosensing technique of surface-plasmon resonance (SPR), the Luminex method and the MBio Biosensor [28–30]. However, more antibodies with high specificity to CYN are required for further development to increase the sensitivity and applicability of the different immunoassays.

Time-resolved fluorescence immunoassay (TRFIA) uses lanthanide chelates and has been widely used for clinical screening and diagnostics. Lanthanide chelates have unique luminescent properties, such as a long Stokes' shift and exceptional decay times, which allow for efficient temporal discrimination of background interferences in the assays [31]. The technique TRFIA is characterized by a long storage time, high sensitivity and specificity, good repeatability and wide detection range with no radioactive contamination [32,33]. In this study, specific monoclonal antibodies to CYN were produced and used to improve on the TRFIA method to develop a sensitive immunological technique for quantification of CYN.

2. Results and Discussion

Cylindrospermopsin is becoming one of the most commonly studied cyanotoxins because of its wide distribution and multiorgan toxicity, and its ecological role and triggering environmental factors have not fully been understood; however, there are relatively few methods available for its analysis [34–36]. In this study, we developed a novel direct competitive TRFIA technique to measure CYN at trace levels (Figure 1). Our TRFIA provides a fast and highly sensitive screening method and may act as an early warning detection tool for CYN monitoring.

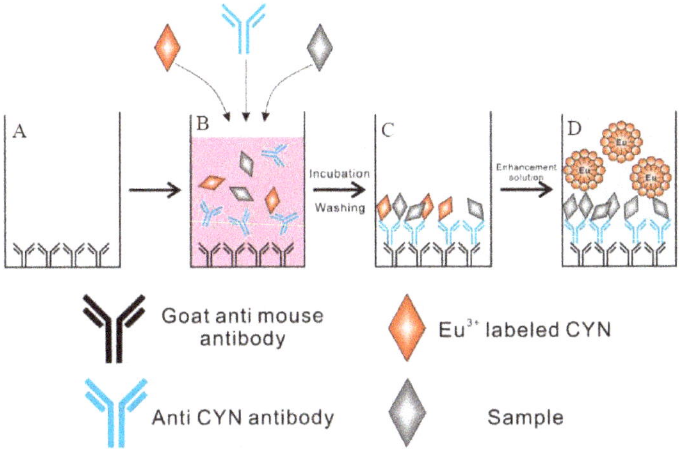

Figure 1. Example of the newly developed TRFIA employing a europium chelate label. (**A**) adsorption of goat anti-mouse antibody; (**B**) a direct competitive reaction; (**C**) formation of CYN/antibody complex; (**D**) measurement of fluorescence intensity.

2.1. CYN Conjugate Preparation

Like other cyanotoxins, CYN is a low-molecular-weight nonimmunogenic toxin. To become immunogenic, these toxins need be conjugated to carrier proteins by means of different chemical approaches [28,37]. The Mannich reaction, a conjugation between one active hydrogen and primary amines in the presence of formaldehyde, has been used to link another alkaloid toxin-saxitoxin (STX) to BSA, while polyclonal antibody R895 was generated using the conjugate as an immunogen [38]. Among five chemical approaches to synthesizing CYN immunogens, the modified Mannich reaction was considered the most effective method for antibody production [28]. Although many carrier proteins have been used in the coupling of cyanotoxins, BSA and KLH are the most common ones. In the present study, we used BSA and KLH to synthesize CYN conjugates via the Mannich reaction and then obtained an immunogen (KLH-CYN) and coating antigen (BSA-CYN).

2.2. Monoclonal Antibody Production

Hybridized cells were plated on six 96-well plates and screened by ELISA for monoclonal antibody production. The first screening with BSA-CYN yielded 64 antibody-producing clones, while the second screening yielded 22. Next, 22 cell lines were re-cloned twice, after which nine clones remained. The subclasses of these nine positive clones determined by the Pierce® Rapid Isotyping Kit were N1 (IgG1 λ), N2 (IgG1κ), N3 (IgG2a λ), N4 (IgG1 λ), N5 (IgG1κ), N6 (IgG2a λ), N8 (IgG1 λ), N10 (IgG1 λ) and N16 (IgG1 λ).

The binding capacity of different MAbs to CYN was examined by a direct competitive TRFIA: 0.1 ng mL^{-1} CYN standard, an appropriate dilution of Eu^{3+}-labeled BSA-CYN, and addition of each MAb to the plate coating with goat anti-mouse IgG. After incubation, washing and fluorescence measurement was determined as follows:

$$\text{binding \%} = (F/F_0) \times 100$$

where F corresponds to the fluorescence value of wells in the presence of 0.1 ng mL^{-1} CYN, and F_0 is the fluorescence in the absence of CYN. The binding of all nine MAbs to BSA-CYN can be inhibited by low concentration of free CYN, but the inhibition varied significantly and N8 had higher affinity to CYN than the other eight antibodies (Figure 2).

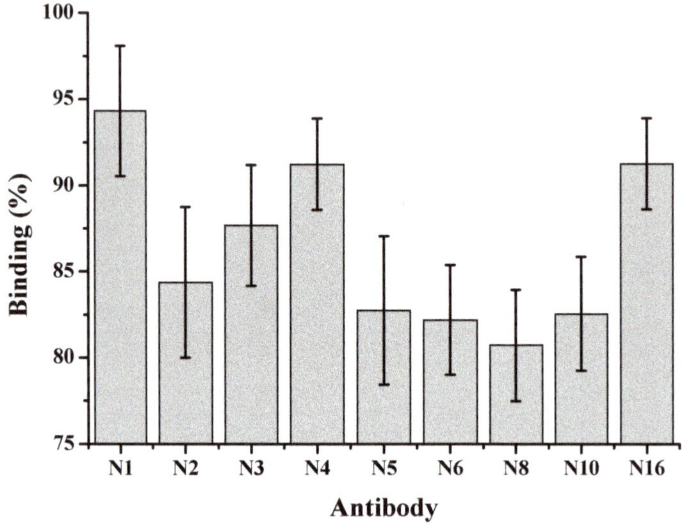

Figure 2. Percentage binding of nine monoclonal antibodies to BSA-CYN in the presence of 0.1 ng mL^{-1} CYN.

To the best of our knowledge, this was the second detailed investigation of CYN antibody production after Elliott et al. [28]. Furthermore, rabbit anti-CYN polyclonal antibodies have been employed in the immunoassays according to the instructions of commercial ELISA kits, but no information regarding antibody production is available. Suitable antibodies are essential to the establishment of immunoassays. In this regard, MCs have received a great deal of attention, and several types of antibodies including conventional polyclonal or monoclonal antibodies and novel recombinant antibodies have been developed [37,39–41]. Moreover, antibodies have been raised specifically against some MC variants [40,42,43]. Weller [44] pointed out that the production of high quality antibodies would be highly desirable for all other cyanotoxins, a situation that caused delays to the development of multianalyte immunoassays and biosensors. Apparently a gap exists in both antibody production to CYN and its immunological techniques. Therefore, our novel monoclonal antibody N8 has wide application prospects.

2.3. Establishment of Standard Curves

The direct competitive TRFIA curve established with antibody N8 showed a typical sigmoidal response for CYN at concentrations of 0.01–100 ng mL^{-1} (Figure 3A). In addition, the fluorescence signal was evaluated following our immunoassay design with a serial dilution of standards obtained from 10 separate assays. The coefficient of variation of each standard was less than 10%, indicating high reproducibility of the TRFIA curve. The logit-log method was used for curve fitting, and the best-fit calibration fell into the concentration range of 0.1–50 ng mL^{-1} (Figure 3B). Concentration of CYN in the sample is quantitatively determined by the logit-log model:

$$\text{In}\left\{\frac{\frac{F}{F_0}}{1-\frac{F}{F_0}}\right\} = -2.08 \times \text{LogX} - 0.516$$

where F is the fluorescence value of wells with the unknown sample, F_0 is the fluorescence value of zero concentration, and X is the CYN concentration of the unknown sample.

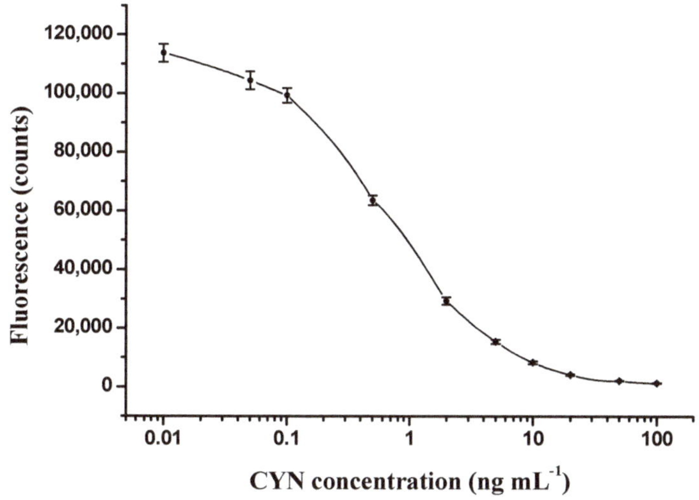

(A)

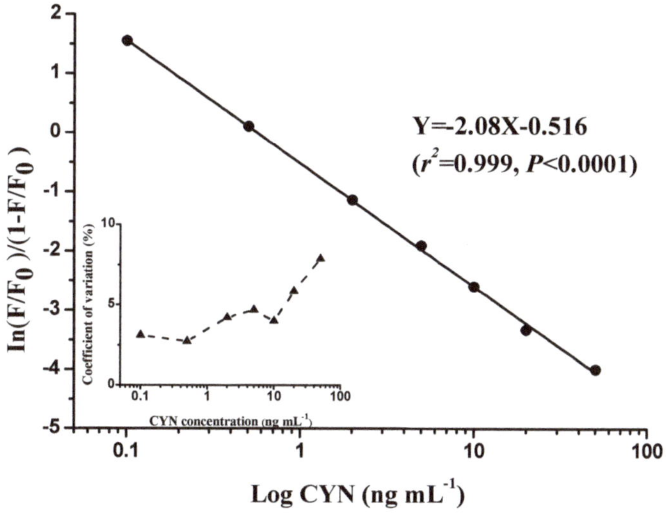

(B)

Figure 3. Typical standard curve showing the normalized fluorescence signal as a function of CYN concentration in the range of 0 to 100 ng mL^{-1} (**A**). The corresponding logit-log linear calibration curve and intra-assay precision profile (each point was based on 10 replicates) are shown in (**B**).

In this way, a TRFIA was established to detect CYN in environmental samples. This immunoassay has been applied previously to detect microcystins and nodularin, and some advantages such as a wide detection range and high sensitivity have been demonstrated [33,45,46]. In our study, the TRFIA also exhibited a wide quantitative range between 0.1 and 50 ng mL^{-1} and maintained good reliability. We did not test the cross-reactivity of N8 to other CYN variants because no commercial standard was available. According to a previous study [28], polyclonal and monoclonal antibodies raised against

CYN showed relatively low cross-reactivities with deoxy-CYN. However, the CYN ELISA kit from Abraxis LLC appeared to provide good quantification for the assessment of CYN and its variants [27]. The difference in performance may be attributed to antibody specificity, and some antibodies with broad specificity are capable of recognizing several cyanotoxins with similar structures [37].

2.4. Assay Validation

2.4.1. Analytical Sensitivity of the Present Method

Under the optimized conditions, the detection limit of the assay was 0.02 ng mL^{-1} CYN. Sensitivity of the newly developed TRFIA method is similar to that of the monoclonal antibody-based ELISA of Elliott et al. [28] and the Abraxis ELISA kit, but higher than that of the monoclonal antibody-based SPR of Elliott et al. [28] and the MBio assay of McNamee et al. [30]. Both TRFIA and commercial ELISA kits are capable of detecting CYN within the guideline value of 1 μg L^{-1} proposed by Humpage and Falconer [35].

2.4.2. Precision and Accuracy of the Present Method

Table 1 shows the precision of the developed TRFIA for the quantitative detection of CYN. Coefficients of variation for both intra- and inter-assays ranged from 4.1% to 6.5%. Precision of the present assay was excellent, and none of the coefficients of variation were significant (≥10%). The general analytical recovery of the assay was in the range of 90–110%, indicating the high accuracy of the measurements.

Table 1. Precision and accuracy test of the newly developed TRFIA.

Type	Samples	Nominal Value (ng mL^{-1})	Mean ± SD (ng mL^{-1})	CV (%)	Recovery (%)
Intra-assay (n = 8)	A	2.0	1.91 ± 0.078	4.1	95.9
	B	5.0	5.16 ± 0.24	4.6	103.2
	C	20.0	20.3 ± 1.18	5.8	101.5
Inter-assay (n = 12)	A	2.0	1.96 ± 0.096	4.9	98.1
	B	5.0	5.08 ± 0.24	4.8	101.7
	C	20.0	19.5 ± 12.7	6.5	97.4

CV: coefficient of variation. SD: standard deviation.

2.4.3. Recovery of the Developed Method

The recoveries of CYN-spiked water samples ranged from 92.6% to 108.8%, with coefficients of variation <16.38% (Table 2). The results showed that our TRFIA was not affected by the matrix of the natural environment when detecting CYN in water samples.

Table 2. Recovery and coefficient of variation of CYN-spiked samples.

Spiked Value (ng mL^{-1})	Measured Value ± SD (ng mL^{-1})	Recovery (%)	CV (%)
0.25	0.246 ± 0.04	98.6	16.38
5.0	5.44 ± 0.54	108.8	9.99
25	23.15 ± 1.91	92.6	8.24

CV: coefficient of variation. SD: standard deviation.

2.4.4. Dilution Linearity for the Present Method

Table 3 shows the dilution linearity results of the assay when we used positive samples serially diluted with our assay buffer. The expected values were derived from the initial value of potency in the undiluted samples. We found that the expected values were almost identical to the measured values, as evidenced by the high recoveries (94.5–108.7%). These results confirmed that linearity was good over a wide range of dilution and detection would be unaffected if the sample was diluted with assay buffer. Therefore, the TRFIA method provides flexibility to assay samples with distinct levels of CYN.

Table 3. Dilution Linearity test for the newly developed TRFIA.

Sample	Dilution	Expected Value (ng mL^{-1})	Observed Value (ng mL^{-1}, n = 3)	Recovery (%)
A	NA		10.0	
	1:2	5.0	4.86	97.2
	1:4	2.5	2.41	96.4
	1:8	1.25	1.29	103.2
	1:16	0.62	0.66	106.5
B	NA		50.0	
	1:2	25.0	25.4	101.6
	1:4	12.5	13.0	104.6
	1:8	6.25	5.91	94.5
	1:16	3.12	3.39	108.7

NA, not applicable.

2.5. Comparison of Assay Results and Performance of the Developed TRFIA and ELISA

The assay performance of the developed TRFIA and the performance data provided in the instruction manuals of the commercial ELISA kits are compared in Table 4. It should be noted that the TRFIA method was faster, more precise, and had a wider detection range than the ELISA kits. To investigate reliability of the developed TRFIA for measurement of CYN in different samples, a total of 91 samples (76 water samples collected from Guangdong reservoirs in South China and 15 *Cylindrospermopsis* cultures) were assayed by TRFIA and ELISA. There were 84 positive samples identified by TRFIA and 83 by ELISA, indicating good agreement between the two methods. The results of our method are compared to those of the Beacon ELISA kit in Figure 4. The estimated contents obtained from the present TRFIA method and the Beacon ELISA kit showed very high correlation (r^2 = 0.963, $p < 0.0001$, n = 91); thus, the novel immunoassay developed by our group can be considered a useful tool for detection of CYN in water and algal samples.

Table 4. Comparison of assay performance for the newly developed TRFIA reagent and commercial ELISA kits.

Method	Recovery	Imprecision	Operating Time	Maximum Quantitative Value
TRFIA	95.9–103.2%	4.1–6.5%	1 h	50 ng mL^{-1}
ELISA (Beacon)	80–120%	<20%	1.5 h	2 ng mL^{-1}
ELISA (Abraxis)	98–108%	4.3–8.3%	1.25–1.5 h	2 ng mL^{-1}

We can infer from the large positive correlation coefficient that examination of CYN samples by ELISA resulted in slightly higher reported concentrations than TRFIA. Previous studies suggest that ELISA may overestimate the CYN concentration [20,22,47]. Moreover, Metcalf et al. [27] found that the non-cyanotoxin-producing green alga, *Chlorella* sp., also gave positive responses in ELISA. The qualitative difference might result from undesired cross-reactivity of antibodies used in ELISA [47]. The present TRFIA was developed using N8 monoclonal antibody; however, commercial ELISA kits typically use rabbit anti-CYN polyclonal antibodies. For the polyclonal sera raised against CYN-OVA, a high level of non-specific background response was observed when evaluated by ELISA [28]. The high specificity of the monoclonal antibody may reduce the probability of cross reactivity and non-specific binding [48]. Future work is needed to evaluate the cross-reactivity of N8 monoclonal antibody to CYN analogues namely deoxy-CYN and 7-epi-CYN. Additionally, LC-MS/MS has been developed as the ideal confirmation technique for trace CYN in environmental samples, and comparison of the present TRFIA method with LC-MS/MS is highly desirable for further confirming the accuracy and reliability of the TRFIA.

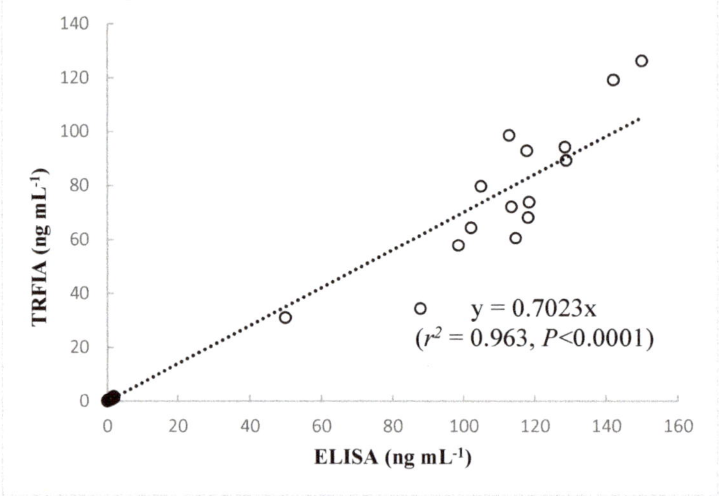

Figure 4. Correlations analysis between CYN concentrations measured by the newly developed TRFIA and by ELISA in 91 samples. Data represent the means of three determinations.

3. Materials and Method

3.1. Chemicals and Solutions

Pure CYN (purity > 95%) was obtained from Enzo Life Sciences (Farmingdale, NY, USA). Keyhole limpet hemocyanin (KLH), bovine serum albumin (BSA), 2-morpholinoethanesulfonic acid (MES), Jeffamine, EDC, N-hydroxysuccinimide (NHS) and other reagents were purchased from Sigma–Aldrich (St. Louis, MO, USA). Goat anti-Mouse IgG secondary antibody was acquired from Biodesign International (Saco, ME, USA) and Eu^{3+} labeled kits were obtained from PerkinElmer (Turku, Finland).

The MES buffer consisted of 50 mM 2-morpholinoethanesulfonic acid with 500 mM NaCl (pH 5), while the coating buffer was 50 mM Na_2CO_3-$NaHCO_3$ buffer (pH 9.6) and the blocking solution was 50 mM Na_2CO_3-$NaHCO_3$ buffer (pH 9.6) containing 1% BSA. The labeling buffer was 50 mM Na_2CO_3-$NaHCO_3$ (pH 8.5) with 155 mM NaCl, elution buffer was 50 mM Tris-HCl (pH 7.4) with 0.2% BSA and 0.9% NaCl, standard buffer was 50 mM Tris-HCl (pH 7.8) containing 0.1% NaN_3 and 0.2% BSA, and assay buffer was 50 mM Tris-HCl (pH 7.8) with 0.02% BSA, 0.05% Tween-20 and 0.05% NaN_3. The enhancement solution was 100 mM acetate-phthalate buffer (pH 3.2) containing 15 μM β-naphthoyltrifluoroacetate, 50 μM tri-n-octylphosphine oxide and 0.1% triton X-100. The washing buffer was 25 mM Tris-HCl (pH 7.8) with 0.9% NaCl and 0.06% Tween-20.

3.2. Preparation of Protein Conjugates KLH-CYN and BSA-CYN

The KLH-CYN protein conjugate was prepared using a modification of the Mannich reaction as described by Elliott et al. [28]. Cylindrospermopsin (250 μg) was added to KLH (1.5 mg) dissolved in 200 μL of phosphate buffer. Formaldehyde (6 μL) was then added, after which the mixture was stirred in the dark at room temperature for 50 h. The conjugate was subsequently purified by dialysis in 0.15 M saline solution.

The BSA-CYN protein conjugate was prepared using a modification of the Mannich reaction described by Compbell et al. [38]. An aliquot (250 μL) of EDC (20 mg) and NHS (8 mg) dissolved in MES buffer was added to the BSA (10 mg dissolved in MES buffer) and mixed for 5 min at room temperature. Jeffamine (50 μL, 1 M) was then added and the mixture was allowed to react for 3 h at room temperature. The Jeffamine-BSA conjugate was subsequently purified using a PD-10 column (GE Healthcare, Little Chalfont, UK). The freeze-dried Jeffamine-BSA (2 mg) was resuspended and both CYN (550 μg) and formaldehyde (12 μL) were added. The resultant mixture was allowed to react for 50 h, followed by dialysis over 24 h in 0.15 M saline solution.

3.3. Monoclonal Antibody Production

Hybridomas-producing anti-CYN monoclonal antibodies (MAbs) were prepared with a standard method for immunization and cell fusion. BALB/c mice were first immunized with 100 μg KLH-CYN in 0.1 mL sterile saline and 0.1 mL of complete Freund's adjuvant. Then, the mice were boosted with KLH-CYN in incomplete Freund's adjuvant with an internal of three weeks. When a high antibody titer was observed, a final immunization was performed intraperitoneally without adjuvant. The spleen cells of BALB/c mice immunized with KLH-CYN were fused with Sp2/0 cells. The resulting hybridomas were screened by indirect ELISA to select antibody-producing clones specifically reacting with BSA-CYN. Positive hybridomas were ultimately propagated in vitro, and then used for the preparation of ascites. This study has been approved and registered by the laboratory animal welfare and ethics committee of Southern Medical University (Guangzhou, China), and the care and use of the animals was conducted according to the Institutional Animal Ethics Committee guidelines.

Monoclonal antibodies were purified using a protein G column (ThermoFisher Scientific, Shanghai, China) according to manufacturer's instructions. Verification of the purified antibody subclass was accomplished using a Pierce® Rapid Isotyping Kit with Kappa and Lambda-Mouse from Thermo Scientific (Rockford, IL, USA).

3.4. Eu^{3+}-Labeled BSA-CYN

To prepare the Eu^{3+}-labeled BSA-CYN, 0.1 mg DTTA-Eu (N1-[P-isothiocyanato-benzyl] -diethylene-triamine-N1, N2, N3-tetraacetate-Eu^{3+}) was added to 0.5 mg of BSA-CYN in 100 μL labeling buffer. The mixture was stirred and kept overnight at room temperature. The Eu^{3+}-labeled BSA-CYN was then separated from unreacted chelates and aggregated proteins by Sephadex G-50 gel filtration using elution buffer. Fluorescence of the labeled BSA-CYN was measured at 615 nm and then aliquots of the labeled protein were kept at −20 °C.

3.5. Coating with Secondary Antibody

Goat anti-mouse IgG secondary antibody (200 μL) diluted to final concentration of 3 μg mL^{-1} with coating buffer was pipetted into each well, after which the plates were incubated at 4 °C overnight and then washed three times with washing buffer. Next, 250 μL of blocking buffer was added to each well and the samples were maintained at 4 °C overnight. Following removal of the blocking buffer, the plates were vacuum dried and stored with a desiccant at −20 °C.

3.6. Development of the Direct Competitive TRFIA

The assay was performed using the one-step procedure and the direct competitive protocol. Following immobilization and antibody labeling protocols described above, 100 μL of CYN standards (0, 0.01, 0.05, 0.1.0.5, 2, 5, 10, 20, 50 and 100 ng mL^{-1}) or samples, 25 μL of assay buffer containing 400 ng N8 MAb, and 25 μL of assay buffer containing 20 ng of Eu^{3+}-labeled BSA-CYN were added into each well and then coated with goat anti-mouse IgG antibody. A competition model in wells was generated subsequent to one hour of incubation with continuous slow shaking. The wells were then washed six times and filled with 100 μL of enhancement solution. The plates were then shaken for 5 min at room temperature and fluorescence intensity measured using a Victor3 1420 Multilabel Counter equipped with filters for Eu^{3+} (excitation wavelength, 340 nm; emission wavelength, 613 nm; delay time, 0.40 ms; window time, 0.40 ms; cycling time, 1.0 ms). Fluorescence intensities were corrected for Eu^{3+}-labeled BSA-CYN binding without the presence of CYN by dividing the signal of the sample or standard solution (F) by that of the zero-concentration calibrator (F_0). The logit-log method was used to generate a linear calibration curve.

3.7. Validation

3.7.1. Analytical Sensitivity

Analytical sensitivity was determined by subtracting two standard deviations (SD) to the mean fluorescence value of 20 zero standard replicates and calculating the corresponding concentration.

3.7.2. Precision and Accuracy of the Assay

To assess repeatability (intra-assay) and reproducibility (inter-assay), three CYN standards (2, 5 and 20 ng mL^{-1}) were analyzed with the same batch of reagents on separate days.

3.7.3. Spiked Sample Analysis

Negative water samples were filtered through Whatman GF/C filters, after which the filtrates were spiked with a 1 μg mL^{-1} CYN solution to achieve levels of 0.25, 5 and 25 ng mL^{-1}. Each spiked sample was then analyzed in triplicate and the CYN recoveries were calculated by comparing measured and spiked values.

3.7.4. Dilution Linearity Test

Serial dilutions of the CYN positive samples (10.0 and 50.0 ng mL^{-1}) were made in assay buffer and the potency of each dilution was determined in triplicate using the developed TRFIA. Assay recovery was assessed by comparing observed and expected values.

3.8. Comparison of Sample Analysis with ELISA Kit

Water samples collected from the reservoirs in Guangdong province, South China, were first filtered through Whatman GF/C filters, with the filtrates subjected to CYN determination. The cultured *Cylindrospermopsis* samples were frozen at −20 °C; the cells were then lysed by freeze-thaw prior to measurement. Insoluble cell debris was removed by centrifugation for 10 min, and the supernatant was used for CYN detection. The samples were simultaneously analyzed using the TRFIA protocol as described above and a Beacon ELISA kit following manufacturer's instructions.

3.9. Data Analysis

Statistical analysis of the data was performed using the Statistical Product and Service Solutions (SPSS) software (version 20.0, SPSS Inc., Chicago, IL, USA, 2011). A two-tailed test was applied for statistical analysis in all tests. A p value < 0.05 was considered statistically significant.

4. Conclusions

We successfully generated nine monoclonal antibodies against CYN, among which N8 had higher affinity to CYN than the other antibodies in relation to binding capacity. A direct competitive TRFIA based on this antibody was developed and validated. This is the first report employing this novel immunoassay to CYN detection. Along with the sensitivity and reliability, this makes the developed TRFIA an efficient tool for the monitoring of CYN in water and algal samples.

Author Contributions: L.L. and B.-p.H. conceived of the study and wrote the paper, L.P. and Y.Y. assisted in analyzing the data.

Acknowledgments: This work was funded by a National Natural Science Foundation of China (NSFC) grant (No. 31770507) and the Water Resource Science and Technology Innovation Program of Guangdong Province (Grant No. 2016-29).

Conflicts of Interest: The authors have no financial or personal conflicts of interest to declare.

References

1. O'Neil, J.M.; Davis, T.W.; Burford, M.A.; Gobler, C.J. The rise of harmful cyanobacteria blooms: The potential roles of eutrophication and climate change. *Harmful Algae* **2012**, *14*, 313–334. [CrossRef]
2. Dittmann, E.; Fewer, D.P.; Neilan, B.A. Cyanobacterial toxins: Biosynthetic routes and evolutionary roots. *FEMS Microbiol. Rev.* **2013**, *37*, 23–43. [CrossRef] [PubMed]
3. Moreira, C.; Azevedo, J.; Antunes, A.; Vasconcelos, V. Cylindrospermopsin: Occurrence, methods of detection and toxicology. *J. Appl. Microbiol.* **2013**, *114*, 605–620. [CrossRef] [PubMed]
4. Rzymski, P.; Poniedziałek, B. In search of environmental role of cylindrospermopsin: A review on global distribution and ecology of its producers. *Water Res.* **2014**, *66*, 320–337. [CrossRef] [PubMed]
5. Ohtani, I.; Moore, R.E.; Runnegar, M.T.C. Cylindrospermopsin, a potent hepatotoxin from the blue-green alga *Cylindrospermopsis raciborskii*. *J. Am. Chem. Soc.* **1992**, *114*, 7942–7944. [CrossRef]
6. Banker, R.; Carmeli, S.; Werman, M.; Teltsch, B.; Porat, R.; Sukenik, A. Uracil moiety is required for toxicity of the cyanobacterial hepatotoxin cylindrospermopsin. *J. Toxicol. Environ. Health Part A* **2001**, *62*, 281–288. [CrossRef] [PubMed]
7. Li, R.; Carmichael, W.W.; Brittain, S.; Eaglesham, G.K.; Shaw, G.R.; Liu, Y.D.; Watanabe, M.M. First report of the cyanotoxins cylindrospermopsin and deoxycylindrospermopsin from *Rhaphidiopsis curvata* (Cyanobacteria). *J. Phycol.* **2001**, *37*, 1121–1126. [CrossRef]
8. Poniedziałek, B.; Rzymski, P.; Kokociński, M. Cylindrospermopsin: Water-linked potential threat to human health in Europe. *Environm. Toxicol. Pharmacol.* **2012**, *34*, 651–660. [CrossRef] [PubMed]
9. Pichardo, S.; Cameán, A.M.; Jos, A. In Vitro Toxicological Assessment of Cylindrospermopsin: A Review. *Toxins (Basel)* **2017**, *9*, E402. [CrossRef] [PubMed]
10. Poniedzialek, B.; Rzymski, P.; Karczewski, J. The role of the enzymatic antioxidant system in cylindrospermopsin-induced toxicity in human lymphocytes. *Toxicol. In Vitro* **2015**, *29*, 926–932. [CrossRef] [PubMed]
11. Poniedzialek, B.; Rzymski, P.; Wiktotowicz, K. First report of cylindrospermopsin effect on human peripheral blood lymphocytes proliferation in vitro. *Cent. Eur. J. Immunol.* **2012**, *37*, 314–317. [CrossRef]
12. Poniedzialek, B.; Rzymski, P.; Wiktorowicz, K. Toxocity of cylindrospermopsin in human lymphocytes: Proliferation, viability and cell cycle studies. *Toxicol. In Vitro* **2014**, *28*, 968–974. [CrossRef] [PubMed]
13. Kinnear, S. Cylindrospermopsin: A Decade of Progress on Bioaccumulation Research. *Mar. Drugs* **2010**, *8*, 542–564. [CrossRef] [PubMed]
14. Adamski, M.; Żmudzki, P.; Chrapusta, E.; Bober, B.; Kaminski, A.; Zabaglo, K.; Latkowska, E.; Bialczyk, J. Effect of pH and temperature on the stability of cylindrospermopsin. Characterization of decomposition products. *Algal Res.* **2016**, *15*, 129–134. [CrossRef]
15. Moreira, C.; Ramos, V.; Azevedo, J.; Vasconcelos, V. Methods to detect cyanobacteria and their toxins in the environment. *Appl. Microbiol. Biotechnol.* **2014**, *98*, 8073–8082. [CrossRef] [PubMed]
16. Eaglesham, G.K.; Norris, R.L.; Shaw, G.R.; Smith, M.J.; Chiswell, R.K.; Davis, B.C.; Neville, G.R.; Seawright, A.A.; Moore, M.R. Use of HPLC-MS/MS to monitor cylindrospermopsin, a blue–green algal toxin, for public health purposes. *Environ. Toxicol.* **1999**, *14*, 151–154. [CrossRef]
17. Welker, M.; Bickel, H.; Fastner, J. HPLC-PDA detection of cylindrospermopsin—Opportunities and limits. *Water Res.* **2002**, *36*, 4659–4663. [CrossRef]
18. Rasmussen, J.P.; Giglio, S.; Monis, P.T.; Campbell, R.J.; Saint, C.P. Development and field testing of a real-time PCR assay for cylindrospermopsin-producing cyanobacteria. *J. Appl. Microbiol.* **2008**, *104*, 1503–1515. [CrossRef] [PubMed]
19. Yilmaz, M.; Philips, E.D.; Szabo, N.J.; Badylak, S. A comparative study of Florida strains of *Cylindrospermopsis* and *Aphanizomenon* for cylindrospermopsin production. *Toxicon* **2008**, *51*, 130–139. [CrossRef] [PubMed]
20. Bláhová, L.; Oravec, M.; Maršálek, B.; Šejnohová, L.; Šimek, Z.; Bláha, L. The first occurrence of the cyanobacterial alkaloid toxin cylindrospermopsin in the Czech Republic as determined by immunochemical and LC/MS methods. *Toxicon* **2009**, *53*, 519–524. [CrossRef] [PubMed]

21. Davis, T.W.; Orr, P.T.; Boyer, G.L.; Burford, M.A. Investigating the production and release of cylindrospermopsin and deoxy-cylindrospermopsin by *Cylindrospermopsis raciborskii* over a natural growth cycle. *Harmful Algae* **2014**, *31*, 18–25. [CrossRef] [PubMed]
22. Berry, J.P.; Lind, O. First evidence of "paralytic shellfish toxins" and cylindrospermopsin in a Mexican freshwater system, Lago Catemaco., and apparent bioaccumulation of the toxins in "tegogolo" snails (*Pomacea patula catemacensis*). *Toxicon* **2010**, *55*, 930–938. [CrossRef] [PubMed]
23. Al-Tebrineh, J.; Merrick, C.; Ryan, D.; Humpage, A.; Bowling, L.; Neilan, B.A. Community composition, toxigenicity, and environmental conditions during a cyanobacterial bloom occurring along 1,100 kilometers of the Murray River. *Appl. Environ. Microbiol.* **2012**, *78*, 263–272. [CrossRef] [PubMed]
24. Mohamed, Z.A.; Al-Shehri, A.M. Assessment of cylindrospermopsin toxin in an arid Saudi lake containing dense cyanobacterial bloom. *Environ. Monit. Assess.* **2013**, *185*, 2157–2166. [CrossRef] [PubMed]
25. Somdee, T.; Kaewsan, T.; Somdee, A. Monitoring toxic cyanobacteria and cyanotoxins (microcystins and cylindrospermopsins) in four recreational reservoirs (Khon Kaen, Thailand). *Environ. Monit. Assess.* **2013**, *185*, 9521–9529. [CrossRef] [PubMed]
26. Lei, L.; Peng, L.; Huang, X.; Han, B.-P. Occurrence and dominance of *Cylindrospermopsis raciborskii* and dissolved cylindrospermopsin in urban reservoirs used for drinking water supply, South China. *Environ. Monit. Assess.* **2014**, *186*, 3079–3090. [CrossRef] [PubMed]
27. Metcalf, J.S.; Young, F.M.; Codd, G.A. Performance assessment of a cylindrospermopsin ELISA with purified compounds and cyanobacterial extracts. *Environ. Forensics* **2017**, *18*, 147–152. [CrossRef]
28. Elliott, C.T.; Redshaw, C.H.; George, S.E.; Campbell, K. First development and characterisation of polyclonal and monoclonal antibodies to the emerging fresh water toxin cylindrospermopsin. *Harmful Algae* **2013**, *24*, 10–19. [CrossRef]
29. Fraga, M.; Vilariño, N.; Louzao, M.C.; Rodríguez, L.P.; Alfonso, A.; Campbell, K.; Elliott, C.T.; Taylor, P.; Ramos, V.; Vasconcelos, V.; et al. Multi-detection method for five common microalgal toxins based on the use of microspheres coupled to a flow-cytometry system. *Anal. Chim. Acta* **2014**, *850*, 57–64. [CrossRef] [PubMed]
30. McNamee, S.E.; Elliott, C.T.; Greer, B.; Lochhead, M.; Campbell, K. Development of a planar waveguide microarray for the monitoring and early detection of five harmful algal toxins in water and cultures. *Environ. Sci. Technol.* **2014**, *48*, 13340–13349. [CrossRef] [PubMed]
31. Dickson, E.F.; Pollak, A.; Diamandis, E.P. Time-resolved detection of lanthanide luminescence for ultrasensitive bioanalytical assays. *J. Photochem. Photobiol. B* **1995**, *27*, 3–19. [CrossRef]
32. Wang, K.; Huang, B.; Zhang, J.; Zhou, B.; Gao, L.; Zhu, L.; Jin, J. A novel and sensitive method for the detection of deoxynivalenol in food by time-resolved fluoroimmunoassay. *Toxicol. Mech. Methods* **2009**, *19*, 559–564. [CrossRef] [PubMed]
33. Qin, X.; Wang, Y.; Song, B.; Wang, X.; Ma, H.; Yuan, J. Homogeneous time-resolved fluoroimmunoassay of microcystin-LR using layered WS2 nanosheets as a transducer. *Methods Appl. Fluoresc.* **2017**, *5*, 024007. [CrossRef] [PubMed]
34. Hawkins, P.R.; Chandrasena, N.R.; Jones, G.J.; Humpage, A.R.; Falconer, I.R. Isolation and toxicity of *Cylindrospermopsis raciborskii* from an ornamental lake. *Toxicon* **1997**, *35*, 341–346. [CrossRef]
35. Humpage, A.R.; Falconer, I.R. Oral toxicity of the cyanobacterial toxin cylindrospermopsin in male Swiss albino mice, determination of no observed adverse effect level for deriving a drinking water guideline value. *Environ. Toxicol.* **2003**, *18*, 94–103. [CrossRef] [PubMed]
36. De la Cruz, A.A.; Hiskia, A.; Kaloudis, T.; Chernoff, N.; Hill, D.; Antoniou, M.G.; He, X.; Loftin, K.; O'Shea, K.; Zhao, C.; et al. A review on cylindrospermopsin: The global occurrence, detection, toxicity and degradation of a potent cyanotoxin. *Environ. Sci. Process Impacts* **2013**, *15*, 1979–2003. [CrossRef] [PubMed]
37. Yang, H.; Dai, R.; Zhang, H.; Li, C.; Zhang, X.; Shen, J.; Wen, K.; Wang, Z. Production of monoclonal antibodies with broad specificity and development of an immunoassay for microcystins and nodularin in water. *Anal. Bioanal. Chem.* **2016**, *408*, 6037–6044. [CrossRef] [PubMed]
38. Campbell, K.; Stewart, L.D.; Doucette, G.J.; Fodey, T.L.; Haughey, S.A.; Vilariño, N.; Kawatsu, K.; Elliott, C.T. Assessment of specific binding proteins suitable for the detection of paralytic shellfish poisons using optical biosensor technology. *Anal. Chem.* **2007**, *79*, 5906–5914. [CrossRef] [PubMed]

39. Nagata, S.; Soutome, H.; Tsutsumi, T.; Hasegawa, A.; Sekijima, M.; Sugamata, M.; Harada, K.; Suganuma, M.; Ueno, Y. Novel monoclonal antibodies against microcystin and their protective activity for hepatotoxicity. *Nat. Toxins* **1995**, *3*, 78–86. [CrossRef] [PubMed]
40. Young, F.M.; Metcalf, J.S.; Meriluoto, J.A.; Spoof, L.; Morrison, L.F.; Codd, G.A. Production of antibodies against microcystin-RR for the assessment of purified microcystins and cyanobacterial environmental samples. *Toxicon* **2006**, *48*, 295–306. [CrossRef] [PubMed]
41. Murphy, C.; Stack, E.; Krivelo, S.; McPartlin, D.A.; Byrne, B.; Greef, C.; Lochhead, M.J.; Husar, G.; Devlin, S.; Elliott, C.T.; et al. Detection of the cyanobacterial toxin, microcystin-LR, using a novel recombinant antibody-based optical-planar waveguide platform. *Bios. Bioelectron.* **2015**, *67*, 708–714. [CrossRef] [PubMed]
42. Zeck, A.; Eikenberg, A.; Weller, M.G.; Niessner, R. Highly sensitive immunoassay based on amonoclonal antibody specific for [4-arginine] microcystins. *Anal. Chim. Acta* **2001**, *441*, 1–13. [CrossRef]
43. Sheng, J.W.; He, M.; Shi, H.C. A highly specific immunoassay for microcystin-LR detection based on a monoclonal antibody. *Anal. Chim. Acta* **2007**, *603*, 111–118. [CrossRef] [PubMed]
44. Weller, M.G. Immunoassays and biosensors for the detection of cyanobacterial toxins in water. *Sensors (Basel)* **2013**, *13*, 15085–15112. [CrossRef] [PubMed]
45. Lei, L.M.; Wu, Y.S.; Gan, N.Q.; Song, L.R. An ELISA-like time-resolved fluorescence immunoassay for microcystin detection. *Clin. Chim. Acta* **2004**, *348*, 177–180. [CrossRef] [PubMed]
46. Akter, S.; Vehniäinen, M.; Kankaanpää, H.T.; Lamminmäki, U. Rapid and highly sensitive non-competitive immunoassay for specific detection of nodularin. *Microorganisms* **2017**, *5*, E58. [CrossRef] [PubMed]
47. Kokociński, M.; Mankiewicz-Boczek, J.; Jurczak, T.; Spoof, L.; Meriluoto, J.; Rejmonczyk, E.; Hautala, H.; Vehniäinen, M.; Pawełczyk, J.; Soininen, J. *Aphanizomenon gracile* (Nostocales), a cylindrospermopsin-producing cyanobacterium in Polish lakes. *Environ. Sci. Pollut. Res. Int.* **2013**, *20*, 5243–5264. [CrossRef] [PubMed]
48. Lipman, N.S.; Jackson, L.R.; Trudel, L.J.; Weis-Garcia, F. Monoclonal versus polyclonal antibodies: Distinguishing characteristics, applications, and information resources. *ILAR J.* **2005**, *46*, 258–268. [CrossRef] [PubMed]

toxins

MDPI

Article

Resveratrol Ameliorates Microcystin-LR-Induced Testis Germ Cell Apoptosis in Rats via SIRT1 Signaling Pathway Activation

Haohao Liu †, Shenshen Zhang †, Chuanrui Liu, Jinxia Wu, Yueqin Wang, Le Yuan, Xingde Du, Rui Wang, Phelisters Wegesa Marwa, Donggang Zhuang, Xuemin Cheng and Huizhen Zhang *

College of Public Health, Zhengzhou University, Zhengzhou 450001, China; Liuhlin2018@126.com (H.L.); zsslb2005@163.com (S.Z.); lcr0624@126.com (C.L.); wjxsir@126.com (J.W.); wyq2018@stu.zzu.edu.cn (Y.W.); yl19920215@126.com (L.Y.); dxd1993@163.com (X.D.); wr935314032@163.com (R.W.); wegesalee@gmail.com (P.W.M.); zdg@zzu.edu.cn (D.Z.); cxm@zzu.edu.cn (X.C.)

* Correspondence: huizhenzhang@zzu.edu.cn; Tel.: +86-151-8835-7252

† These two authors contribute equally to this work.

Received: 22 May 2018; Accepted: 5 June 2018; Published: 9 June 2018

Abstract: Microcystin-leucine arginine (MC-LR), a cyclic heptapeptide produced by cyanobacteria, is a strong reproductive toxin. Studies performed in rat Sertoli cells and Chinese hamster ovary cells have demonstrated typical apoptosis after MC-LR exposure. However, little is known on how to protect against the reproductive toxicity induced by MC-LR. The present study aimed to explore the possible molecular mechanism underlying the anti-apoptosis and protective effects of resveratrol (RES) on the co-culture of Sertoli–germ cells and rat testes. The results demonstrated that MC-LR treatment inhibited the proliferation of Sertoli–germ cells and induced apoptosis. Furthermore, sirtuin 1 (SIRT1) and Bcl-2 were inhibited, while p53 and Ku70 acetylation, Bax expression, and cleaved caspase-3 were upregulated by MC-LR. However, RES pretreatment ameliorated MC-LR-induced apoptosis and SIRT1 inhibition, and downregulated the MC-LR-induced increase in p53 and Ku70 acetylation, Bax expression, and caspase-3 activation. In addition, RES reversed the MC-LR-mediated reduction in Ku70 binding to Bax. The present study indicated that the administration of RES could ameliorate MC-LR-induced Sertoli–germ cell apoptosis and protect against reproductive toxicity in rats by stimulating the SIRT1/p53 pathway, suppressing p53 and Ku70 acetylation and enhancing the binding of Ku70 to Bax.

Keywords: apoptosis; microcystin-LR (MC-LR); reproductive toxicity; resveratrol; sirtuin 1 (SIRT1)

Key Contribution: Resveratrol is particularly effective in ameliorating MC-LR-induced Sertoli–germ cell apoptosis signaling pathways associated with: (1) increasing the expression of SIRT1; (2) attenuating the MC-LR-induced acetylation of p53 and Ku 70; and (3) reversing the MC-LR-mediated reduction in Ku70 and Bax binding.

1. Introduction

Cyanobacterial blooms caused by water eutrophication represent a health hazard to humans and animals, evoking global concerns [1,2]. Microcystins (MCs) are a family of over 100 different structural analogue compounds with seven stable cyclic heptapeptide structures, and are produced by cyanobacteria such as *Microcystis* [3]. Microcystin-leucine arginine (MC-LR) is the most abundant and most toxic MC found in natural water, causing growing environmental and public health issues [4]. Humans are most likely exposed to MC-LR through the consumption of contaminated water and food resources, and dermal exposure/inhalation during recreational activities in contaminated surface water. Thus, a safety limit (1.0 μg/L) of MC-LR has been set by World Health Organization (WHO) in drinking water. However, the concentration is usually much higher in natural water. Chen et al. considered that further studies are needed to determine whether the present WHO provisional MC-LR guideline for drinking water is protective for humans [5].

MC-LR can accumulate in several tissues such as the liver, brain, ovary, intestine, kidney, and muscle [6–10]. The liver is the most affected organ in humans, followed by the gonads [11]. Accordingly, MC-LR has been shown to induce sperm abnormalities by downregulating miR-96 and altering deleted-in azoospermia-associated protein 2 (DAZAP2) expressions [12]. Chen et al. found that MC-LR was cytotoxic to Sertoli cells by altering the expression of miRNAs and mRNAs [13]. In a previous study conducted by the investigators, it was demonstrated that Chinese hamster ovary (CHO) cell apoptosis after MC-LR treatment may be associated with the activation of endoplasmic reticulum stress (ERs) and autophagy [14].

Sirtuin 1, which is a member of the sirtuin family of proteins encoded by the *SIRT1* gene and is also a NAD-dependent deacetylase protein [15], is associated with the regulatory control of diverse cellular process including cell survival, apoptosis, DNA repair, autophagy, and cell migration, through deacetylating histones and non-histones proteins [16,17]. SIRT1 could regulate p53 activity through deacetylation modification [18]. Acetylation plays a vital role in the activation of p53. Acetylated p53 induces the expression of many genes, causing either cell cycle arrest or apoptosis [19]. The study conducted by Vaziri et al. [18] demonstrated that SIRT1 downregulated the acetylated p53 levels, reduced *p53* transcriptional activity, and prevented p53-dependent apoptosis. P53 is a central stress sensor that responds to apoptosis, cell death, oxidative stress, and autophagy, which can stimulate the expression of Bax and suppress Bcl-2 protein expression, and thereby induce apoptosis through the mitochondria-dependent pathway [20,21]. Recent studies showed that the enhanced expression of SIRT1 could decrease p53 acetylation, thereby inhibiting mitochondria apoptosis [22,23]. Similarly, the potent SIRT1 activator resveratrol (RES) enhances cell survival and inhibits apoptosis by stimulating SIRT1 activation and the deacetylation of p53 [17,24,25].

Ku70, a key factor of the non-homologous end joining (NHEJ), is one of the crucial downstream mediators of SIRT1. It is an evolutionarily conserved protein that regulates cell death by binding to the proapoptotic factor Bax in the cytoplasm [26]. Cohen et al. have shown that increased acetylation of Ku70 could induce disruption of the Ku70–Bax interaction, which blocks Bax-mediated apoptosis [27]. The acetylation of Ku70 can trigger Bax release and activation, leading to Bax-mediated cell death [28,29]. In addition, the SIRT1 protein can directly interact with Ku70 to physically form a complex that controls the acetylation status of Ku70 protein. Furthermore, Ku70 deacetylation by SIRT1 can promote DNA repair, thereby extending its life span [30,31].

Sertoli cells are scaffolds of germ cells that can form a blood–testis barrier through tight junctions, which protect sperm formation and provide a high concentration of androgen environment for sperm maturation. Germ cells acquire nutrients through Sertoli cells, and the structural changes of Sertoli cells play a vital role in the apoptosis of germ cells. In this study, Sertoli cells were used as a feeder layer for germ cells to stimulate the reproductive environment in vivo, and investigate the unexplored SIRT1/p53 pathway-mediated apoptosis. The Sertoli cells and germ cells co-cultured in a model were insufficient in the past single Sertoli cell culture system, but have scientific and practical significance for the study of the reproductive toxicity of MC-LR.

RES is a potent activator of SIRT1, but little is known about its effects on the acetylation of Ku70 and p53, and eventually, the MC-LR-induced testis germ cell apoptosis. Therefore, the present study was designed: (1) to investigate the expression of SIRT1 and the acetylation of Ku70 and p53 in vitro and in vivo following treatment with MC-LR or RES; (2) explore its related signaling pathway and underlying mechanism; and (3) reveal the effects of RES on MC-LR-induced germ cell apoptosis.

2. Results

2.1. The Identification and Viability of Co-Cultured Sertoli–Germ Cells

Sertoli cells are shuttle-like polygonal membranous epithelioid cells that grow by adhering to the plate, and proliferate in vitro. Germ cells are round and are mainly attached to Sertoli cells for growth. Sertoli cells begin to adhere and germ cells begin to attach to differentiating Sertoli cells after 24 h of incubation (Figure 1A). Sertoli cells flatten out to form a monolayer and germ cell clusters remain attached to the monolayer after 48 h of incubation (Figure 1B). The hematoxylin and eosin (H&E) staining results revealed that there were many vacuoles in the cytoplasm of Sertoli cells, which were degenerated germ cells (red arrow). Furthermore, the Sertoli cell nucleus was large and elliptic (green arrow), while germ cells were round with a deeply stained nuclei (yellow arrow) (Figure 1C).

Cell Counting Kit-8 (CCK8) kits were used to assess the viability of Sertoli–germ cells exposed to MC-LR or RES for 24 h. As shown in Figure 1D, cell viability gradually decreased with the increase in concentration of MC-LR (1–60 μg/mL). The IC_{50} dose of MC-LR for Sertoli–germ cells was 36 μg/mL by calculating cell inhibition rate. Hence, $IC_{50}/4$, $IC_{50}/2$, and IC_{50} were used for the subsequent experiments. Cell viability slightly increased after treatment with RES (1–20 μM), but declined significantly when the concentration ranged from 30 μM to 60 μM (Figure 1E). Hence, the selected dose of RES in subsequent experiments was 20 μM.

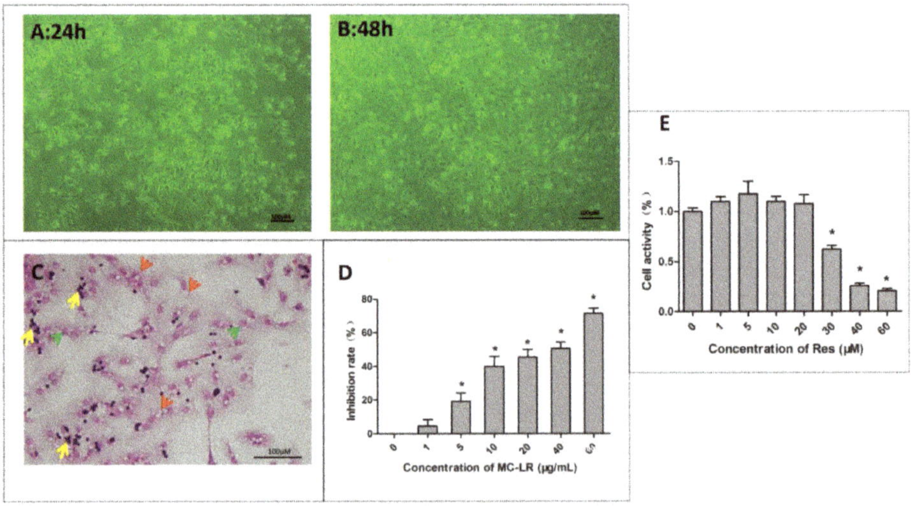

Figure 1. The identification and viability of co-cultured Sertoli–germ cells. Hematoxylin and eosin (H&E) staining was used to identify the co-cultured Sertoli–germ cells. (**A**,**B**) The appearance of co-cultured Sertoli–germ cells after culture for 24 or 48 h by light microscopy (×100); (**C**) H&E staining used to identify the co-cultured Sertoli–germ cells (×200), showing degenerated germ cells (red arrow), Sertoli cell nucleus (green arrow), and germ cells (yellow arrow). (**D**,**E**) The viability of co-cultured Sertoli–germ cells after 24 h of treatment with microcystin-leucine arginine (MC-LR; 0–60 μg/mL) or resveratrol (RES; 0–60 μM) was tested by Cell Counting Kit-8 (CCK8) kits. The calculated IC_{50} dose of MC-LR was 36 μg/mL. The dose used for RES was 20 μM. The results are expressed as mean ± SEM ($n = 3$); * $p < 0.05$ vs. the control group. Scale bar: 100 μm.

2.2. Protective Effect of RES on MC-LR-Induced Testicular Cell Apoptosis

After 24 h of exposure to MC-LR in vitro, the co-cultured Sertoli–germ cell apoptosis rates which included early apoptosis and late apoptosis were examined by Annexin V–fluoresceine isothiocyanate/propidium iodide (FITC/PI) apoptosis detection kits. As shown in Figure 2A–G, the apoptosis significantly increased at 9 μg/mL or higher concentrations of MC-LR. However, the apoptosis rate was remarkably decreased when pretreatment with RES (20 μM) for two hours. Terminal deoxynucleotidyl transferase dUTP nick end labeling (TUNEL) tests and confocal microscopy were performed in vivo to detect the induction of apoptosis in the testicular tissues of rats exposed to MC-LR. As shown in Figure 2H,I, the expression of testicular apoptotic (TUNEL-positive) cells increased more than in the control group. However, pretreatment with RES for two hours followed by MC-LR injection dramatically alleviated the apoptotic cells in the testis when compared to the MC-LR group.

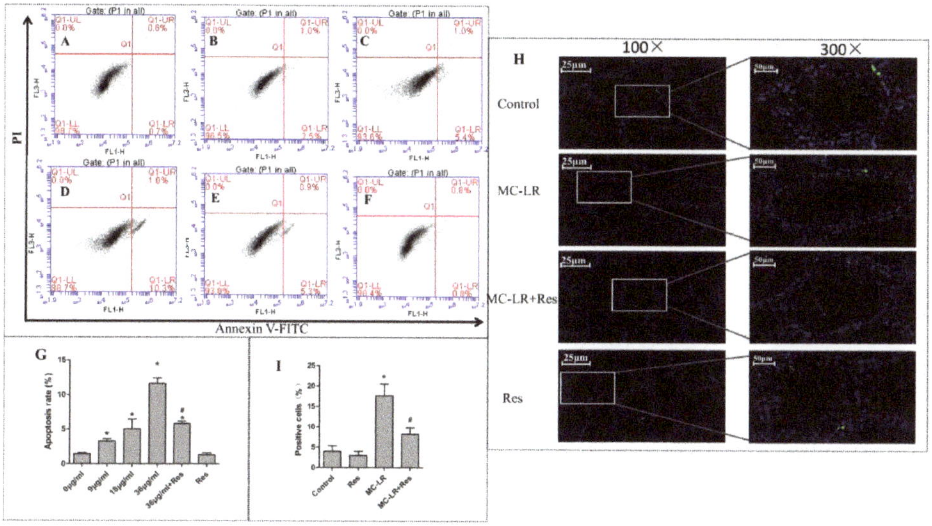

Figure 2. The protective effect of RES on MC-LR-induced testicular cell apoptosis. (**A–G**) The apoptotic cells were quantified by flow cytometric analysis. Sertoli–germ cells were stained with fluoresceine isothiocyanate-coupled annexin V and propidium iodide: (**A**) Control; (**B**) 9 μg/mL; (**C**) 18 μg/mL; (**D**) 36 μg/mL; (**E**) RES + 36 μg/mL; and (**F**) RES. Q1-LL represents normal cells, Q1-LR represents early apoptotic cells, Q1-UR represents late stage apoptotic cells, and Q1-UL represents necrotic cells. (**G**) Results are expressed as mean ± SEM (n = 3). * $p < 0.05$ vs. the control group, # $p < 0.05$ vs. 36 μg/mL of MC-LR group. (**H**) The effects of apoptosis on the testicular tissues of rats were tested by terminal deoxynucleotidyl transferase dUTP nick end labeling (TUNEL) assay and confocal microscopy. TUNEL assay was used to examine the apoptosis rate of testicular tissues in rats exposed to MC-LR with or without RES: green, TUNEL-positive cell; blue, nuclear. Left panel: at least 10 tubules in each random field and three fields randomly selected were evaluated for each testis (100×), Scale bar: 25 μm. Right panel: magnified images (300×) of the boxes in the left panel, scale bar: 50 μm. Three testes from different mice were tested in each group. (**I**) The results are expressed as mean ± SEM (n = 9); * $p < 0.05$ vs. the control group, # $p < 0.05$ vs. 40 μg/kg of the MC-LR group. PI: propidium iodide.

2.3. *Effect of RES on MC-LR-Induced Mitochondrial Membrane Collapse*

The early disruption of mitochondrial membrane potential (ΔΨm) is the most critical event during apoptosis, representing a prerequisite in drug-induced cell apoptosis. Therefore, in order to evaluate the effect of RES and MC-LR on the intrinsic pathway, the co-cultures of Sertoli–germ cells were labeled with JC-1 dye. MC-LR (9, 18, and 36 μg/mL) was found to induce mitochondrial membrane depolarization, characterized by the decrease in ΔΨ, as compared to the control group (Figure 3). Compared to the 36 μg/mL used in the MC-LR group, RES pretreated cells rescued the

MC-LR-induced ΔΨ collapse. Thus, the observation confirmed that RES (20 μM) effectively mitigated the MC-LR-induced disruption of mitochondrial membrane potential (Figure 3).

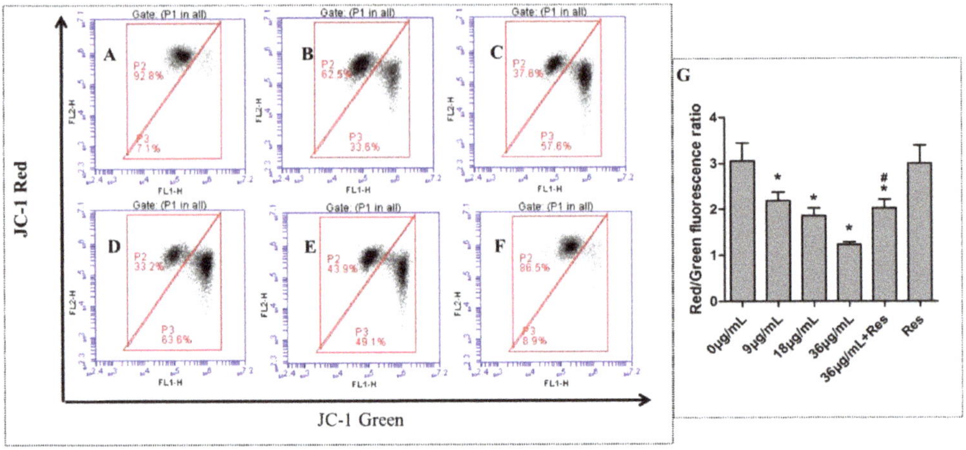

Figure 3. Effects of RES and MC-LR on the mitochondrial membrane potential in Sertoli–germ cells. The mitochondrial membrane potential of cells was detected by JC-1 staining and flow cytometry: (**A**) 0 μg/mL; (**B**) 9 μg/mL; (**C**) 18 μg/mL; (**D**) 36 μg/mL; (**E**) 36 μg/mL + RES; and (**F**) RES. (**G**) Quantitative analysis of green/red fluorescence in Sertoli–germ cells. Data are expressed as mean ± SEM (n = 3); * $p < 0.05$ vs. the control group, # $p < 0.05$ vs. the MC-LR group (36 μg/mL). Red/Green fluorescence ratio was calculated via mean red fluorescence value (mean FL2-H value) and mean green fluorescence value (mean FL1-H value) which were given by flow cytometer. P2: red fluorescent cells; P3: green fluorescent cells.

2.4. mRNA Levels of SIRT1/p53 Pathway Markers

In order to examine whether MC-LR induced apoptosis through the *SIRT1/p53* pathway, and evaluate the effect of RES on the target genes, the effects of MC-LR on the mRNA levels of the *SIRT1/p53* pathway markers were tested. The co-culture of Sertoli–germ cells exposed to MC-LR (9, 18 and 36 μg/mL) with or without RES for 24 h and the mRNA expression of *SIRT1*, *p53*, *Bcl-2*, *caspase-3* and *Bax* were tested in vitro by qPCR analysis. MC-LR exposure significantly increased mRNA expression for *caspase-3* and *Bax* at 18 and 36 μg/mL in Figure 4A. A slight upregulation in *p53* mRNA expression was found in the MC-LR group, but there was no significant difference, when compared to the control group. Furthermore, *Bcl-2* and *SIRT1* mRNA expression was downregulated after exposure to 36 μg/mL of MC-LR. In the presence of 20 μM of RES, the MC-LR-induced upregulation of *caspase-3* and *Bax* was blunted. The suppression of *Bcl-2* and *SIRT1* mRNA expression by MC-LR was rescued by RES pretreatment (Figure 4A). MC-LR exposure in vivo induced the mRNA expression of *p53*, *Bax* and *caspase-3* to increase in testicular tissues (Figure 4B). Conversely, *SIRT1* and *Bcl-2* mRNA expression was decreased by MC-LR treatment. Compared to the MC-LR group, RES pretreatment suppressed the MC-LR-induced *p53*, *Bax*, and *caspase-3* mRNA expression, and increased *SIRT1* and *Bcl-2* mRNA levels, which was similar to the untreated group (Figure 4B). Taken together, these results indicate that RES could protect against MC-LR-induced germ cell apoptosis through the *SIRT1/p53* pathway.

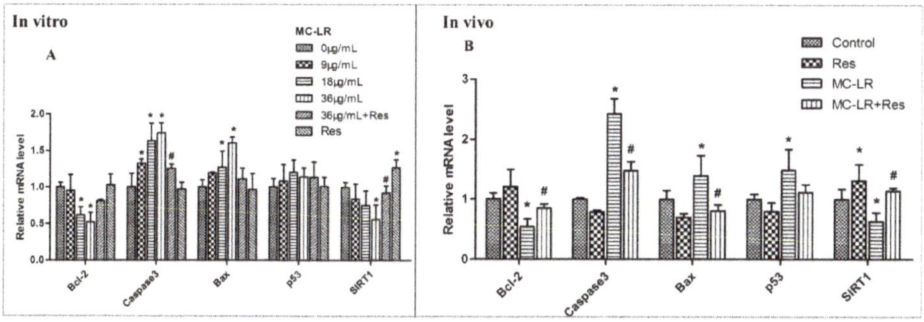

Figure 4. Apoptosis-related gene expression levels in co-cultured Sertoli–germ cells and testicular tissues exposed to MC-LR with or without RES. (**A**) Relative mRNA level in vitro; (**B**) Relative mRNA level in vivo. Data are presented as mean ± SEM for each group; * $p < 0.05$ vs. the control group, # $p < 0.05$ vs. the MC-LR group (in vitro: 36 μg/mL; in vivo: 40 μg/kg).

2.5. RES Increased SIRT1 Protein Expression and Mediated p53-Related Apoptotic Protein Expression

SIRT1 plays a number of crucial physiological roles in various cellular functions, such as gene silencing, cell cycle, apoptosis, and energy homeostasis [32]. Most notably, SIRT1 could deactivate p53 and attenuate its ability as a transcription factor, thereby ameliorating apoptosis [33]. In order to further explore the mechanisms of MC-LR-induced apoptosis, the protein expression of SIRT1, Acetyl-p53, Bax, Bcl-2, and cleaved caspase-3 were examined in the co-culture of Sertoli–germ cells and testicular tissues by western blot. The expression levels of p53, Acetyl-p53, cleaved caspase-3, and Bax in vitro exhibited an increase within the dose range of 9-36 μg/mL, and these levels peaked at the highest concentration of MC-LR (36 μg/mL) (Figure 5A). Conversely, with increasing doses of MC-LR, the protein levels of Bcl-2 and SIRT1 were significantly inhibited. Hence, 36 μg/mL of MC-LR had the most significant effect on Sertoli–germ cells, and was thereby used for subsequent experiments. The pretreatment with RES (20 μM) markedly rescued MC-LR-suppressed SIRT1 and Bcl-2 expression. In contrast, MC-LR (36 μg/mL) exposure dramatically increased p53, Acetyl-p53, cleaved caspase-3, and Bax protein expression, while RES pretreatment inhibited the induction effect of MC-LR (Figure 5C). The related protein expression levels in vivo were also examined to verify the mechanism of MC-LR-induced apoptosis. The SIRT1 protein level decreased in the MC-LR group as compared to the normal group, but a marked increase in the RES + MC-LR group was observed when compared to the former group. The acetylated p53, p53, cleaved caspase-3, and Bax protein levels were all significantly upregulated in the MC-LR-treated group, while RES pretreatment attenuated the MC-LR-induced protein changes. Similarly, the Bcl-2 protein level was reduced by MC-LR treatment, and was rescued by RES pretreatment (Figure 5E). In addition, the testicular SIRT1 expression decreased to approximately 50% in the MC-LR group, as compared to the control group (Figure 5G). However, the protein expression increased in RES-treated rats when compared to that in untreated rats. Furthermore, RES pretreatment with MC-LR injection improved the MC-LR-induced downregulation of SIRT1 expression. In summarize, the protective effect of RES on MC-LR-induced apoptosis may involve stimulation of SIRT1 followed by deactivation of p53 and regulation of apoptosis-related protein expression.

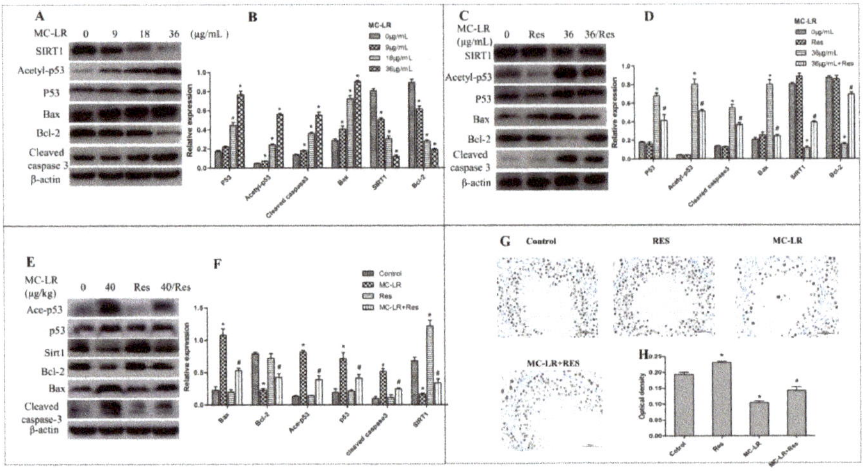

Figure 5. Apoptosis-related protein expression levels in co-cultured Sertoli–germ cells and testicular tissues exposed to MC-LR in the presence or absence of RES. (**A–D**) Western blot of sirtuin 1 (SIRT1), p53, acetylated p53, Bax, Bcl-2, and cleaved caspase-3 in co-cultured Sertoli–germ cells exposed to MC-LR in the presence or absence of RES; and (**E**,**F**) Western blot of apoptosis-related protein expression levels in testicular tissues of rats exposed to MC-LR with or without RES. The expression levels were quantified with Quantity One. β-actin was used as a loading control. Data are presented as mean ± SEM for each group; * $p < 0.05$ vs. the control group, # $p < 0.05$ vs. the MC-LR group (in vitro: 36 μg/mL; in vivo: 40 μg/kg). (**G**) Immunohistochemistry was used to examine the effect of MC-LR and RES on SIRT1 protein expression in rat testes (×400). Scale bar: 50 μm. (**H**) The quantitative analysis of SIRT1 expressed in each of the groups, * $p < 0.05$ vs. the control group, # $p < 0.05$ vs. the MC-LR group (40 μg/kg).

2.6. The Level of Acetylated Ku70 and Ku70–Bax Binding in the Co-Culture of Sertoli–Germ Cells and Testicular Tissues

Ku70 is known to bind to Bax and isolate it from the mitochondria to regulate apoptosis [27]. Acetylation of Ku70 causes the isolation of Bax from Ku70 and accelerates cell death through Bax-mediated apoptosis. With the above results showing that the deacetylase SIRT1 was regulated after MC-LR or RES treatment, it was examined whether the change in SIRT1 could mediate Ku70 deacetylation in Sertoli–germ cells. As shown in Figure 6A,B, MC-LR significantly enhanced Ku70 acetylation in Sertoli–germ cells. Interestingly, pretreatment with RES lowered the content of acetylated Ku70, when compared to the MC-LR group. Moreover, co-immunoprecipitation was used to confirm the interaction between Bax and Ku70 (Figure 6E). These results show that even though the Ku70–Bax binding complex was disrupted by MC-LR, RES could ameliorate the disruption effects of MC-LR on the Ku70–Bax binding. This suggests that MC-LR disrupted the interaction of Ku70 with Bax, induced Ku70 acetylation and stimulated Bax dissociation from Ku70. Then, the dissociated Bax enters the mitochondria to trigger apoptosis and Sertoli–germ cell death.

Rats were exposed to MC-LR for 14 days in vivo, after which the acetylated Ku70 was found to be upregulated. The effect of RES combined with MC-LR treatment was reductive, decreasing total acetylation (Figure 6C). In order to determine whether Ku70 interacted with Bax induced by MC-LR in the testis, co-immunoprecipitation was used to assess the interaction of Ku70–Bax. As shown in Figure 6G,H, the treatment with MC-LR alone significantly decreased the content of Bax, which was connected to Ku70. In reverse-IP experiments, the amount of Ku70 was also lower than the control group. These results indicate that the Ku70–Bax interaction is disrupted by MC-LR treatment. RES-pretreated rats improved the disruptive effects of MC-LR on the interaction. Based on the above results, it was speculated that MC-LR disrupts the interaction of Ku70–Bax, stimulates the acetylation of Ku70 and releases Bax, allowing it to translocate to the mitochondria and thereby trigger apoptosis.

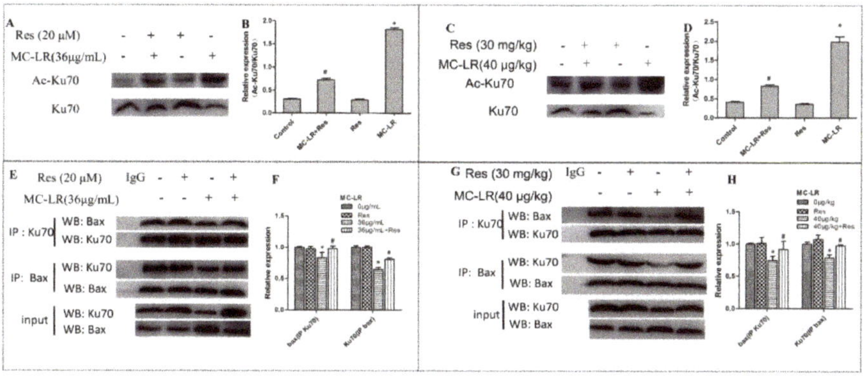

Figure 6. The western blot of acetylated Ku70 protein levels and the immunoprecipitation analysis of the Ku70–Bax binding state in co-cultures of Sertoli–germ cells and testicular tissues. (**A**,**B**) The western blot of acetylated Ku70 in co-cultures of Sertoli–germ cells after treatment with MC-LR with or without RES; (**C**,**D**) The western blot of acetylated Ku70 in the testicular tissues of rats exposed to MC-LR with or without RES; (**E**,**F**) The change in Ku70–Bax binding state in co-cultured Sertoli–germ cells exposed to MC-LR with or without RES by immunoprecipitation analysis; (**G**,**H**) The change in Ku70–Bax binding state in testicular tissues of rats exposed to MC-LR with or without RES by immunoprecipitation analysis. The band intensities of the immunoblots were quantified using Quantity One software, and are statistically presented in graphs. Data are presented as mean ± SEM for each group; * $p < 0.05$ vs. the control group, # $p < 0.05$ vs. the MC-LR group (in vitro: 36 μg/mL; in vivo: 40 μg/kg).

2.7. *Effect of RES on MC-LR-Induced Pathological Change*

In order to assess the effects of MC-LR and RES on the testis, the testicular histomorphology of SD rats was performed after MC-LR treatment for 14 days. The histopathological changes in the testis and seminiferous tubules were observed under a light microscope in different groups. MC-LR made the testis structure loose, and caused seminiferous tubule degeneration, structural shrinkage, and vacuolation in the mesenchyme. In RES-pretreated SD rats, Sertoli–germ cells were arranged in regular seminiferous tubules and compared to those of MC-LR-treated rats (Figure 7).

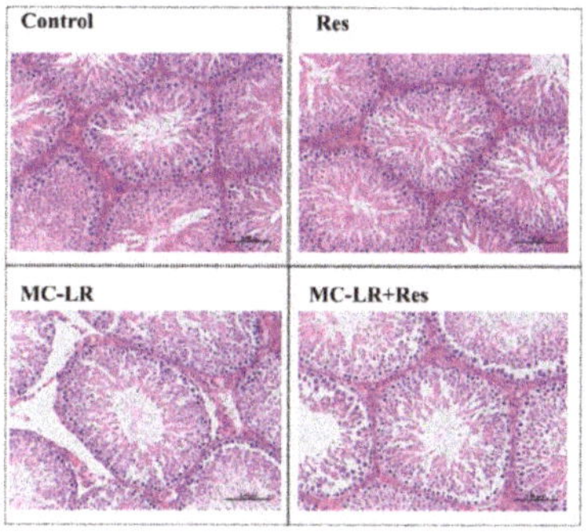

Figure 7. Effect of RES on MC-LR-induced pathological changes in the testis. The morphologic changes in testes exposed to MC-LR with or without RES (×200). The morphological changes of rat testes were examined by H&E staining. Scale bar: 100 μm.

3. Discussion

There is increasing evidence that infertility in animals and humans is potentially linked to environment exposure to reproductive toxins. Although we used a higher dose than that suggested by the WHO (1 μg/L), the concentration is usually much higher in natural water (10 μg/L). The highest microcystins contents in fish intestines were 85.67 μg/g dry weight (DW) [34]. The microcystins contents in cyanobacteria of water bloom can reach 7300 μg/g DW [35]. In addition, humans and animals can be exposed to microcystins in a variety of ways, such as through the digestive tract, respiratory tract, skin, and food chain [36]. Therefore, it is difficult for humans and animals to avoid the risk caused by higher doses of microcystins. The previous study conducted by the investigators demonstrated that MC-LR induced the apoptotic death of Sertoli cells and CHO cells through the activation of the mitochondrial caspase cascade [37], as well as ERs and autophagy [14], respectively. In the present study, a combination of in vitro and in vivo studies were used to investigate the mechanistic basis of the toxic effects of MC-LR on Sertoli–germ cells, and determine whether RES could rescue cells from apoptosis.

The interactions between germ cells and Sertoli cells are crucial for the successful production of male gametes [38]. Interference in the normal interaction between Sertoli and germ cells may cause testicular dysfunction, which could result in the improper release of mature sperm [39]. The inhibited cell proliferation and upregulated cell apoptosis in Sertoli–germ cells isolated from rat testes by MC-LR indicated a detrimental effect on the male reproductive system. The polyphenol RES possesses comprehensive biochemical and physiological properties, ranging from antiplatelet, anti-inflammatory, and neuroprotective activity to anti-apoptosis [40]. In this study, RES pretreatment attenuated MC-LR-induced Sertoli–germ cell apoptosis and stimulated cell proliferation. The loss of mitochondrial membrane potential plays a vital role in the mitochondria-mediated apoptotic pathway. This study showed that MC-LR induced the depolarization of mitochondrial membrane potential, and caused $\Delta\Psi$ collapse. However, RES pretreatment prevented the MC-LR-induced disruption of $\Delta\Psi$. Such data suggests that RES protected against MC-LR-induced $\Delta\Psi$ collapse.

MC-LR induces reproductive toxicity by being transported into testicular tissues and stimulating cell apoptosis by targeting the spermatogonia and Sertoli cells [41]. SIRT1 is a nicotinamide adenine dinucleotide (NAD+)-dependent class III histone deacetylase, and has important physiological roles in regulating cell survival, and protecting against apoptosis [42]. In this study, SIRT1 expression was inhibited both in MC-LR-treated Sertoli–germ cells and co-cultured cells obtained from rats treated with MC-LR for 14 days. Moreover, MC-LR treatment promoted cellular apoptosis. RES is a SIRT1 activator [43] that can mitigate type-1 diabetes mellitus-induced sperm abnormalities and DNA damage by activating SIRT1 [44]. RES inhibited H_2O_2-induced apoptosis by activating SIRT1 and inhibiting p53 acetylation and caspase-3 activation [45]. In the present study, there was a remarkable recovery in SIRT1 level in RES-pretreated Sertoli–germ cells, as well as in the in vivo rat model. In addition, the protein level of Bcl-2 was significantly suppressed by the sole treatment with MC-LR for 24 h, while the p53 level, particularly acetylated p53, Bax and cleaved caspase-3, was promoted by the treatment with MC-LR. This indicates the regulatory effect of SIRT1 and mitochondria-dependence of MC-LR-induced apoptosis in vitro and in vivo. Interestingly, these present results also revealed that the pretreatment with SIRT1 activator RES inhibited the acetylation of p53 and alleviated the suppression of Bcl-2 in both Sertoli–germ cells and rat testes. Furthermore, MC-LR-induced caspase cleavage and Bax expression were inhibited, while Bcl-2 protein expression was upregulated by RES treatment. These results indicate that RES protected Sertoli–germ cells from MC-LR-induced apoptosis in vitro and in vivo through the upregulation of SIRT1 and deregulation of p53 acetylation.

Ku70 is an essential factor in the non-homologous end joining (NHEJ) pathway of DNA repair factor. Numerous previous studies have shown that Ku70 could bind to the apoptotic protein Bax to regulate cell death in the cytoplasm [26]. More importantly, SIRT1 can enhance DNA repair capacity by physically forming a complex with Ku70 and deacetylasing Ku70 in Q293A cells [30]. The study conducted by Anekonda et al. [46] revealed that RES can protect retinal cells from apoptosis through the downregulation of Bax, the upregulation of SIRT1 and Ku70 activity, and the inhibition of caspase-3 activity. The investigators suspected that SIRT1 activator RES might prevent against MC-LR-induced reproductive toxicity by deacetylation of Ku70. The present results revealed that the acetylated Ku70 and SIRT1 protein levels in the RES+MC-LR group were lower than those in the MC-LR group, both in vitro and in vivo. These results indicate that RES suppressed the acetylation of Ku70. In addition, Bax is member of the Bcl-2 protein family that promotes apoptosis. It can translocate to the mitochondria and promote mitochondrial membrane potential loss, subsequently leading to apoptosis. Ku70 can bind to the apoptotic protein Bax in the cytoplasm and block Bax-mediated cell death [28], while the acetylated form of Ku70 releases Bax, allowing it to translocate to the mitochondria, and trigger caspase-dependent apoptosis [29]. In another study, the treatment of A549 cells with epigallocatechin gallate-upregulated Ku70 acetylation blocked the combination of Ku70 and Bax, and subsequently triggered lung cancer cell apoptosis [47]. The present study revealed that Ku70 and the acetylated Ku70 may play a crucial role in the Bax-mediated mitochondrial apoptosis pathway. Furthermore, the present study found that after treating Sertoli–germ cells and rats with MC-LR, the interaction of Bax–Ku70 markedly decreased. Surprisingly, the combined treatment of RES and MC-LR reduced the acetylation of Ku70, resulting in the effective increase in the interaction of Bax–Ku70, ultimately protecting against MC-LR-induced Sertoli–germ cell apoptosis due to RES. The above results showed that RES protected MC-LR-induced Sertoli–germ cells and rat testes from apoptosis by promoting the interaction of Bax–Ku70.

In conclusion, the present study revealed that MC-LR exposure downregulated SIRT1 levels in primary co-cultured Sertoli–germ cells and rat testes, which was accompanied by the upregulation of p53 and Ku70 acetylation, Bax expression, and cleaved caspase-3, and a decrease in Bcl-2 expression. RES pretreatment promoted the activation of SIRT1 and the interaction of Ku70 and Bax, and downregulated the acetylation of p53 and Ku70, Bax expression, and caspase-3 activation (Figure 8). This data suggests that the administration of RES ameliorates MC-LR-induced testis germ cell apoptosis, and protects against reproductive toxicity in rats by stimulating the SIRT1/p53 pathway.

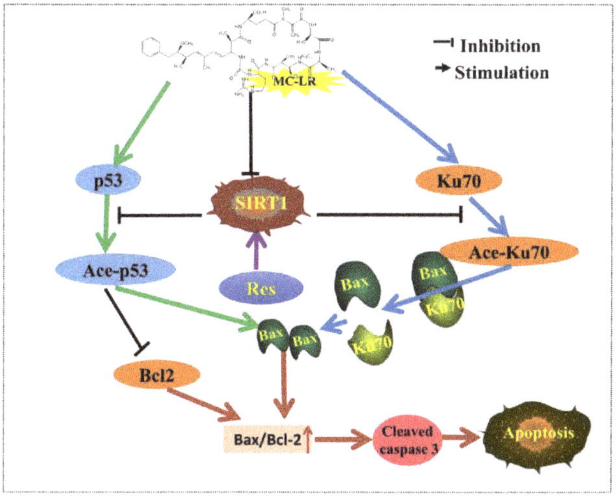

Figure 8. Proposed model for MC-LR-induced testis germ cell apoptosis in rats via SIRT1 signaling pathway activation. MC-LR exposure downregulated SIRT1 levels in primary co-cultured Sertoli–germ cells and rat testes, subsequently increasing p53 and Ku70 acetylation, and Bax/Bcl-2 and cleaved caspase-3 expression. RES pretreatment could promote the activation of SIRT1, which could resist MC-LR-induced apoptosis via reducing the acetylation of p53 and Ku70, as well as Bax/Bcl-2 and caspase-3 activation.

4. Materials and Methods

4.1. Chemicals

MC-LR (purity > 96%) was purchased from Beijing Express Technology Co. (Beijing, China). An institutional safety procedure was used to carry out the experiment according to the textbook of the "*Experimental methods and techniques of toxicology*". Dulbecco's modified eagle medium/nutrient mixture F-12 (DMEM/F-12), fetal bovine serum (FBS), penicillin–streptomycin, 0.25% trypsin, and collagenase type I were purchased from GIBCO (Rockville, MD, USA). RES was purchased from Abcam (Cambridge, UK). The annexin V–FITC/PI apoptosis detection kit and mitochondrial membrane potential assay kit were purchased from Beyotime Institute of Biotechnology (Shanghai, China). Cell Counting Kit-8 (CCK8) was purchased from Dojindo Laboratories (Kyushu Island, Japan).

4.2. Isolation and Identification of Sertoli–Germ Cells

The isolation and identification of co-cultured Sertoli–germ cells were performed as previously other investigators described [48], with some improvements. Specific pathogen-free (SPF) male Sprague-Dawley (SD) rats (20–22 days old) were obtained from the Experimental Animal Center. Sertoli–germ cells were separated from these SD rats. Briefly, the testes were removed membrane, cut into pieces, washed two times with a pre-chilled PBS, and digested with 0.25% trypsin in an incubator for 30 min at 37 °C. Next, 0.1% collagenase was used to digest testicular fragments for another 30 min. Then, the homogenate was filtered by a 200-mesh stainless steel filter, and cells were washed with PBS for two times after collection. After centrifugation for five minutes at 1000 rpm, the cell pellets, which included the Sertoli cells and germ cells, were resuspended with DMEM/F-12 medium contained 10% FBS, and cultured in an incubator for 24 h at 37 °C (95% air and 5% CO_2). Hematoxylin and eosin (H&E) staining was used to identify the co-cultured Sertoli–germ cells. The resuspended cells climbed to the carry sheet glass (20 mm), and were cultured for 24 h. The slides of these cells were cleaned using cold PBS, and fixed 15 min by 4% paraformaldehyd. Then, the cell specimens were assigned for H&E staining for routine histological examinations. The slides of these cells were examined under a microscope (Nikon Eclipse E100, Tokyo, Japan).

4.3. Cell Viability Assay

Sertoli–germ cells (density: 1×10^5 cells per mL) were inoculated to a 96-well plate with 200 μL culture medium. After 24 h, cells were used with MC-LR for final concentrations of 0, 1, 5, 10, 20, 40, and 60 μg/mL for 24 h, and RES at final concentrations of 0, 1, 5, 10, 20, 30, 40, and 60 μmol/L for another 24 h. Next, CCK8 reagents were added to each well and incubated at 37 °C for four hours. Automated microplate reader (BioTek, Winooski, VT, USA) was used to measure absorbance at 450 nm. Cell inhibition rate and viability were calculated, and the IC50 dose of MC-LR for 24 h was determined. Cell viability = $[(As - Ab)/(Ac - Ab)] \times 100\%$, and inhibition rate = $[(Ac - As)/(Ac - Ab)] \times 100\%$. As: experimental hole absorbance (including medium, cells, CCK8, MC-LR, or RES), Ac: control hole absorbance (including medium, cells, CCK8, non-MC-LR, or RES), Ab: blank hole absorbance (including medium and CCK8, non-cells, non-MC-LR, or RES).

4.4. Apoptosis Assay

Flow cytometry was used to examine the apoptosis rate of Sertoli–germ cells. Briefly, cells were transplanted to a 6-well plate and exposed to different concentrations of MC-LR (0, 9, 18 and 36 μg/mL) with or without RES (20 μM). After 24 h, cells were collected, washed with cold PBS for two times and centrifuged at 1000 rpm for five minutes. Then, 5×10^5 cells were selected for resuspension in 500 μL binding buffer, 5 μL annexin V-FITC, and 5 μL PI. These cells were kept in the dark at room temperature for 15 min. Flow cytometry was used to examined cells by a FACS Calibur flow cytometer (BD Accuri C6, Franklin Lakes, NJ, USA), and analyzed with the software.

4.5. Measurement of Mitochondrial Membrane Potential

JC-1, a cationic dye that accumulates in energized mitochondria, is an indicator of mitochondrial potential in a variety of cell types. Briefly, cells exposed to MC-LR with or without RES were collected and washed. Next, cells were loaded with 500 μL of JC-1 working solution. Then, these cells were incubated at 37 °C for 25 min. The fluorescence was detected with a FACS Calibur flow cytometer (BD Accuri C6, Franklin Lakes, NJ, USA).

4.6. Western Blot

Total protein was extracted from the testes of rats and co-cultured Sertoli–germ cells exposed to different concentrations of MC-LR with or without RES. Samples containing 30 μg of protein underwent electrophoresis with a Bio-Rad electrophoresis apparatus (Bio-Rad, Hercules, CA, USA), were separated by 12% SDS-PAGE, and subsequently transferred onto a polyvinylidene fluoride (PVDF) membrane (Millipore, Bedford, MA, USA). The membranes were blocked in tris buffered saline with tween (TBST) containing 5% BSA at 23 °C for two hours, and immunoblotted using primary anti-Ku70 (sc-17789, Santa Cruz Biotechnology, Boston, CA, USA), anti-SIRT1 (ab110304), anti-p53 (ab131442), anti-p53 (acetyl K381, ab61241), anti-cleaved-caspase-3 (ab2302), anti-Bax (ab32503), anti-Bcl-2 (ab7973), and anti-β-actin (ab6276) (Abcam, Cambridge, UK). An enhanced chemiluminescence detection kit (Beijing ComWin Biotech Co., Ltd., Beijing, China) was used to analyze the protein bands. The intensity of the bands was quantified using the Bio-Rad Quantity One software.

4.7. Real-Time PCR

Total RNA was isolated from cells and the testes of rats by TRIZOL reagent (TaKaRa, Dalian, China). The purity of the RNA was tested using the quotient of the optical density (OD) at 260/280 nm. Then, the purified total RNA (1 mg) was reverse-transcribed using an EasyScript First-Strand cDNA synthesis super mix kit (TaKaRa, Dalian, China). The qPCR analysis was performed using SYBR Premix Ex Taq II. The PCR reaction was performed at 95 °C (5 min), followed by 45 cycles of denaturation for 95 °C (15 s), 60 °C (20 s), and 72 °C (20 s). The 2-ΔΔCt method was used to calculate the transcriptional levels of genes, with glyceraldehyde-3phosphate dehydrogenase (GAPDH) as the internal reference. Each sample was run in triplicate for qPCR. The sequences of primer pairs used in this assay were in the Table 1.

Table 1. The sequences of primers used in Real-Time PCR

Gene	Forward Primer	Reverse Primer
SIRT1	5′-TCATTCTGACTGTGATGACGA-3′	5′-CTGCCACAGTGTCATATCCAA-3′
P53	5′-CCCCTGAAGACTGGATAACTGTC-3′	5′-AACTCTGCAACATCCTGGGG-3′
Bax	5′-GAACCATCATGGGCTGGACA-3′	5′-GTGAGTGAGGCAGTGAGGAC-3′
Bcl-2	5′-CTGAACCGGCATCTGCACAC-3′	5′-GCAGGTCTGCTGACCTCACT-3′
Caspase-3	5′-GACTGCGGTATTGAGACAGA-3′	5′-CGAGTGAGGATGTGCATGAA-3′
GAPDH	5′-GGCACAGTCAAGGCTGAGAATG-3′	5′-ATGGTGGTGAAGACGCCAGTA-3′

4.8. Animal Treatment

SPF male SD rats were purchased from the Henan Province Experimental Animal Center (license number: SCXK (YU) 2015-0004) and were fed at the barrier environment animal laboratory of colleague of public health in Zhengzhou University (license number: SYXK (YU) 2012-0007). The animals were fed with standard rodent pellet diet, provided with water ad libitum, and kept on a 12-h light–dark cycle. Rats were handled according to the guidelines for the care and use of laboratory animals published by the Ministry of Health of the People's Republic of China. All studies were approved by the Animal Study Committee of Zhengzhou University (Date: March, 2014). Twenty-four male SD rats were randomly divided into four groups: control, RES (30 mg/kg bw) [49], MC-LR (40 μg/kg·bw), and RES+MC-LR groups. Rats were treated daily with MC-LR or vehicle by i.p. injection for 14 days. The rats in the RES+MC-LR group were pretreated with RES for two hours prior to MC-LR injection. At 24 h after the last injection, the testes of rats were excised for analysis.

4.9. Hematoxylin-Eosin Staining

The testes were quickly collected from the SD rats, cleaned using cold PBS, and fixed with 4% paraformaldehyde for 24 h. Then, 30% phosphate-buffered sucrose solution was used to equilibrate the testes for 2 h. Testes were embedded in paraffin and cut into 6-μm sections. Next, the sections were dehydrated by xylene and 100% alcohol, subsequently conducted hematoxylin staining and eosin staining. Finally, a microscope (Nikon Eclipse E100, Tokyo, Japan) was used to observe the morphological changes of testes.

4.10. TUNEL Assay

The terminal deoxynucleotidyl transferase dUTP nick end labeling (TUNEL) (Roche, Switzerland) was used to perform the detection of apoptosis. Briefly, the testes were fixed in 4% paraformaldehyde for 24 h at 23 °C, permeabilized with 0.1% Triton X-100, and washed twice. Then, the terminal deoxynucleotidyl transferase (TdT)-labelled nucleotide mix was added to each slide, incubated for one hour at 37 °C, and observed using a fluorescent microscope (Olympus, Tokyo, Japan) at 488-nm excitation and 530-nm emission. Image-Pro Plus 6.0 (Media Cybernetics, Inc., Rockville, MD, USA) was used to select the labeled green fluorescent nuclei as a unified standard for judging positive cells in all photographs. DAPI blue nuclei with the same markers were selected as the total cells. The percentage of positive cells (number of positive cells / number of total cell × 100) was determined as the apoptosis rate (%).

4.11. Histology and Immunohistochemistry

The testes obtained from rats were fixed in buffered 4% formaldehyde and embedded in paraffin. Immunohistochemistry was performed with paraffin sections (5 μM) obtained from the testes specimen. The slides were incubated with a diluted primary antibody, anti-SIRT1 (1:1000, Abcam), while negative control was incubated with antibody diluents. Horse radish peroxidase (HRP)-labeled secondary antibody (Beijing ComWin Biotech, Beijing, China) was added to the specimens, and incubated for 30 min at 23 °C. Then, the sections were counterstained with hematoxylin, subsequently dehydrated, and observed under a light microscope (XSP-C204, CIC). Image-Pro Plus 6.0 (Media Cybernetics, Inc., Rockville, MD, USA, 2006) was used to select the same brown color as a unified standard for judging the positive of all photographs. Each photograph was analyzed to quantify the integral optical density of the immunohistochemical staining of each photograph.

4.12. Co-Immunoprecipitation Analysis

The testes were rapidly homogenized in Radio-Immunoprecipitation Assay (RIPA) buffer. The protein was precleared with Protein A agarose suspension (Abcam, Cambridge, UK), and 500 μg of protein were incubated with 5 μg of the antibody (Ku70/Bax) on a benchtop shaker for two hours at room temperature. After incubation, the mixture was rotated on a benchtop shaker by adding 100 μL of Protein A agarose suspension at 4 °C overnight. Beads were collected and washed with RIPA buffer. After that, beads were boiled for five minutes in 5 × SDS sample loading buffer, subsequently separated by 12% SDS-PAGE, and transferred onto PVDF membranes for western blot.

4.13. Statistical Analysis

Data are expressed as mean ± standard error of the mean (SEM). One-way analysis of variance (ANOVA, Birmingham, UK) was used to analyze the significant differences between groups, followed by Student–Newman–Keuls test. $p < 0.05$ was considered statistically significant. All statistical analyses were carried out using SPSS 21.0 (Armonk, NY, USA, 2012).

Author Contributions: Conceptualization, H.Z.Z. and H.H.L.; Methodology, H.H.L. and S.S.Z.; Software, L.Y., Y.Q.W., C.R.L., and R.W.; Validation, C.R.L. and J.X.W.; Investigation, H.H.L. and S.S.Z., Resources, H.Z.Z.; Data Curation, X.D.D., P.W.M., X.M.C., and D.G.Z.; Writing—Original Draft Preparation, H.H.L.; Writing—Review and Editing, S.S.Z.; Visualization, H.H.L and S.S.Z.; Supervision, H.Z.Z.; Funding Acquisition, H.Z.Z. All authors have read and approved the final manuscript.

Funding: This research was funded by The National Nature Science Foundation of China (grant numbers 81472948 and 81773384), The Scientific and Technological Project of Henan Province (grant number 142102310344), and the Program of Science and Technology Development of Henan province (grant number 122102310208).

Acknowledgments: We would like to thank Mustapha U. Imam for his help in improving the manuscript.

Conflicts of Interest: The authors declare no conflict of interest.

References

1. Codd, G.A.; Morrison, L.F.; Metcalf, J.S. Cyanobacterial toxins: Risk management for health protection. *Toxicol. Appl. Pharmacol.* **2005**, *203*, 264–272. [CrossRef] [PubMed]
2. Merel, S.; Walker, D.; Chicana, R.; Snyder, S.; Baures, E.; Thomas, O. State of knowledge and concerns on cyanobacterial blooms and cyanotoxins. *Environ. Int.* **2013**, *59*, 303–327. [CrossRef] [PubMed]
3. Puddick, J.; Prinsep, M.R.; Wood, S.A.; Kaufononga, S.A.; Cary, S.C.; Hamilton, D.P. High levels of structural diversity observed in microcystins from microcystis cawbg11 and characterization of six new microcystin congeners. *Mar. Drugs* **2014**, *12*, 5372–5395. [CrossRef] [PubMed]
4. Hoeger, S.J.; Hitzfeld, B.C.; Dietrich, D.R. Occurrence and elimination of cyanobacterial toxins in drinking water treatment plants. *Toxicol. Appl. Pharmacol.* **2005**, *203*, 231–242. [CrossRef] [PubMed]
5. Chen, L.; Giesy, J.P.; Xie, P. The dose makes the poison. *Sci. Total Environ.* **2017**, *621*, 649–653. [CrossRef] [PubMed]
6. Woolbright, B.L.; Williams, C.D.; Ni, H.; Kumer, S.C.; Schmitt, T.; Kane, B.; Jaeschke, H. Microcystin-LR induced liver injury in mice and in primary human hepatocytes is caused by oncotic necrosis. *Toxicon* **2017**, *125*, 99–109. [CrossRef] [PubMed]
7. Milutinovic, A.; Zorc-Pleskovic, R.; Petrovic, D.; Zorc, M.; Suput, D. Microcystin-LR induces alterations in heart muscle. *Folia Biol.* **2006**, *52*, 116–118.
8. Chen, C.; Liu, W.; Wang, L.; Li, J.; Chen, Y.; Jin, J.; Kawan, A.; Zhang, X. Pathological damage and immunomodulatory effects of zebrafish exposed to microcystin-LR. *Toxicon* **2016**, *118*, 13–20. [CrossRef] [PubMed]
9. Qin, W.; Xu, L.; Zhang, X.; Wang, Y.; Meng, X.; Miao, A.; Yang, L. Endoplasmic reticulum stress in murine liver and kidney exposed to microcystin-LR. *Toxicon* **2010**, *56*, 1334–1341. [CrossRef] [PubMed]
10. Wu, J.; Shao, S.; Zhou, F.; Wen, S.; Chen, F.; Han, X. Reproductive toxicity on female mice induced by microcystin-LR. *Environ. Toxicol. Pharmacol.* **2014**, *37*, 1–6. [CrossRef] [PubMed]
11. Zhou, Y.; Yuan, J.; Wu, J.; Han, X. The toxic effects of microcystin-LR on rat spermatogonia in vitro. *Toxicol. Lett.* **2012**, *212*, 48–56. [CrossRef] [PubMed]
12. Zhou, Y.; Xiang, Z.; Li, D.; Han, X. Regulation of microcystin-LR-induced toxicity in mouse spermatogonia by mir-96. *Environ. Sci. Technol.* **2014**, *48*, 6383–6390. [CrossRef] [PubMed]
13. Chen, Y.; Zhou, Y.; Wang, J.; Wang, L.; Xiang, Z.; Li, D.; Han, X. Microcystin-leucine arginine causes cytotoxic effects in sertoli cells resulting in reproductive dysfunction in male mice. *Sci. Rep.* **2016**, *6*, 39238. [CrossRef] [PubMed]
14. Zhang, S.; Liu, C.; Li, Y.; Imam, M.U.; Huang, H.; Liu, H.; Xin, Y.; Zhang, H. Novel role of ER stress and autophagy in microcystin-LR induced apoptosis in Chinese hamster ovary cells. *Front. Phys.* **2016**, *7*, 527. [CrossRef] [PubMed]
15. Yi, J.; Luo, J. Sirt1 and p53, effect on cancer, senescence and beyond. *Biochim. Biophys. Acta* **2010**, *1804*, 1684–1689. [CrossRef] [PubMed]
16. Ben Salem, I.; Boussabbeh, M.; Pires Da Silva, J.; Guilbert, A.; Bacha, H.; Abid-Essefi, S.; Lemaire, C. Sirt1 protects cardiac cells against apoptosis induced by zearalenone or its metabolites alpha- and beta-zearalenol through an autophagy-dependent pathway. *Toxicol. Appl. Pharmacol.* **2017**, *314*, 82–90. [CrossRef] [PubMed]

17. Wang, X.L.; Wu, L.Y.; Zhao, L.; Sun, L.N.; Liu, H.Y.; Liu, G.; Guan, G.J. Sirt1 activator ameliorates the renal tubular injury induced by hyperglycemia in vivo and in vitro via inhibiting apoptosis. *Biomed. Pharmacother.* **2016**, *83*, 41–50. [CrossRef] [PubMed]
18. Vaziri, H.; Dessain, S.K.; Ng Eaton, E.; Imai, S.I.; Frye, R.A.; Pandita, T.K.; Guarente, L.; Weinberg, R.A. Hsir2(sirt1) functions as an nad-dependent p53 deacetylase. *Cell* **2001**, *107*, 149–159. [CrossRef]
19. Reed, S.M.; Quelle, D.E. P53 acetylation: Regulation and consequences. *Cancers* **2014**, *7*, 30–69. [CrossRef] [PubMed]
20. Kannan, K.; Amariglio, N.; Rechavi, G.; Jakob-Hirsch, J.; Kela, I.; Kaminski, N.; Getz, G.; Domany, E.; Givol, D. DNA microarrays identification of primary and secondary target genes regulated by p53. *Oncogene* **2001**, *20*, 2225–2234. [CrossRef] [PubMed]
21. Miyashita, T.; Krajewski, S.; Krajewska, M.; Wang, H.G.; Lin, H.K.; Liebermann, D.A.; Hoffman, B.; Reed, J.C. Tumor suppressor p53 is a regulator of bcl-2 and bax gene expression in vitro and in vivo. *Oncogene* **1994**, *9*, 1799–1805. [PubMed]
22. Yan, S.; Wang, M.; Zhao, J.; Zhang, H.; Zhou, C.; Jin, L.; Zhang, Y.; Qiu, X.; Ma, B.; Fan, Q. Microrna-34a affects chondrocyte apoptosis and proliferation by targeting the sirt1/p53 signaling pathway during the pathogenesis of osteoarthritis. *Int. J. Mol. Med.* **2016**, *38*, 201–209. [CrossRef] [PubMed]
23. Zhang, F.; Zhang, M.; Wang, A.; Xu, M.; Wang, C.; Xu, G.; Zhang, B.; Zou, X.; Zhuge, Y. Tweak increases sirt1 expression and promotes p53 deacetylation affecting human hepatic stellate cell senescence. *Cell Biol. Int.* **2017**, *41*, 147–154. [CrossRef] [PubMed]
24. Howitz, K.T.; Bitterman, K.J.; Cohen, H.Y.; Lamming, D.W.; Lavu, S.; Wood, J.G.; Zipkin, R.E.; Chung, P.; Kisielewski, A.; Zhang, L.L.; et al. Small molecule activators of sirtuins extend saccharomyces cerevisiae lifespan. *Nature* **2003**, *425*, 191–196. [CrossRef] [PubMed]
25. He, N.; Zhu, X.; He, W.; Zhao, S.; Zhao, W.; Zhu, C. Resveratrol inhibits the hydrogen dioxide-induced apoptosis via sirt 1 activation in osteoblast cells. *Biosci. Biotechnol. Biochem.* **2015**, *79*, 1779–1786. [CrossRef] [PubMed]
26. Lieber, M.R. The mechanism of double-strand DNA break repair by the nonhomologous DNA end-joining pathway. *Annu. Rev. Biochem.* **2010**, *79*, 181–211. [CrossRef] [PubMed]
27. Cohen, H.Y.; Lavu, S.; Bitterman, K.J.; Hekking, B.; Imahiyerobo, T.A.; Miller, C.; Frye, R.; Ploegh, H.; Kessler, B.M.; Sinclair, D.A. Acetylation of the c terminus of ku70 by cbp and pcaf controls bax-mediated apoptosis. *Mol. Cell* **2004**, *13*, 627–638. [CrossRef]
28. Hada, M.; Kwok, R.P. Regulation of ku70-bax complex in cells. *J. Cell Death* **2014**, *7*, 11–13. [CrossRef] [PubMed]
29. Subramanian, C.; Opipari, A.W., Jr.; Bian, X.; Castle, V.P.; Kwok, R.P. Ku70 acetylation mediates neuroblastoma cell death induced by histone deacetylase inhibitors. *Proc. Natl. Acad. Sci. USA* **2005**, *102*, 4842–4847. [CrossRef] [PubMed]
30. Jeong, J.; Juhn, K.; Lee, H.; Kim, S.H.; Min, B.H.; Lee, K.M.; Cho, M.H.; Park, G.H.; Lee, K.H. Sirt1 promotes DNA repair activity and deacetylation of ku70. *Exp. Mol. Med.* **2007**, *39*, 8–13. [CrossRef] [PubMed]
31. Oberdoerffer, P.; Michan, S.; McVay, M.; Mostoslavsky, R.; Vann, J.; Park, S.K.; Hartlerode, A.; Stegmuller, J.; Hafner, A.; Loerch, P.; et al. Sirt1 redistribution on chromatin promotes genomic stability but alters gene expression during aging. *Cell* **2008**, *135*, 907–918. [CrossRef] [PubMed]
32. Yamamoto, H.; Schoonjans, K.; Auwerx, J. Sirtuin functions in health and disease. *Mol. Endocrinol.* **2007**, *21*, 1745–1755. [CrossRef] [PubMed]
33. Luo, J.; Nikolaev, A.Y.; Imai, S.; Chen, D.; Su, F.; Shiloh, A.; Guarente, L.; Gu, W. Negative control of p53 by sir2alpha promotes cell survival under stress. *Cell* **2001**, *107*, 137–148. [CrossRef]
34. Chen, J.; Xie, P.; Zhang, D.; Lei, H. In situ studies on the distribution patterns and dynamics of microcystins in a biomanipulation fish–bighead carp (aristichthys nobilis). *Environ. Pollut.* **2007**, *147*, 150–157. [CrossRef] [PubMed]
35. Zhang, Q.; Carmichael, W.; Yu, M.; Li, S. Cyclic peptide hepatotoxins from freshwater cyanobacterial (biue-green algae) waterblooms collected in central China. *Environ. Toxicol. Chem.* **1991**, *10*, 313–321. [CrossRef]
36. Zhang, D.; Xie, P.; Liu, Y.; Qiu, T. Transfer, distribution and bioaccumulation of microcystins in the aquatic food web in lake taihu, china, with potential risks to human health. *Sci. Total Environ.* **2009**, *407*, 2191–2199. [CrossRef] [PubMed]

37. Huang, H.; Liu, C.; Fu, X.; Zhang, S.; Xin, Y.; Li, Y.; Xue, L.; Cheng, X.; Zhang, H. Microcystin-LR induced apoptosis in rat sertoli cells via the mitochondrial caspase-dependent pathway: Role of reactive oxygen species. *Front. Phys.* **2016**, *7*, 397. [CrossRef] [PubMed]
38. Sinha, N.; Adhikari, N.; Saxena, D.K. Effect of endosulfan on the enzymes of polyol pathway in rat sertoli-germ cell coculture. *Bull. Environ. Contam. Toxicol.* **2001**, *67*, 821–827. [CrossRef] [PubMed]
39. Mishra, V.; Saxena, D.K.; Das, M. Effect of argemone oil and argemone alkaloid, sanguinarine on sertoli-germ cell coculture. *Toxicol. Lett.* **2009**, *186*, 104–110. [CrossRef] [PubMed]
40. Danz, E.D.; Skramsted, J.; Henry, N.; Bennett, J.A.; Keller, R.S. Resveratrol prevents doxorubicin cardiotoxicity through mitochondrial stabilization and the sirt1 pathway. *Free Radic. Boil. Med.* **2009**, *46*, 1589–1597. [CrossRef] [PubMed]
41. Wang, L.; Wang, X.; Geng, Z.; Zhou, Y.; Chen, Y.; Wu, J.; Han, X. Distribution of microcystin-LR to testis of male sprague-dawley rats. *Ecotoxicology* **2013**, *22*, 1555–1563. [CrossRef] [PubMed]
42. Jang, J.; Huh, Y.J.; Cho, H.J.; Lee, B.; Park, J.; Hwang, D.Y.; Kim, D.W. Sirt1 enhances the survival of human embryonic stem cells by promoting DNA repair. *Stem Cell Rep.* **2017**, *9*, 629–641. [CrossRef] [PubMed]
43. Coussens, M.; Maresh, J.G.; Yanagimachi, R.; Maeda, G.; Allsopp, R. Sirt1 deficiency attenuates spermatogenesis and germ cell function. *PLoS ONE* **2008**, *3*, e1571. [CrossRef] [PubMed]
44. Park, S.J.; Ahmad, F.; Philp, A.; Baar, K.; Williams, T.; Luo, H.; Ke, H.; Rehmann, H.; Taussig, R.; Brown, A.L.; et al. Resveratrol ameliorates aging-related metabolic phenotypes by inhibiting camp phosphodiesterases. *Cell* **2012**, *148*, 421–433. [CrossRef] [PubMed]
45. Abdelali, A.; Al-Bader, M.; Kilarkaje, N. Effects of trans-resveratrol on hyperglycemia-induced abnormal spermatogenesis, DNA damage and alterations in poly (adp-ribose) polymerase signaling in rat testis. *Toxicol. Appl. Pharmacol.* **2016**, *311*, 61–73. [CrossRef] [PubMed]
46. Anekonda, T.S.; Adamus, G. Resveratrol prevents antibody-induced apoptotic death of retinal cells through upregulation of sirt1 and ku70. *BMC Res. Notes* **2008**, *1*, 122. [CrossRef] [PubMed]
47. Li, M.; Li, J.J.; Gu, Q.H.; An, J.; Cao, L.M.; Yang, H.P.; Hu, C.P. Egcg induces lung cancer a549 cell apoptosis by regulating ku70 acetylation. *Oncol. Rep.* **2016**, *35*, 2339–2347. [CrossRef] [PubMed]
48. Rastogi, D.; Narayan, R.; Saxena, D.K.; Chowdhuri, D.K. Endosulfan induced cell death in sertoli-germ cells of male wistar rat follows intrinsic mode of cell death. *Chemosphere* **2014**, *94*, 104–115. [CrossRef] [PubMed]
49. Uguralp, S.; Usta, U.; Mizrak, B. Resveratrol may reduce apoptosis of rat testicular germ cells after experimental testicular torsion. *Eur. J. Pediatr. Surg.* **2005**, *15*, 333–336. [CrossRef] [PubMed]

toxins

MDPI

Article

Toxic Cyanobacteria in Svalbard: Chemical Diversity of Microcystins Detected Using a Liquid Chromatography Mass Spectrometry Precursor Ion Screening Method

Julia Kleinteich [1,2,*], Jonathan Puddick [3], Susanna A. Wood [3], Falk Hildebrand [4], H. Dail Laughinghouse IV [5], David A. Pearce [6,7], Daniel R. Dietrich [8] and Annick Wilmotte [2]

1 Center for Applied Geosciences, Eberhard Karls Universität Tübingen, Hölderlinstr. 12, 72074 Tübingen, Germany

2 BCCM/ULC, University of Liege, In-Bios Centre for Protein Engineering, B6, 4000 Liege, Belgium; awilmotte@uliege.be

3 Cawthron Institute, Halifax Street East, Nelson 7010, New Zealand; jonathan.puddick@cawthron.org.nz (J.P.), susie.wood@cawthron.org.nz (S.A.W.)

4 Structural and Computational Biology, European Molecular Biology Laboratory, Meyerhofstrasse 1, 69117 Heidelberg, Germany; falk.hildebrand@gmail.com

5 Fort Lauderdale Research and Education Center, University of Florida, Davie, FL 33314, USA; hlaughinghouse@ufl.edu

6 Department of Applied Sciences, Faculty of Health and Life Sciences, University of Northumbria at Newcastle, Newcastle NE1 8ST, UK; david.pearce@northumbria.ac.uk

7 British Antarctic Survey, Cambridge CB3 0ET, UK

8 Human and Environmental Toxicology, University of Konstanz, 78457 Konstanz, Germany; daniel.dietrich@uni-konstanz.de

* Correspondence: Julia.Kleinteich@gmail.com; Tel.: +49-7071-29-72495

Received: 9 February 2018; Accepted: 29 March 2018; Published: 3 April 2018

Abstract: Cyanobacteria synthesize a large variety of secondary metabolites including toxins. Microcystins (MCs) with hepato- and neurotoxic potential are well studied in bloom-forming planktonic species of temperate and tropical regions. Cyanobacterial biofilms thriving in the polar regions have recently emerged as a rich source for cyanobacterial secondary metabolites including previously undescribed congeners of microcystin. However, detection and detailed identification of these compounds is difficult due to unusual sample matrices and structural congeners produced. We here report a time-efficient liquid chromatography-mass spectrometry (LC-MS) precursor ion screening method that facilitates microcystin detection and identification. We applied this method to detect six different MC congeners in 8 out of 26 microbial mat samples of the Svalbard Archipelago in the Arctic. The congeners, of which [Asp^3, ADMAdda^5, Dhb^7] MC-LR was most abundant, were similar to those reported in other polar habitats. Microcystins were also determined using an Adda-specific enzyme-linked immunosorbent assay (Adda-ELISA). *Nostoc* sp. was identified as a putative toxin producer using molecular methods that targeted 16S rRNA genes and genes involved in microcystin production. The *mcy* genes detected showed highest similarities to other Arctic or Antarctic sequences. The LC-MS precursor ion screening method could be useful for microcystin detection in unusual matrices such as benthic biofilms or lichen.

Keywords: arctic; benthic mats; cyanotoxins; ELISA; 16S rRNA gene

key Contribution: A time-efficient LC-MS precursor ion screening method was applied to detect unusual microcystin congeners in complex sample matrices of microbial biofilms on the Svalbard archipelago in the Arctic.

---

1. Introduction

Cyanobacteria are phototrophic prokaryotes that occur in a diverse range of terrestrial and aquatic ecosystems worldwide. They are most infamously known for their mass occurrence (blooms) in tropical and temperate freshwaters [1,2]. These blooms are becoming progressively more problematic as they are reinforced by increasing nutrient loads and elevated water temperature mediated by climate change [3,4]. Many of the bloom-forming cyanobacterial species produce toxic secondary metabolites that pose a threat to human and animal health [5,6]. The compounds include heptapeptides with hepato- and neurotoxic potential, as well as neurotoxic and cytotoxic alkaloids [5–8]. Planktonic cyanobacterial blooms and the associated toxins have a direct impact on drinking water quality, the usability of water for recreational activities and receive significant attention from the scientific community, media and the general public [6,9]. It is important to understand the potential chemical diversity of cyanobacterial toxins and identify cyanobacterial species producing them to assist in management and risk assessment of cyanobacterial blooms. This knowledge may also help in understanding the evolution and the ecological function of the secondary metabolites.

In contrast to planktonic species, far less scientific and public attention has been devoted to non-planktonic habitats. Cyanobacteria growing in benthic mats, however, may also provide a source for novel secondary metabolites. Quite recently, it has been shown that cyanobacterial species in benthic mats, lichen-associations or epilithic biofilms produce toxins [8–10] as well as novel toxin congeners [11,12] previously undescribed from typical planktonic blooms.

Freshwater habitats of the polar regions are inhabited by a large taxonomic diversity of benthic cyanobacterial species [13,14]. Recent studies show that 20–96% of screened polar samples contain cyanobacterial toxins [15–20]. Benthic cyanobacterial mats in polar meltwater ponds, cryoconite holes, wet soil and marshy moss cushions are therefore suitable candidates to discover potentially toxic cyanobacteria and new secondary metabolites [20]. The neurotoxic saxitoxin was detected in a benthic cyanobacterial community from the Arctic [18] and the cytotoxic cylindrospermopsin was found in a similar habitat in the Antarctic. However, no known toxin-producing organisms were observed in polar samples and these have yet to be identified [19]. Microcystins (MCs), the most commonly identified and widely distributed cyanotoxins, have also been detected in cyanobacterial mats from the Arctic and the Antarctic. Microcystins are cyclic heptapeptides (Figure 1) composed of seven D- and L- amino acids, including uncommon amino acids such as 3*S*-amino-9*S*-methoxy-2*S*,6,8*S*-trimethyl-10-phenyldeca-4*E*,6*E*-dienoic acid (Adda) or *N*-methyl dehydroalanine (Mdha). The number of known MC variants currently exceeds 250 [21]. This variety is mainly based on two variable amino acids and modifications of the amino acids, such as methylation [12]. Microcystins found in the polar regions include a range of uncommon or previously unknown variants [11,12,17,22]; e.g., congeners that contained the rare substitution of the position one amino acid, the usual D-alanine, to glycine [12,17]. Microcystins act as protein phosphatase inhibitors in eukaryotic cells, inducing a breakdown of the cellular cytoskeleton and eventually leading to cell death, but they require active transport/uptake into the cell via organic anion transporting polypeptides (OATPs) [23]. The structure of the MC congeners affects their protein phosphatase inhibition and cellular uptake characteristics, and thus their final toxicity [8]. Accurate identification of MC congeners is therefore vital for risk assessment and freshwater management.

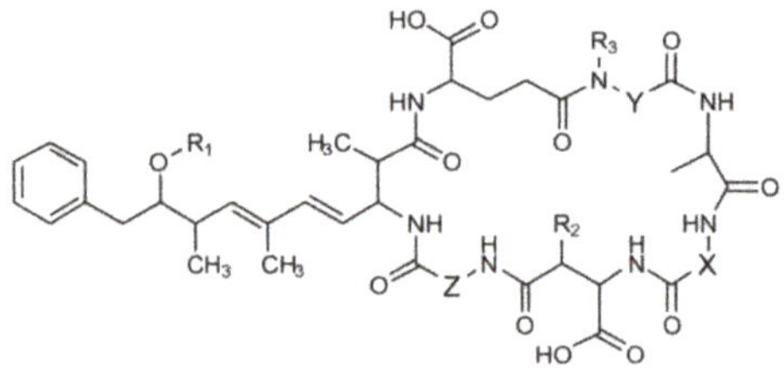

	R_1	R_2	R_3	X	Y	Z
MC-LR	$-CH_3$	$-CH_3$	$-CH_3$	Leu	Dha	Arg
MC-LA	$-CH_3$	$-CH_3$	$-CH_3$	Leu	Dha	Ala
[Asp³, Ser⁷] MC-RR	$-CH_3$	-H	-H	Arg	Ser	Arg
[Asp³, Dha⁷] MC-RR	$-CH_3$	-H	-H	Arg	Dha	Arg
[Asp³, ADMAdda⁵, Thr⁷] MC-LR	$-COCH_3$	-H	-H	Leu	Thr	Arg
[Asp³, ADMAdda⁵, Dhb⁷] MC-LR	$-COCH_3$	-H	-H	Leu	Dhb	Arg
[Asp³, ADMAdda⁵, Dhb⁷] MC-RR	$-COCH_3$	-H	-H	Arg	Dhb	Arg

Figure 1. Structure of microcystin-LR and the 6 microcystin (MC) congeners identified in this study (ADMAdda = *O*-acetyl-*O*-demethyl 3-amino-9-methoxy-2,6,8-trimethyl-10-phenyldeca-4,6-dienoic acid, Ala = alanine, Arg = arginine, Asp = aspartic acid, Dha = dehydroalanine, Dhb = dehydrobutyrine, Leu = leucine, Ser = serine and Thr = threonine).

However, the increasing number of structural toxin congeners complicates the identification of MCs. Additionally, unusual sample types, such as polar benthic microbial mats, have a complex matrix containing pigments, polysaccharides and secondary metabolites [18,24]. These and other compounds may interfere with certain detection methods, for example by cross-reactivity of antibodies in an ELISA [25,26]. Analytical tools such as high-performance liquid chromatography (HPLC) and liquid chromatography-tandem mass spectrometry (LC-MS/MS) are often used for MC detection in complex matrices. Identification of unusual MCs using standard HPLC and LC-MS/MS methods, though, requires comparison with costly reference standards or time-consuming identification and structural characterization by experienced personnel.

The aims of this study were to (1) develop an LC-MS precursor ion scanning method that would simplify the identification and characterization procedure by reducing the number of candidate compounds that need to be characterized, and (2) to use this methodology to screen environmental microbial mat samples with a complex sample matrix collected from the Arctic.

2. Results

For this study, 26 cyanobacteria dominated microbial mat samples from Svalbard were available (Supplementary Table S1 and Figures S1 and S2). Twenty of the samples were analyzed using an Adda-specific ELISA (Table 1). Of those 20 samples, 18 showed a signal above the detection limit in the ELISA, ranging between 2 and 54 μg of microcystin per liter extract. In three of these samples (SV-54, -74, and -75), the measured MC concentration exceeded the range of the standard curve despite several dilution steps and the MC concentration was therefore estimated to be above 50 μg/L. Twelve of the samples that were positive in the Adda-ELISA and six additional samples (SV-A, -B, -C, -D, -E and -81, not analyzed by ELISA) were analyzed using the MC congener precursor ion scanning method developed in this study (Table 1 and Figure 2).

When the chromatograms from the precursor ion screen were compared to those acquired by collecting positive ion scan data (*m/z* 450-1,150; Figure 2), the data collected using the LC-MS precursor ion scanning method contained fewer candidate ions as expected. For the 18 samples analyzed during this study, there were 55% fewer peaks to further investigate when using the precursor ion scan searching for Adda product ions (Supplementary Table S2).

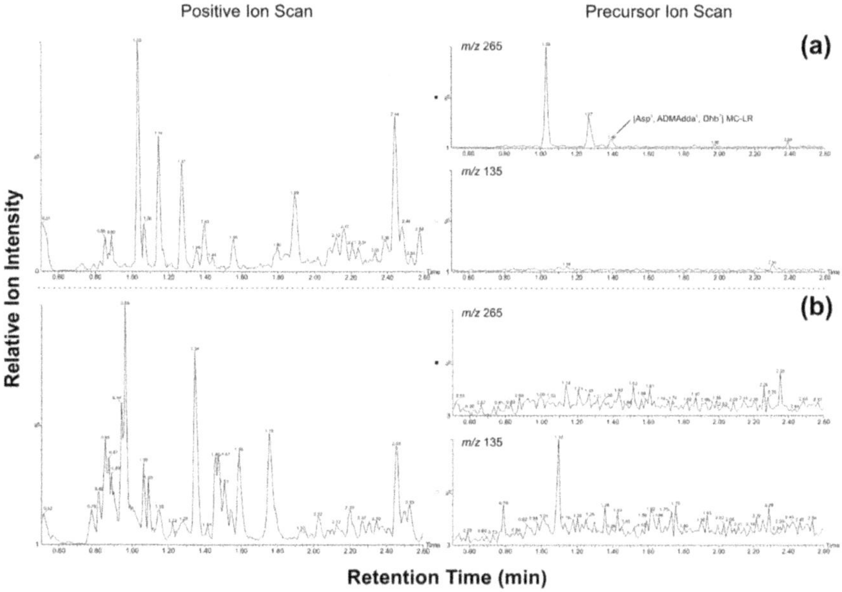

Figure 2. Base-peak chromatograms of positive ion scans (left; m/z 450-1150) and precursor ion scans (right; m/z 135 for Adda-containing compounds or m/z 265 for ADMAdda-containing compounds) for (**a**) SV-81 (Category 1) and (**b**) SV-02 (Category 3). See Table 1 for definition of categories.

Table 1. Microbial mat samples from Svalbard, the extracted mass of lyophilized material, their toxin content as determined by enzyme-linked immune sorbent assays (ELISAs), the liquid chromatography mass spectrometry (LC-MS) precursor ion screening method as well as a detailed LC-MS/MS analysis and the detection of genes involved in toxin production. Categories of the LC-MS precursor ion scan: (1) Microcystin (MC) likely to be present in the sample, (2) MC possibly present in the sample, and (3) MC absent from the sample. Genes: Non-ribosomal peptide synthetase (NRPS), polyketide synthase (PKS), microcystin gene E (*mcyE*), microcystin gene B (*mcyB*), and saxitoxin gene A (*sxtA*). Structures of MC congeners can be found in Figure 1.

Sample	Extracted Mass [g]	ELISA [μg/L]	LC-MS Precursor Ion Category (1–3)	LC-MS/MS Characterization	Genes
SV-A	0.047	n.a.	2	n.d.	NRPS, PKS
SV-B	0.078	n.a.	3	n.a.	NRPS, PKS
SV-C	0.089	n.a.	3	n.a.	-
SV-D ^{N}	0.093	n.a.	1	[Asp^{3}, ADMAdda^{5}, Dhb^{7}] MC-LR	NRPS, PKS, *mcyE*, *mcyB*
SV-E ^{N}	0.068	n.a.	3	n.a.	NRPS, PKS, *mcyE*
SV-2	2.017	8	3	n.a.	NRPS, PKS
SV-8	0.679	9	3	n.a.	PKS
SV-13	0.109	14	3	n.a.	-

Table 1. *Cont.*

Sample	Extracted Mass [g]	ELISA [µg/L]	LC-MS Precursor Ion Category (1–3)	LC-MS/MS Characterization	Genes
SV-14	0.025	0	n.a.	n.a.	NRPS, PKS
SV-16	0.105	19	3	n.a.	NRPS, PKS
SV-17	0.262	18	2	n.d.	NRPS, PKS
SV-24	1.010	2	n.a.	n.a.	-
SV-28	0.893	8	n.a.	n.a.	-
SV-39	0.020	0	n.a.	n.a.	NRPS, PKS
SV-40 N	0.815	2	n.a.	n.a.	-
SV-46	2.438	6	n.a.	n.a.	-
SV-49 N	0.663	54	1	[Asp^3, ADMAdda^5, Dhb^7] MC-LR	-*
SV-54	1.267	>STD	3	n.a.	-
SV-56 N	0.569	37	3	n.a.	-
SV-65	1.805	3	n.a.	n.a.	-
SV-74 N	0.101	>STD	1	MC-LA	-*
SV-75 N	0.056	>STD	1	[Asp^3, ADMAdda^5, Dhb^7] MC-RR [Asp^3, ADMAdda^5, Dhb^7] MC-LR	NRPS, PKS, *mcyE*, *mcyB*, *sxtA*
SV-77	0.256	2	n.a.	n.a.	-
SV-80 H	0.104	25	1	[Asp^3, Ser^7] MC-RR [Asp^3, Dha^7] MC-RR [Asp^3, ADMAdda^5, Dhb^7] MC-RR [Asp^3, ADMAdda^5, Thr^7] MC-LR [Asp^3, ADMAdda^5, Dhb^7] MC-LR	NRPS, PKS, *mcyE*, *mcyB*
SV-81 H,E	11.309	n.a.	1	[Asp^3, ADMAdda^5, Dhb^7] MC-LR	NRPS, PKS, *mcyE*, *mcyB*
SV-83 H	0.126	2	1	[Asp^3, ADMAdda^5, Dhb^7] MC-RR [Asp^3, ADMAdda^5, Dhb^7] MC-LR Unidentified microcystin	-*

n.a. = not analyzed; n.d. = MCs not detected; >STD = above standard curve; N = sample dominated by *Nostoc*; H = Hotspring; E = endolithic; * = low DNA quality.

For the purpose of identifying candidate MCs, compounds that eluted between 1–1.25 min with *m/z* 500–575 were assumed to be doubly-protonated ions of MCs containing two arginine residues. This premise was further strengthened by the presence of the corresponding singly-protonated ion between *m/z* 1000–1150. Compounds that eluted between 1.25–1.55 min with *m/z* 850–1150 were assumed to be MCs containing one arginine residue in position two (Figure 1). Compounds in the same mass region that eluted between 1.55–1.9 min were assumed to be MCs that contained one arginine residue in position four [27]. Finally, compounds that eluted between 1.9–2.35 min with an *m/z* 850–1150 were assumed to be hydrophobic MC congeners containing no arginine residues. These retention times were determined using available microcystin reference standards such as MCs -RR, -YR, -LR and -LF, and using an extract *Microcystis* CAWBG11 that produces a wide array of microcystins and has been well characterized in our laboratory [27]. From this analysis, the LC-MS precursor ion screens were classified in three categories: (1) MCs likely to be present in the sample; (2) MCs possibly present in the sample, and; (3) MCs absent from the sample. The classification between Categories 1 and 2 took into account whether MCs with the same precursor ion mass had been reported in the past [28]. Seven of the 18 samples analyzed using the precursor ion screen were classified in Category 1, two in Category 2 and the remaining nine in Category 3 (Table 1).

When the Category 1 and 2 samples were investigated further by MS/MS, all seven category 1 samples were positive for MC congeners. Six known MCs and an unidentified MC congener were detected (Figure 1, Table 1). The unidentified congener, detected in sample SV-83, had a mass of 1052 Da and contained dehydrobutyrine (Dhb), but its structure could not be elucidated at this point in time due to insufficient structural information from the product ions acquired. The most commonly observed MC congener in the samples was [Asp^3, $ADMAdda^5$, Dhb^7] MC-LR (Figure 1), identified in SV-D, SV-49, SV-75, SV-80, SV-81 and SV-83.

The genes for non-ribosomal peptide synthesis (NRPS) and polyketide synthetases (PKS), involved in general secondary metabolite production were detected in 13 samples, irrespective of their category in the pre-cursor scan (Table 1). The *mcyB* and *mcy*E genes were shown to be present in four samples (SV-D, SV-75, SV-80 and SV-81), whereas only *mcyE* was detected in sample SV-E (Table 1). All samples containing *mcyE* or *mcyB* genes, except sample SV-E, were Category 1 in the LC-MS precursor ion scanning method and contained MC congeners as detected using detailed LC-MS/MS analysis. The partial sequences of the *mcyE* genes were most similar to those of the genus *Nostoc* sp. 152, with a pairwise similarity between 93% and 99% as well as to an uncultured *Nostoc* clone MVMG1 from Antarctica (98–99%) in SV-D, E, 75 and 80 and to an uncultured cyanobacterium isolate *nda*F gene from the Gulf of Finland (93%, Supplementary Table S3) in SV-81. The putative *mcyB* genes were related to non-ribosomal peptide synthetase gene cluster sequences in *Microcoleus* PCC-8701 with a maximum pairwise similarity of 73% as well as *Cylindrospermum* sp. NIES-4074 and *Scytonema* NIES-4073 whole genomes. Similarities of up to 76% were recorded to *Nostoc* strains, but with a lower sequence coverage. These similarities appear quite high as polar toxin gene sequence similarities to the sequences in GenBank from other habitats are generally low [18,19]. Moreover, genetic similarities for protein-coding genes are usually lower than for the highly conserved ribosomal subunit sequences. In order to focus on the functionality of the protein, we translated the DNA sequence into the corresponding amino acid sequence. Using the blastp search tool of GenBank, the *mcyE* amino acid sequences appeared most similar (>97%) to various Arctic and Antarctic *Nostoc* strains (Supplementary Table S3). The putative *mcyB* amino acid sequences were most similar (>83%) to a non-ribosomal peptide synthetase of the Antarctic cyanobacterium *Phormidesmis priestleyi* ULC007 (#WP_073072318.1).

In addition to MCs, two samples (SV-13 and SV-83) were positive in the cylindrospermopsin ELISA (data not shown). However, this could not be confirmed by HPLC or molecular analysis. A saxitoxin ELISA was negative for all samples assessed (data not shown), but a fragment of the *sxtA* gene was amplified in sample SV-75. The sequence matched with 100% similarity to a *sxt* sequence detected in a *Nostoc* dominated cyanobacterial mat from Baffin Island [18] and to *Scytonema* sp. UCFS15 from New Zealand with 98% similarity (Supplementary Table S3).

To elucidate the cyanobacterial diversity of the samples and to identify potential toxin producers, five samples were analyzed by 454 next generation sequencing. The genomic library obtained from the partial 16S rRNA gene revealed a high homogeneity between the samples (Figure 3). The dominant OTU (OTU1) in three samples belonged to the family Oscillatoriaceae, with the highest match on GenBank being an uncultured bacterium clone from the Antarctic Peninsula. Similarities above 98% to cultured strains were observed for *Phormidium* sp. JR20 later identified as *Microcoleus favosus* JR20 [29] and *Phormidium* sp. Ant-Orange, both isolated from Antarctica. Representative sequences and putative identities of the 6 most abundant OTUs can be found in Supplemental Table S4. One sample, SV-D, was distinctly different from the other samples and not dominated by OTU1. SV-D had a higher taxonomic richness and was composed by five OTUs assigned to the *Oscillatoriales* family (OTU2 and 6), *Tolypothrix* (OTU3) and *Nostoc* (OTU4 and 5). OTU4 had highest similarities to a sequence from Baffin Island in the Canadian Arctic [18]. Upon macroscopic and microscopic examination, *Nostoc* was clearly dominant in sample SV-D and SV-E; however, it was only a minor component in the

454 sequencing analysis. Although SV-C, SV-D, and SV-E contained OTUs that could be assigned to the *Nostoc* genus, the *Nostoc* specific OTUs did not exceed 20% of the sequences in any of the samples.

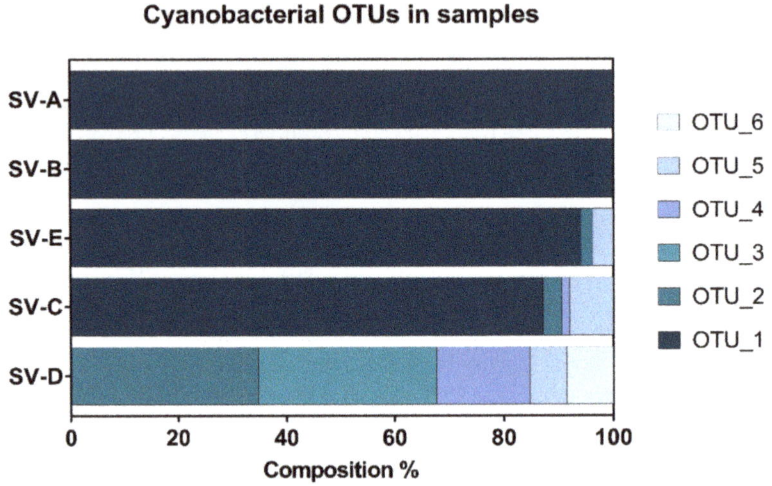

Figure 3. Phylogenetic composition of Svalbard samples based on 16S ribosomal RNA gene amplification using 454 sequencing and filtered for solely cyanobacterial reads. See text and Supplementary Figure S4 for identification of operational taxonomic units (OTUs).

3. Discussion

In the present study, we show that the cyanobacterial toxin MC is widely distributed in benthic cyanobacterial mats in the Svalbard archipelago. The identified MC congeners were chemically diverse and structurally similar to MC congeners detected previously in polar habitats [11,12,18,19]. The presence of MCs in our samples as well as the detection of toxins and other compounds with bioactive potential in other studies [11,12,18,19,30–32] indicate that cyanobacteria in polar regions could be a rich source of secondary metabolites.

Cyanobacterial toxins are commonly known from warm habitats but have been reported repeatedly in the polar regions over the last two decades [11,15–20,33]. Previous records of cyanotoxins in Svalbard reported the presence of MC variants in samples of cyanolichens [11] and MC-LR in biocrusts [33], without an exact determination of the MC variant in the latter study. In the present study, 90% of the samples were positive using the Adda-MC ELISA. The presence of specific MC variants was confirmed by LC-MS/MS analysis in 20% of the samples. A clear signal in the LC-MS precursor ion screen was confirmed by further MS/MS characterization in seven samples. The detection of the *mcyE* and *mcyB* genes confirmed the presence of a toxin producer in five samples.

More samples tested positive for MC in the ELISA than in the LC-MS analysis. This difference may be explained by the lower detection limit and the cumulative signal of all Adda-containing molecules in the ELISA. In contrast, LC-MS targets single compounds at a higher detection limit. However, the LC-MS precursor ion screen provided more definitive information than the Adda-ELISA. The ELISA seemed to have more problems with the complex sample matrix of Arctic environmental samples containing pigments, polysaccharides and other secondary metabolites [24] that may interfere with antibodies and other reagents [25,26].

The LC-MS precursor ion scanning method allowed candidate MCs to be identified more easily than traditional LC-MS identification methods. In our study, the precursor ion scan provided 57% fewer peaks to assess when compared to conducting a preliminary assessment using a conventional positive ion scan of the appropriate mass range. Furthermore, because the precursor ion scan assesses for product ions associated with MCs, there is more confidence that further investigation will result in the detection of MCs. Conventional dereplication strategies were employed to reduce the number of candidate ions by a further 68% (i.e., comparing the retention time and *m/z* with that which would be expected for MCs). Traditional MS/MS characterization techniques were still required to confirm the identity of candidate ions, but the workload was reduced significantly by decreasing the number of candidate compounds through the MC precursor ion screen.

Whilst the MC congeners detected most frequently during this study are rarely detected in planktonic cyanobacterial blooms, they are commonly observed in Arctic and Antarctic habitats [11,12,17–19]. This includes the desmethyl Asp^3 modification, the substitution of *N*-methyl dehydroalanine (Mdha) to dehydrobutyrine (Dhb) at position seven (Figure 1) as well as the acetyl desmethyl modification of the position five Adda group (ADMAdda), which are commonly observed in MCs produced by *Nostoc* species [27]. However, the SV-74 sample contained the more conventional MC-LA, which is observed in planktonic cyanobacterial blooms relatively frequently. In SV-80, Ser^7 and Thr^7 congeners were present. These amino acids are the precursor compounds for Dha^7 and Dhb^7, commonly observed in position seven of MCs. Their presence indicates that either the dehydrogenase enzyme in the non-ribosomal peptide synthetase (NRPS) module was not functioning effectively [34] in the dominant cyanobacterial strain or that multiple MC-producing strains were present and the dehydrogenase gene was not functional in a subset of the population. Isolation of cultured strains would be required to further understand the MC diversity observed in these samples, highlighting some of the limitations of working on environmental material.

The chemical diversity of MCs and geographic connection was also reflected in the MC synthetase (*mcy*) genes and translated amino acid sequences amplified here. These sequences were most similar to *mcyE* sequences of an uncultured *Nostoc* detected in Antarctic habitats [17] and to an *ndaF* gene of an uncultured cyanobacterium in the Gulf of Finland [35]. The closest cultured match, *Nostoc* sp.152, was originally isolated in Finland [36]. For the putative *mcyB* gene, no close similarities (>90%) to gene sequences from environmental sources could be detected, indicating that the amplified product may be an NRPS but not located on a *mcy* operon. Nevertheless, more than 83% similarity to a non-ribosomal peptide synthetase amino acid sequence of the Antarctic cyanobacterium *Phormidesmis priestleyi* ULC007 suggests a geographic connection to polar organisms. The here detected nucleotide and translated amino acid sequences differ from those of species usually detected in planktonic blooms of warmer climate such as *Microcystis aeruginosa* (e.g., NIES-843, accession #NC_010296) or *Planktothrix rubescens* (e.g., NIVA-CYA 407 accession #NZ_AVFW00000000). Unfortunately, no known Arctic or Antarctic MC-producing strain has been isolated or cultivated to date and few sequences of the *mcy* genes are available from polar habitats, so that a comprehensive geographic distribution analysis of toxin genes is not possible. The full genome sequencing of an Arctic/Antarctic MC producer, or more *mcy* sequences from polar regions, could help to understand the detected MC diversity.

During this study, the cyanobacterial genus *Nostoc* was identified as a likely candidate for producing MCs on Svalbard. This genus has previously been suspected to be a producer of MCs in the polar regions [17,18,20] and the results of this study support this suggestion. *Nostoc* was predominant, identified macroscopically and microscopically, in samples containing the highest concentrations of MC (Samples SV-D, -E, -49, -56, -74 and -75). Moreover, the *mcyE* genes detected were most similar to *mcyE* genes of the *Nostoc* genus as described above. In addition, Kaasalainen et al. [11] detected MCs in lichen-associated *Nostoc* on Svalbard. The *Nostoc* genus is widely distributed on the Svalbard archipelago and may make a significant contribution to local nitrogen cycling [37].

Even though *Nostoc* clearly predominated some of the samples, the genus was underrepresented in the pyrosequencing results. This underrepresentation of *Nostoc* is likely an artifact of insufficient DNA extraction due to the extensive exopolysaccharidic sheaths of the Nostocalean family possibly in combination with primer biases [38]. This highlights the limits of next generation sequencing techniques, which can only be regarded as semi-quantitative [39,40]. The dominant OTUs detected in the five samples analyzed were similar to other Arctic or Antarctic cyanobacterial sequences (Supplemental Table S4). A circumpolar cold-climate biogeography of cyanobacterial species has been suggested before [41,42] and needs to be studied in detail in the future. Unfortunately, no cyanobacterial sequences from the Russian Arctic are available, but they would be a valuable addition for polar cyanobacterial biogeography studies.

Interestingly, the three samples from a polar geothermal spring with temperatures of around 20 °C contained relatively more congeners than samples from other habitats on the Svalbard archipelago. This would suggest that Arctic MC producers increase production in warmer temperatures or that warmer temperatures support different strains of cyanobacteria. It has been shown previously that polar cyanobacterial communities, when growing at warmer temperatures, changed species composition and increased MC production [43]. Another hypothesis is that MC producers are adapted to warmer habitats and thrive in the hot springs, though they are normally able to survive in the cold polar climate. Geothermal sites in the polar regions could thus serve as a refuge for organisms generally adapted to a warmer climate as was inferred by Fraser et al. [44] for Antarctica. As a possible consequence, a temperature rise due to climate change, like the one already ongoing in some parts of the Arctic and the Antarctic [45], could lead to an increased abundance of toxic cyanobacteria in Svalbard.

In summary, benthic cyanobacteria from Svalbard proved to be a rich source for structurally interesting MC congeners. The precursor ion screening tool facilitated the identification of six different MC congeners by specifically assessing product ions diagnostic of MCs. The MC congeners produced were structurally dissimilar to those usually observed in planktonic blooms in warmer regions, but were similar to those produced by the genus *Nostoc*. It is possible that the 'unusual' MC variants detected in extreme environments are not unusual, but have been overlooked in temperate and other bioregions to date. It has been suggested that "Given the ecological plasticity of cyanobacteria [. . .] potentially toxic cyanobacteria are much more widely distributed than currently thought." [20]. Consequently, cyanobacteria worldwide and in all habitat types (e.g., terrestrial and aquatic, benthic and planktonic) may contain a yet unknown diversity of MC variants as well as other secondary metabolites. Benthic cyanobacteria in polar, temperate and tropical environments remain an interesting source for the identification of bioactive compounds including MCs. The precursor ion screening methodology described here will assist in the discovery of these MC congeners as it simplifies and speeds up the identification of non-conventional congeners.

4. Materials and Methods

Svalbard is an archipelago in the North Atlantic Ocean at 77°50′ N and 19°50′ E. Samples were collected during two field seasons (Supplementary Table S1). Five cyanobacterial samples were collected at various locations in the vicinity of Longyearbyen, Björndalen and Colesbukta in June 2012 and 21 samples in a diversity of biotopes across the entire archipelago in July 2013 (Supplementary Figure S1). Cyanobacterial biofilms, mats and crusts were collected from streams, wet soil, moss cushions and in 'hot springs' with water temperature up to 25 °C (Supplementary Figure S2, Supplementary Table S1). Biofilm material was directly sampled using a sterile spatula and stored in sterile tubes or bags and were stored at −20 °C within 8 h from collection until further analysis. Biofilm material was examined microscopically and macroscopically in the laboratory.

For MC extraction, the frozen material was thawed, homogenized with a sterile spatula and lyophilized. In a methanolic extraction step, organic compounds were extracted as follows: one mL of 80% aqueous methanol (v/v) with 0.1% formic acid (v/v) was added to 0.02–2.4 mg of lyophilized material (Table 1), incubated for 2 h at room temperature, vortexed vigorously, and ultra-sonicated for 15 min. The organic material was pelleted by centrifugation (11,400× g, 10 min) and the extraction repeated twice on the pellet. The resulting supernatants were combined and dried at 37 °C under continuous nitrogen flow or in a speed-vac (Savant SPD111V, Thermo Fisher Scientific, Waltham, MA, USA). The dried extracts were re-solubilized (200 µL, 80% v/v methanol with 0.1% formic acid; v/v), centrifuged (13,000× *g*, 15 min) to remove residual particles and stored at −20 °C until LC-MS analysis. The samples collected during 2013 were further purified using C_{18} cartridges (Sep-Pak, Waters, Dublin, Ireland) as described previously [19] to decrease matrix effects in the subsequent ELISA.

Twenty samples collected during 2013 were screened using ELISA for MCs (MC-ADDA ELISA), cylindrospermopsins (CYNs) and saxitoxins (STXs) according to the recommended protocol (ABRAXIS, Warminster, PA, USA). The assays have an LOD of 0.15 ng MC/mL, 0.05 ng CYN/mL and 0.02 ng STX/mL, respectively.

Based on the preliminary results, 12 samples that generated a strong MC signal in the ELISA and five additional extracts from the 2012 field trip plus one from hot spring were selected for analysis using LC-MS applying the precursor ion screening method. Clarified extracts were analyzed using an Acquity I-Class ultra-performance liquid chromatography system (UPLC; Waters Ltd., Borehamwood, UK) coupled to a Xevo-TQS triple quadrupole mass spectrometer (Waters Ltd.). Compound separation was achieved using an Acquity BEH-C_{18} UPLC column (Waters Ltd.; 1.7-μm; 50 × 2.1 mm) at 40 °C. Sample components were eluted at 0.4 mL/min with a gradient from 10% acetonitrile (Solvent A; *v*/*v*) to 90% acetonitrile (Solvent B; *v*/*v*), each containing 100 mM formic acid and 4 mM ammonia. The sample extracts (5 μL) were injected at 5% Solvent B (*v*/*v*) and held for 12 s before a linear gradient up to 35% Solvent B (*v*/*v*) over 24 s, to 50% Solvent B (*v*/*v*) over a further 72 s and to 65% Solvent B (*v*/*v*) over a final 42 s, before flushing with 100% Solvent B for 60 s and returning to the initial column conditions to equilibrate for 60 s. The electrospray ionization source was operated in positive-ion mode (150 °C; capillary 1.5 kV; nitrogen desolvation gas 1000 L/h at 500 °C; cone gas 150 L/h) with the mass spectrometer conducting precursor ion scans.

In precursor ion scanning mode, the first quadrupole was set to scan between m/z 450–1150 before the ions were introduced to a collision cell (the second quadrupole) and fragmented with argon gas at a collision energy of 40 V. The third quadrupole was set to filter specific product ions, in this case m/z 135 for Adda-containing MCs and m/z 265 for ADMAdda-containing MCs. The precursor ions, which resulted in the specified product ions, were then determined by the MassLynx software (Version 4.1, Waters Ltd.). Following this, the samples were further de-replicated by assessing the observed retention time in comparison to the molecular weight of the precursor ions (e.g., microcystins with two arginine residues elute in an earlier retention region than other microcystin congeners and generally have masses >1000 Da) in order to compile a list of candidate MCs to be identified using conventional structural characterisation methods (described below).

The candidate MCs identified using the precursor ion scanning method were further investigated by generating MS/MS spectra for each ion of interest. Tandem MS spectra were collected in positive ion mode over an m/z range of 100–1200. Compounds that were presumed to be MCs containing two arginine residues (e.g., MC-RR) were fragmented using a collision energy of 25–30 V and compounds presumed to be MCs containing one or no arginine residues (e.g., MC-LR or MC-LA respectively) were fragmented using a collision energy of 40–45 V. The spectra were primarily annotated with the assistance of the software mMass [46]. When discrepancies were apparent, the spectra were annotated manually using previously published MS/MS investigations of MCs as a guide [12,27,36,47,48]. To confirm whether Dha, Mdha or Dhb were present in the MCs identified, a micro-scale thiol derivatization was performed [49]. Microcystin solutions were reacted with β-mercaptoethanol as described in Puddick et al. [50] but using the precursor ion scan described above. If no reaction had occurred within 2 h at 30 °C, the MC was classified as containing Dhb^7. A control reaction containing MC-RR, MC-LR and MC-LA was run in parallel to confirm the reaction rate for Mdha.

DNA was extracted from each sample using the PowerSoil® DNA Isolation Kit (former MO BIO laboratories, Carlsbad, CA, USA, now Qiagen, Germantown, MA, USA) for the samples collected in 2012 and the PowerBiofilm® DNA Isolation Kit (former MO BIO laboratories, Carlsbad, CA, US, now Qiagen, Germantown, MA, USA) for the samples from 2013. Between 5 and 10 mg of frozen cyanobacterial material was extracted following the manufacturer's recommendations. The resulting DNA was eluted in sterile DNAse-free water and stored frozen at −20 °C until further use.

The DNA quality was checked by the amplification of the 16S rRNA gene using the primer set 27F/809R [22]. Several PCRs were performed on the extracted DNA, targeting genes of the *mcy*, *sxt*, and *cyr* operon involved in MC, STX, and CYN synthesis, respectively. For all reactions the Phusion™ polymerase Master Mix (NEB, Ipswich, MA, USA) or the iTaq™ PCR Master Mix solution (iNtRON Biotechnology, Sangdaewon-dong, South-Korea) was used. Primers, references and PCR conditions are listed in Supplementary Table S5. Bands of interest were excised from TAE buffered 1.5% agarose-gels (*w*/*v*) using a sterile scalpel, purified with a Gel Extraction Kit (Fermentas, St. Leon-Rot, Germany) and bi-directionally sequenced using the same primer combination as for amplification on a Sanger Sequencer (3730 DNA Analyzer) at the sequencing facility GIGA (http://www.giga.uliege.be) of the University of Liège. *Microcystis aeruginosa* UAM501 served as a positive control for *mcy* genes, *Aphanizomenon ovalisporum* UAM290 for *cyr* genes, and *Aphanizomenon gracile* UAM531 for *sxt* genes [51]. The sequences obtained were aligned and manually edited using Geneious™ software (Geneious Pro 7.1.1., Biomatters Ltd., Auckland, New Zealand). Translation into the corresponding amino acid code was also done using Geneious™ by aligning the amplified fragments to published and annotated *mcy* genes (of KC699835 *Nostoc* sp. 152 and AY768451 *Microcoleus* sp. PCC 8701). The closest cultured and non-cultured phylogenetic hits were identified for each sequence using the megablast and blastn tools of GenBank for nucleotide sequences and the blastp tool for amino acid sequences. Phylogenetic affiliations and accession numbers are given in Supplementary Table S3. The *sxtA* sequence was not submitted since it was represented by a single read only and can thus not be verified.

The primer pair CYA106F 5′-CGGACGGGTGAGTAACGCGTGA-3′ and modified 519-536 5′-GTNTTACNGCGGCKGCTG-3′ were used to amplify a fragment of the 16S rRNA gene including the V2-V3 domains [52,53]. Pyrosequencing using a 454 Sequencing System (Roche 454 Life Sciences, Basel, Switzerland) was performed at the Research and Testing Laboratories (Lubbock, TE, USA) as described previously [19]. The raw 454 data can be downloaded from http://vm-lux.embl.de/~hildebra/Arctic_454/.

We used the LotuS 1.31 pipeline [54] in short amplicon mode with default quality filtering. Raw 16S rRNA gene reads were quality filtered to ensure a minimum length of 250 bp, not more than eight homonucleotides, no ambiguous bases, an average read phred quality equivalent to 25 and an accumulated error below 0.5. Clustering and denoising of OTUs was performed using UPARSE [55], removing chimeric OTUs against the RDP reference database (http://drive5.com/uchime/rdp_gold.fa) with uchime [56], merging reads with FLASH [57] and assigning a taxonomy using an RDP classifier [58]. We could assign on average 5500 $\pm$ 3108 reads to each sample that were of cyanobacterial origin. Further data analysis was conducted with R statistical language Version 3.00 (The R Foundation, https://www.r-project.org/) as described in Hildebrand et al. [59], employing the rtk software [60] for all data normalizations.

Supplementary Materials: The following are available online at http://www.mdpi.com/2072-6651/10/4/147/s1, Supplemental file 1 including Figure S1: Map of Svalbard and sampling locations. Figure S2: In-situ photographs of some cyanobacterial mats sampled on Svalbard, Table S1: Metadata of samples, sampling dates, GPS coordinates and site description. Table S2: Evaluation of the precursor ion screen. Table S3: Amplified *mcyE*, *mcyB*, and *sxtA* sequences from cyanobacterial samples in Svalbard. Table S4: Sequences of most abundant cyanobacterial OTUs. Table S5: PCR conditions and primers. Supplemental File 2: Mass spectrometry data for all samples assessed.

Acknowledgments: This work was funded with a stipend for J.K. by the Carl-ZEISS foundation at the University of Konstanz followed by a Marie Curie COFUND funded post-doctoral position at the University of Liège, Belgium. J.K. and H.D.L. would like to acknowledge the Fonds de la Recherche Scientifique (FNRS) for travel stipends. S.A.W., J.P. and D.R.D. thank the Marsden Fund of the Royal Society of New Zealand (12-UOW-087-Toxic in Crowds and CAW1601-Blooming Buddies) and the Royal Society of New Zealand International Research Staff Exchange Scheme grant (MEAT Agreement 295223) for funding. D.R.D. would like to acknowledge the support of BW-Wassernetzwerk Baden-Württemberg. F.H. was supported by the European Union's Horizon 2020 research and innovation program under the Marie Skłodowska-Curie grant agreement No 600375. J.K. would like to thank UNIS and the Norwegian Government for the Arctic Microbiology course and the teaching staff for the course in 2012. H.D.L. would like to acknowledge the Svalbard Science Forum under the project # RiS-ID: 6206 and the FNRS for a post-doctoral fellowship in the BIPOLE project (2.4.570.09.F). We would like to thank the Antarctic Science Bursary for funding the 454 sequencing and the Federation of European Microbiological Societies (FEMS) conference grant. A.W. is Research Associate of the FRS-FNRS of Belgium. We acknowledge support by the Deutsche Forschungsgemeinschaft and the Open Access Publishing Fund of University of Tübingen.

Author Contributions: J.K., H.D.L. and J.P. designed the study. J.K. and J.P. conceived and designed the experiments and J.K. and H.D.L. collected the samples. F.H. analyzed the genomic data. S.A.W., D.A.P., D.R.D. and A.W. contributed reagents, materials, analysis tools and valuable input in the study design and the manuscript. J.K. and J.P. wrote the paper and were substantially supported by all co-authors.

Conflicts of Interest: The authors declare no conflict of interest.

References

1. Mowe, M.A.D.; Mitrovic, S.M.; Lim, R.P.; Furey, A.; Yeo, D.C.J. Tropical cyanobacterial blooms: A review of prevalence, problem taxa, toxins and influencing environmental factors. *J. Limnol.* **2014**, *73*. [CrossRef]
2. O'Neil, J.M.; Davis, T.W.; Burford, M.A.; Gobler, C.J. The rise of harmful cyanobacteria blooms: The potential roles of eutrophication and climate change. *Harmful Algae* **2012**, *14*, 313–334. [CrossRef]
3. Davis, T.W.; Berry, D.L.; Boyer, G.L.; Gobler, C.J. The effects of temperature and nutrients on the growth and dynamics of toxic and non-toxic strains of *Microcystis* during cyanobacteria blooms. *Harmful Algae* **2009**, *8*, 715–725. [CrossRef]
4. Posch, T.; Köster, O.; Salcher, M.M.; Pernthaler, J. Harmful filamentous cyanobacteria favoured by reduced water turnover with lake warming. *Nat. Clim. Chang.* **2012**, *2*, 809–813. [CrossRef]
5. Chorus, I.; Bartram, J. (Eds.) *Toxic Cyanobacteria in Water: A Guide to Their Public Health Consequences, Monitoring and Management*; E & FN Spon: London, UK; New York, NY, USA, 1999; ISBN 0-419-23930-8.
6. Dietrich, D.R.; Fischer, A.; Michel, C.; Hoeger, S.J. Toxin mixture in cyanobacterial blooms—A critical comparison of reality with current procedures employed in human health risk assessment cyanobacterial metabolites: Health hazards for humans? In *Cyanobacterial Harmful Algal Blooms*; Hudnell, H.K., Ed.; Springer: New York, NY, USA, 2008; pp. 885–912.
7. Feurstein, D.; Kleinteich, J.; Heussner, A.H.A.H.; Stemmer, K.; Dietrich, D.R. Investigation of Microcystin Congener–Dependent Uptake into Primary Murine Neurons. *Environ. Health Perspect.* **2010**, *118*, 1370–1375. [CrossRef] [PubMed]
8. Feurstein, D.; Stemmer, K.; Kleinteich, J.; Speicher, T.; Dietrich, D.R. Microcystin congener- and concentration-dependent induction of murine neuron apoptosis and neurite degeneration. *Toxicol. Sci.* **2011**, *124*, 424–431. [CrossRef] [PubMed]
9. Zimmer, C. Cyanobacteria Are Far from Just Toledo's Problem. *The New York Times*, 2014. Available online: http://www.nytimes.com/2014/08/07/science/cyanobacteria-are-far-from-just-toledos-problem.html (accessed on 15 June 2016).
10. Quiblier, C.; Wood, S.; Echenique-Subiabre, I.; Heath, M.; Villeneuve, A.; Humbert, J.-F. A review of current knowledge on toxic benthic freshwater cyanobacteria—Ecology, toxin production and risk management. *Water Res.* **2013**, *47*, 5464–5479. [CrossRef]
11. Kaasalainen, U.; Fewer, D.P.; Jokela, J.; Wahlsten, M.; Sivonen, K.; Rikkinen, J. Cyanobacteria produce a high variety of hepatotoxic peptides in lichen symbiosis. *Proc. Natl. Acad. Sci. USA* **2012**, *109*, 5886–5891. [CrossRef] [PubMed]
12. Puddick, J.; Prinsep, M.; Wood, S.; Cary, S.; Hamilton, D.; Holland, P. Further Characterization of Glycine-Containing Microcystins from the McMurdo Dry Valleys of Antarctica. *Toxins* **2015**, *7*, 493–515. [CrossRef] [PubMed]
13. Vincent, W.F. Cyanobacterial Dominance in the Polar Regions. In *The Ecology of Cyanobacteria*; Whitton, B.A., Potts, M., Eds.; Springer: Dordrecht, The Netherlands, 2000; pp. 321–340. ISBN 978-0-7923-4735-4.
14. Vincent, W.F.; Quesada, A. Cyanobacteria in High Latitude Lakes, Rivers and Seas. In *Ecology of Cyanobacteria II Their Diversity in Space and Time*; Whitton, B.A., Ed.; Springer: Dordrecht, The Netherlands, 2012; pp. 371–385. ISBN 978-94-007-3855-3.
15. Hitzfeld, B.C.; Lampert, C.S.; Spaeth, N.; Mountfort, D.; Kaspar, H.; Dietrich, D.R. Toxin production in cyanobacterial mats from ponds on the McMurdo Ice Shelf, Antarctica. *Toxicon* **2000**, *38*, 1731–1748. [CrossRef]
16. Jungblut, A.; Hoeger, S.; Mountfort, D. Characterization of microcystin production in an Antarctic cyanobacterial mat community. *Toxicon* **2006**, *47*, 271–278. [CrossRef] [PubMed]

17. Wood, S.A.; Mountfort, D.; Selwood, A.I.; Holland, P.T.; Puddick, J.; Cary, S.C. Widespread distribution and identification of eight novel microcystins in Antarctic cyanobacterial mats. *Appl. Environ. Microbiol.* **2008**, *74*, 7243–7251. [CrossRef] [PubMed]
18. Kleinteich, J.; Wood, S.A.; Puddick, J.; Schleheck, D.; Küpper, F.C.; Dietrich, D. Potent toxins in Arctic environments—Presence of saxitoxins and an unusual microcystin variant in Arctic freshwater ecosystems. *Chem. Biol. Interact.* **2013**, *206*, 423–431. [CrossRef] [PubMed]
19. Kleinteich, J.; Hildebrand, F.; Wood, S.A.; Cirés, S.; Agha, R.; Quesada, A.; Pearce, D.A.; Convey, P.; Küpper, F.C.; Dietrich, D.R. Diversity of toxin and non-toxin containing cyanobacterial mats of meltwater ponds on the Antarctic Peninsula: A pyrosequencing approach. *Antarct. Sci.* **2014**, *26*, 521–532. [CrossRef]
20. Cirés, S.; Casero, M.; Quesada, A. Toxicity at the Edge of Life: A Review on Cyanobacterial Toxins from Extreme Environments. *Mar. Drugs* **2017**, *15*, 233. [CrossRef] [PubMed]
21. Meriluoto, J.; Spoof, L.; Codd, G.A. (Eds.) *Handbook of Cyanobacterial Monitoring and Cyanotoxin Analysis*; John Wiley & Sons: Chichester, UK, 2016; ISBN 9781119068761.
22. Jungblut, A.D.; Neilan, B.A. Molecular identification and evolution of the cyclic peptide hepatotoxins, microcystin and nodularin, synthetase genes in three orders of cyanobacteria. *Arch. Microbiol.* **2006**, *185*, 107–114. [CrossRef] [PubMed]
23. Fischer, A.; Hoeger, S.J.; Stemmer, K.; Feurstein, D.J.; Knobeloch, D.; Nussler, A.; Dietrich, D.R. The role of organic anion transporting polypeptides (OATPs/SLCOs) in the toxicity of different microcystin congeners in vitro: A comparison of primary human hepatocytes and OATP-transfected HEK293 cells. *Toxicol. Appl. Pharmacol.* **2010**, *245*, 9–20. [CrossRef] [PubMed]
24. De los Rios, A.; Ascaso, C.; Wierzchos, J.; Fernández-Valiente, E.; Quesada, A. Microstructural characterization of cyanobacterial mats from the McMurdo Ice Shelf, Antarctica. *Appl. Environ. Microbiol.* **2004**, *70*, 569–580. [CrossRef] [PubMed]
25. Nagata, S.; Tsutsumi, T.; Hasegawa, A.; Yoshida, F.; Ueno, Y. Enzyme Immunoassay for Direct Determination of Microcystins in Environmental Water. *J. AOAC Int.* **1997**, *80*, 408–417.
26. Rivasseau, C.; Racaud, P.; Deguin, A.; Hennion, M.C. Evaluation of an ELISA kit for the monitoring of microcystins (cyanobacterial toxins) in water and algae environmental samples. *Environ. Sci. Technol.* **1999**, *33*, 1520–1527. [CrossRef]
27. Puddick, J.; Prinsep, M.; Wood, S.; Kaufononga, S.; Cary, S.; Hamilton, D. High Levels of Structural Diversity Observed in Microcystins from *Microcystis* CAWBG11 and Characterization of Six New Microcystin Congeners. *Mar. Drugs* **2014**, *12*, 5372–5395. [CrossRef] [PubMed]
28. Puddick, J. Spectroscopic Investigations of Oligopeptides from Aquatic Cyanobacteria: Characterisation of New Oligopeptides, Development of Microcystin Quantification Tools and Investigations into Microcystin Production. Ph.D. Thesis, The University of Waikato, Hamilton, New Zealand, 2013.
29. Strunecký, O.; Komárek, J.; Johansen, J.; Lukešová, A.; Elster, J. Molecular and morphological criteria for revision of the genus *Microcoleus* (Oscillatoriales, Cyanobacteria). *J. Phycol.* **2013**, *49*, 1167–1180. [CrossRef] [PubMed]
30. Biondi, N.; Tredici, M.R.; Taton, A.; Wilmotte, A.; Hodgson, D.A.; Losi, D.; Marinelli, F. Cyanobacteria from benthic mats of Antarctic lakes as a source of new bioactivities. *J. Appl. Microbiol.* **2008**, *105*, 105–115. [CrossRef] [PubMed]
31. Wynn-Williams, D.D.; Edwards, H.G.M.; Garcia-Pichel, F. Functional biomolecules of Antarctic stromatolitic and endolithic cyanobacterial communities. *Eur. J. Phycol.* **1999**, *34*, 381–391. [CrossRef]
32. Asthana, R.K.; Tripathi, M.K.; Srivastava, A.; Singh, A.P.; Singh, S.P.; Nath, G.; Srivastava, R.; Srivastava, B.S. Isolation and identification of a new antibacterial entity from the Antarctic cyanobacterium *Nostoc* CCC 537. *J. Appl. Phycol.* **2009**, *21*, 81–88. [CrossRef]
33. Chrapusta, E.; Węgrzyn, M.; Zabaglo, K.; Kaminski, A.; Adamski, M.; Wietrzyk, P.; Bialczyk, J. Microcystins and anatoxin-a in Arctic biocrust cyanobacterial communities. *Toxicon* **2015**, *101*, 35–40. [CrossRef] [PubMed]
34. Tillett, D.; Dittmann, E.; Erhard, M.; Von Döhren, H.; Börner, T.; Neilan, B.A. Structural organization of microcystin biosynthesis in *Microcystis aeruginosa* PCC7806: An integrated peptide-polyketide synthetase system. *Chem. Biol.* **2000**, *7*, 753–764. [CrossRef]
35. Fewer, D.P.; Köykkä, M.; Halinen, K.; Jokela, J.; Lyra, C.; Sivonen, K. Culture-independent evidence for the persistent presence and genetic diversity of microcystin-producing *Anabaena* (*Cyanobacteria*) in the Gulf of Finland. *Environ. Microbiol.* **2009**, *11*, 855–866. [CrossRef] [PubMed]

36. Oksanen, I.; Jokela, J.; Fewer, D.P.; Wahlsten, M.; Rikkinen, J.; Sivonen, K. Discovery of rare and highly toxic microcystins from lichen-associated cyanobacterium *Nostoc* sp. strain IO-102-I. *Appl. Environ. Microbiol.* **2004**, *70*, 5756–5763. [CrossRef] [PubMed]
37. Solheim, B.; Endal, A.; Vigstad, H. Nitrogen fixation in Arctic vegetation and soils from Svalbard, Norway. *Polar Biol.* **1996**, *16*, 35–40. [CrossRef]
38. Tedersoo, L.; Anslan, S.; Bahram, M.; Põlme, S.; Riit, T.; Liiv, I.; Kõljalg, U.; Kisand, V.; Nilsson, H.; Hildebrand, F.; et al. Shotgun metagenomes and multiple primer pair-barcode combinations of amplicons reveal biases in metabarcoding analyses of fungi. *MycoKeys* **2015**, *10*, 1–43. [CrossRef]
39. Lee, C.K.; Herbold, C.W.; Polson, S.W.; Wommack, K.E.; Williamson, S.J.; McDonald, I.R.; Cary, S.C. Groundtruthing next-gen sequencing for microbial ecology-biases and errors in community structure estimates from PCR amplicon pyrosequencing. *PLoS ONE* **2012**, *7*, e44224. [CrossRef] [PubMed]
40. Pessi, I.S.; de Maalouf, P.C.; Laughinghouse, H.D.; Baurain, D.; Wilmotte, A. On the use of high-throughput sequencing for the study of cyanobacterial diversity in Antarctic aquatic mats. *J. Phycol.* **2016**, *52*, 356–368. [CrossRef] [PubMed]
41. Kleinteich, J.; Hildebrand, F.; Bahram, M.; Voigt, A.Y.; Wood, S.A.; Jungblut, A.D.; Küpper, F.C.; Quesada, A.; Camacho, A.; Pearce, D.A.; et al. Pole-to-Pole Connections: Similarities between Arctic and Antarctic Microbiomes and Their Vulnerability to Environmental Change. *Front. Ecol. Evol.* **2017**, *5*, 137. [CrossRef]
42. Jungblut, A.D.; Lovejoy, C.; Vincent, W.F. Global distribution of cyanobacterial ecotypes in the cold biosphere. *ISME J.* **2010**, *4*, 1–12. [CrossRef] [PubMed]
43. Kleinteich, J.; Wood, S.A.; Küpper, F.C.; Camacho, A.; Quesada, A.; Frickey, T.; Dietrich, D.R. Temperature-related changes in polar cyanobacterial mat diversity and toxin production. *Nat. Clim. Chang.* **2012**, *2*, 356–360. [CrossRef]
44. Fraser, C.I.; Terauds, A.; Smellie, J.; Convey, P.; Chown, S.L. Geothermal activity helps life survive glacial cycles. *Proc. Natl. Acad. Sci. USA* **2014**, *111*, 5634–5639. [CrossRef] [PubMed]
45. Stocker, T.F.; Qin, D.; Plattner, G.-K.; Tignor, M.; Allen, S.K.; Boschung, J.; Nauels, A.; Xia, Y.; Bex, V.; Midgley, P.M. *IPCC, 2013: Climate Change 2013: The Physical Science Basis*; Contribution of Working Group I to the Fifth Assessment Report of the Intergovernmental Panel on Climate Change; IPCC: Cambridge, UK; New York, NY, USA, 2013.
46. Niedermeyer, T.; Strohalm, M. mMass as a software tool for the annotation of cyclic peptide tandem mass spectra. *PLoS ONE* **2012**, *7*, e44913. [CrossRef] [PubMed]
47. Hummert, C.; Dahlmann, J.; Reinhardt, K.; Dang, H.P.H.; Dang, D.K.; Luckas, B. Liquid chromatography-mass spectrometry identification of microcystins in *Microcystis aeruginosa* strain from lake Thanh Cong, Hanoi, Vietnam. *Chromatographia* **2001**, *54*, 569–575. [CrossRef]
48. Welker, M.; Christiansen, G.; von Döhren, H. Diversity of coexisting *Planktothrix* (Cyanobacteria) chemotypes deduced by mass spectral analysis of microcystins and other oligopeptides. *Arch. Microbiol.* **2004**, *182*, 288–298. [CrossRef] [PubMed]
49. Miles, C.O.; Sandvik, M.; Haande, S.; Nonga, H.; Ballot, A. LC-MS Analysis with Thiol Derivatization to Differentiate [Dhb^7]- from [Mdha^7]-Microcystins: Analysis of Cyanobacterial Blooms, *Planktothrix* Cultures and European Crayfish from Lake Steinsfjorden, Norway. *Environ. Sci. Technol.* **2013**, *47*, 4080–4087. [CrossRef] [PubMed]
50. Puddick, J.; Prinsep, M.; Wood, S.; Miles, C.; Rise, F.; Cary, S.; Hamilton, D.; Wilkins, A. Structural Characterization of New Microcystins Containing Tryptophan and Oxidized Tryptophan Residues. *Mar. Drugs* **2013**, *11*, 3025–3045. [CrossRef] [PubMed]
51. Cirés, S.; Wörmer, L.; Ballot, A.; Agha, R.; Wiedner, C.; Velázquez, D.; Casero, M.C.; Quesada, A. Phylogeography of cylindrospermopsin and paralytic shellfish toxin-producing Nostocales cyanobacteria from Mediterranean Europe (Spain). *Appl. Environ. Microbiol.* **2014**, *80*, 1359–1370. [CrossRef] [PubMed]
52. Nübel, U.; Garcia-Pichel, F.; Muyzer, G. PCR primers to amplify 16S rRNA genes from cyanobacteria. *Appl. Environ. Microbiol.* **1997**, *63*, 3327–3332. [PubMed]
53. Weisburg, W.G.; Barns, S.M.; Pelletier, D.A.; Lane, D.J. 16S ribosomal DNA amplification for phylogenetic study. *J. Bacteriol.* **1991**, *173*, 697–703. [CrossRef] [PubMed]
54. Hildebrand, F.; Tadeo, R.; Voigt, A.Y.; Bork, P.; Raes, J. LotuS: An efficient and user-friendly OTU processing pipeline. *Microbiome* **2014**, *2*, 30. [CrossRef] [PubMed]

55. Edgar, R.C. UPARSE: Highly accurate OTU sequences from microbial amplicon reads. *Nat. Methods* **2013**, *10*, 996–998. [CrossRef] [PubMed]
56. Edgar, R.C.; Haas, B.J.; Clemente, J.C.; Quince, C.; Knight, R. UCHIME improves sensitivity and speed of chimera detection. *Bioinformatics* **2011**, *27*, 2194–2200. [CrossRef] [PubMed]
57. Magoč, T.; Salzberg, S.L. FLASH: Fast length adjustment of short reads to improve genome assemblies. *Bioinformatics* **2011**, *27*, 2957–2963. [CrossRef] [PubMed]
58. Wang, Q.; Garrity, G.M.; Tiedje, J.M.; Cole, J.R. Naive Bayesian Classifier for Rapid Assignment of rRNA Sequences into the New Bacterial Taxonomy. *Appl. Environ. Microbiol.* **2007**, *73*, 5261–5267. [CrossRef] [PubMed]
59. Hildebrand, F.; Ebersbach, T.; Nielsen, H.B.; Li, X.; Sonne, S.B.; Bertalan, M.; Dimitrov, P.; Madsen, L.; Qin, J.; Wang, J.; et al. A comparative analysis of the intestinal metagenomes present in guinea pigs (*Cavia porcellus*) and humans (*Homo sapiens*). *BMC Genom.* **2012**, *13*, 514. [CrossRef] [PubMed]
60. Saary, P.; Forslund, K.; Bork, P.; Hildebrand, F. RTK: Efficient rarefaction analysis of large datasets. *Bioinformatics* **2017**, *33*, 2594–2595. [CrossRef] [PubMed]

toxins

MDPI

Article

Co-Occurrence of Microcystins and Taste-and-Odor Compounds in Drinking Water Source and Their Removal in a Full-Scale Drinking Water Treatment Plant

Lixia Shang [1,2], Muhua Feng [1,*], Xiangen Xu [1,3], Feifei Liu [1,4], Fan Ke [1] and Wenchao Li [1]

[1] State Key Laboratory of Lake Science and Environment, Nanjing Institute of Geography and Limnology, Chinese Academy of Sciences, Nanjing 210008, China; lxshangsun@163.com (L.S.); xgxuc@163.com (X.X.); feifei1845@126.com (F.L.); fke@niglas.ac.cn (F.K.); wchli@niglas.ac.cn (W.L.)
[2] Key Laboratory of Marine Ecology and Environmental Sciences, Institute of Oceanology, Chinese Academy of Sciences, Qingdao 266071, China
[3] Changzhou Academy of Environmental Science, Changzhou 213022, China
[4] Nanjing Municipal Design and Research Institute CO., Ltd., Nanjing 210008, China
* Correspondence: mhfeng@niglas.ac.cn; Tel.: +86-258-688-2206

Received: 1 December 2017; Accepted: 28 December 2017; Published: 2 January 2018

Abstract: The co-occurrence of cyanotoxins and taste-and-odor compounds are a growing concern for drinking water treatment plants (DWTPs) suffering cyanobacteria in water resources. The dissolved and cell-bound forms of three microcystin (MC) congeners (MC-LR, MC-RR and MC-YR) and four taste-and-odor compounds (geosmin, 2-methyl isoborneol, β-cyclocitral and β-ionone) were investigated monthly from August 2011 to July 2012 in the eastern drinking water source of Lake Chaohu. The total concentrations of microcystins and taste-and-odor compounds reached 8.86 μg/L and 250.7 ng/L, respectively. The seasonal trends of microcystins were not consistent with those of the taste-and-odor compounds, which were accompanied by dominant species *Microcystis* and *Dolichospermum*. The fate of the cyanobacteria and metabolites were determined simultaneously after the processes of coagulation/flocculation, sedimentation, filtration and chlorination in the associated full-scale DWTP. The dissolved fractions with elevated concentrations were detected after some steps and the breakthrough of cyanobacteria and metabolites were even observed in finished water. Chlorophyll-*a* limits at intake were established for the drinking water source based on our investigation of multiple metabolites, seasonal variations and their elimination rates in the DWTP. Not only microcystins but also taste-and-odor compounds should be taken into account to guide the management in source water and in DWTPs.

Keywords: microcystins; taste-and-odor compounds; water source; drinking water treatment plant; cyanobacterial thresholds

Key Contribution: The co-occurrence of microcystins and taste-and-odor compounds accompanied by dominant species *Microcystis* and *Dolichospermum* were studied in a drinking water source and in the associated full-scale DWTP. Not only microcystins but also taste-and-odor compounds were taken into account to guide the management in source water and in DWTPs.

1. Introduction

Cyanobacteria are well known for their ability to produce diverse secondary metabolites including microcystins (MCs) and taste-and-odor (T&O) compounds [1–3]. Due to the acute and potentially chronic effects of MCs, the World Health Organization [4] issued a guideline value of 1.0 μg/L for microcystin-LR (MC-LR) in drinking water. T&O compounds include geosmin (GEO) and 2-methyl isoborneol (MIB) with an earthy and musty odor, respectively, and β-cyclocitral (CYC) and β-ionone (ION) with a tobacco and violet odor, respectively [3]. These compounds are the primary barriers to drinking water safety. Due to extremely low odor threshold concentrations (OTCs) with 4, 15, 19.3 and 7 ng/L for GEO, MIB, CYC and ION, respectively [3], T&O compounds even at low levels in water are a source of complaints by consumers. Although the effects of T&O compounds on human health are still unclear, they are the primary criteria of drinking water safety considered by consumers [3], which pose challenges for drinking water treatment plant (DWTP) management.

The occurrence and seasonal dynamics of MCs or T&O compounds have been well studied in surface waters worldwide [5–8]. Recently, some researchers began to focus upon the co-occurrence of cyanotoxins and T&O compounds in lakes, reservoirs and rivers. During the 2007 summer's odorous tap water crisis in Wuxi (China), an MC-LR of 7.59 μg/L was detected in the source water at the intake [9]. Graham et al. [10] concluded that microcystin co-occurred with geosmin in 87% of the cynobacterial blooms and MIB in 39% sampled from 23 Midwestern United States lakes. However, seasonal dynamics of MC and T&O compounds may not be so uniform due to their different producers, physicochemical properties, metabolic pathways, and environmental influencing factors [7,11]. Conflicting results in the relationship between MCs and T&O compounds with positive or no correlations were obtained in different waters [10,12], which would complicate their removal by treatment processes in DWTPs.

Scientific literature has reported that conventional water treatment techniques have been ineffective in removing MCs and T&O compounds from water [13,14]. However, most previous studies focused on the removal of algal cells and their associated metabolites only in laboratory experiments or along treatment trains, which did not always conform to reality due to a range of uncertainties in actual situations [13,15,16]. The scant literature involving cyanobacterial metabolites removal by pilot- or full-scale treatment plants present conflicting results where some plants were said to have removed MCs below the guideline of 1.0 μg/L [17], but, in another case, a MC concentration of 2.47 μg/L was detected in the drinking water [18]. Therefore, it was necessary to investigate the fate of cyanobacteria and their metabolites under typical plant operation conditions to assess the potential removal. Moreover, no studies to date have estimated the efficiency of full-scale DWTPs in removing the mixtures of MCs and T&O compounds co-occurring in the same water source, which is needed to comprehensively assess and minimize the hazards caused by cyanobacteria and their metabolites.

To estimate the safety of raw water with cyanobacteria, the Alert Levels Framework (ALF) has been established to define threshold values according to the MC-LR cell quota and the WHO guideline value of MC-LR in drinking water [2]. More than 100 MC variants have been characterized from bloom samples and isolated strains of cyanobacteria [1]; thus, the safety threshold for cyanobacteria was established based on different MC congeners at different cyanobacterial growth periods [19]. To avoid overestimating the risk of cyanobacteria in drinking source water, maximum tolerable (MT) values were calculated involving the toxin removal performance of real treatment trains by Schmidt et al. [20]. In this study, we improved these previous methods by taking account of multiple metabolites with seasonal variations and removal rates of plants and thereby pave the way for a refined management strategy designed to solve the cyanobacterial threat in DWTPs.

The objectives of the present study were to: (1) evaluate the seasonal variations of the co-occurrence of three MC congeners and four T&O compounds associated with two cyanobacterial dominant species, *Microcystis* and *Dolichospermum*, in the eastern drinking water source (EDWS) of Lake Chaohu; (2) assess the fate and elimination of cyanobacteria and the studied metabolites at each treatment step in the full-scale DWTP; and (3) establish the chlorophyll-*a* limits at intake (CLIs) for the drinking water source involving different dominant cyanobacterial species producing more than one kind of metabolites and their removal rates in the DWTP during different seasons. This research is expected to help mitigate the problems associated with cyanobacteria and their metabolites in DWTPs, which will naturally lead to safer drinking water. To the best of our knowledge, this is the first study to report the elimination of the mixtures of MCs and T&O compounds in a full-scale DWTP.

2. Results

2.1. Seasonal Variations of Cyanobacteria, MCs and T&O Compounds in the EDWS

In the EDWS of Lake Chaohu, cyanobacteria and their associated metabolites including MCs and T&O compounds were studied monthly from August 2011 to July 2012 (Table 1). The chlorophyll-*a* concentrations ranged from 5.4 to 68.0 μg/L with the lowest and highest average concentrations detected in the spring (9.0 ± 3.0 μg/L) and in the summer (38.5 ± 14.0 μg/L), respectively. The dominant phytoplankton was cyanobacteria with *Microcystis* and *Dolichospermum* making up more than 98% of the cyanobacterial abundance except in the spring (85%). Cyanobacterial density was lowest in the spring with only *Dolichospermum* detected in most samples. From summer to the middle of autumn, the *Microcystis* abundance increased considerably as the dominant species and the cyanobacteria reached a peak density of 195,000 cells/mL simultaneously. In late autumn and winter, the amount of filamentous species belonging to the genera *Dolichospermum* increased gradually and again became the dominant species with the maximum number of cyanobacteria (200,000 cells/mL) measured.

Co-occurrence of three MC congeners and four T&O compounds with different seasonal trends have been observed in this water source (Table 1). The total concentrations of the three studied MC congeners (MC-LR, MC-RR and MC-YR) ranged from 0.28 to 8.86 μg/L. In the spring, the average concentration of each MC congener was at its lowest value corresponding to the lowest cyanobacterial density of the whole year. In summer and autumn, MCs occurred at higher concentrations with *Microcystis* being dominant. The MC-LR concentration reached its maximum amount in the summer with average values of 2.80 and 3.22 μg/L for EMC-LR and IMC-LR, respectively. By comparison, the highest concentrations of MC-RR (3.73 μg/L) and MC-YR (3.58 μg/L) were both detected in autumn leading to the peak concentration of the total MCs during the study period, while the concentrations of the three other MC variants remained at high levels with a sum of 4.11 μg/L in winter. The concentrations of MCs were detected above the WHO [4] recommended guideline value in summer and autumn, and even maintained a high level of nearly 1.0 μg/L in winter.

Compared to MCs, the four studied T&O compounds showed more complex seasonal dynamics in the source water (Table 1). Furthermore, the seasonal variations of dissolved T&O compounds were inconsistent with those of particle compounds. In the spring, the maximum concentrations of *d*-GEO, *d*-CYC and *d*-ION were obtained at the lowest cyanobacterial density. In the summer, the maximum concentration of MIB was detected at the highest concentrations of both *d*-MIB (193.4 ng/L) and *p*-MIB (34.0 ng/L). Additionally, higher levels of *d*-CYC and *p*-CYC were observed at concentrations around 50 ng/L, while the peak concentration of *p*-ION (152.7 ng/L) was also recorded. In autumn, relatively low concentrations of total T&O compounds were observed but also exceeded their OTCs in some samples. In winter, the maximum concentration of CYC (125.0 ng/L) was observed with the highest *p*-CYC concentration (106.3 ng/L) detected, while the concentration of *p*-ION was up to 115.8 ng/L. The four studied T&O compounds could exceed their OTCs all the year round. During the study period, 69%, 39%, 72% and 86% of the samples for GEO, MIB, CYC and ION exceeded their corresponding OTCs, respectively.

Table 1. Mean and ranges of chlorophyll-*a*, phytoplankton density, extra- (E) and intra-cellular (I) microcystins and dissolved (*d*-) and particle-bound (*p*-) taste-and-odor compounds in the eastern drinking water source of Lake Chaohu (including Intake-A, B and C) during four seasons.

	Spring	**Summer**	**Autumn**	**Winter**
Chlorophyll-*a* (μg/L)	9.0 (5.4–13.7)	38.5 (22.2–68.0)	22.2 (8.0–55.7)	24.6 (5.7–47.9)
Phytoplankton density ($\times 10^4$ cells/mL)	1.7 (0.5–4.0)	19.1 (18.0–20.2)	12.7 (5.7–17.2)	12.2 (7.0–22.1)
Cyanobacterial density ($\times 10^4$ cells/mL)	1.5 (0.3–3.9)	18.5 (18.6–19.5)	12.5 (5.4–17.0)	12.0 (6.7–21.9)
Microcystis density ($\times 10^4$ cells/mL)	0.1 (*n.d.*–0.3)	11.6 (2.8–16.5)	8.8 (3.0–13.1)	2.3 (0.9–4.8)
Dolichospermum density ($\times 10^4$ cells/mL)	1.4 (0.2–3.8)	2.8 (0.6–5.2)	1.6 (1.1–2.2)	9.4 (5.7–15.5)
EMC-LR (μg/L)	0.05 (0.03–0.06)	0.98 (0.07–2.80)	0.65 (0.07–1.04)	0.14 (0.02–0.55)
IMC-LR (μg/L)	0.04 (0.02–0.08)	1.15 (0.12–3.22)	0.11 (0.03–0.18)	0.36 (0.04–1.31)
EMC-RR (μg/L)	0.12 (0.06–0.17)	0.31 (0.21–0.59)	1.42 (0.08–2.45)	0.12 (0.06–0.22)
IMC-RR (μg/L)	0.07 (0.05–0.11)	0.16 (0.11–0.25)	0.56 (0.03–1.47)	0.41 (0.10–1.20)
EMC-YR (μg/L)	0.07 (*n.d.*–0.12)	0.18 (0.11–0.27)	1.34 (0.21–2.83)	0.46 (*n.d.*–0.74)
IMC-YR (μg/L)	0.01 (*n.d.*–0.02)	0.37 (0.02–1.07)	0.82 (0.05–1.35)	0.11 (0.01–0.31)
d-geosmin (ng/L)	16.8 (1.6–47.8)	8.1 (4.8–12.1)	3.7 (1.3–7.2)	7.7 (*n.d.*–17.2)
p-geosmin (ng/L)	0.7 (0.2–1.7)	0.6 (0.2–1.1)	*n.d.*	0.4 (*n.d.*–0.9)
d-2-methyl isoborneol (ng/L)	11.1 (4.5–30.7)	60.2 (*n.d.*–193.4)	14.1 (0.7–27.5)	5.4 (*n.d.*–11.8)
p-2-methyl isoborneol (ng/L)	1.3 (0.4–2.0)	12.6 (0.6–4.0)	0.5 (*n.d.*–1.3)	0.5 (*n.d.*–1.7)
d-*β*-cyclocitral (ng/L)	26.5 (8.0–50.1)	21.9 (8.4–49.2)	14.9 (8.2–33.8)	10.6 (*n.d.*–20.0)
p-*β*-cyclocitral (ng/L)	4.7 (1.0–10.9)	26.3 (5.4–53.4)	4.0 (0.9–8.7)	42.5 (5.4–106.3)
d-*β*-ionone (ng/L)	5.8 (0.4–28.4)	10.8 (3.3–23.2)	7.3 (*n.d.*–21.8)	0.4 (*n.d.*–1.1)
p-*β*-ionone (ng/L)	14.5 (1.8–30.1)	53.4 (2.6–152.7)	4.8 (0.7–11.6)	47.1 (3.5–115.8)

n.d.: below the lower-limit of the calibration range.

2.2. Relationships among Cyanobacteria, MCs and T&O Compounds in the EDWS

Redundancy analysis was used to investigate the relationships among cyanobacteria, MCs and T&O compounds in the eastern drinking water source of Lake Chaohu (Figure 1). No relationship was found between the two predominant species *Microcystis* and *Dolichospermum*, while the density of *Microcystis* but not *Dolichospermum* was positively correlated with cyanobacterial density. Cyanobacterial density together with the *Microcystis* and *Dolichospermum* density correlated positively with the Chl-*a* concentrations. Meanwhile, the concentrations of both the MCs and T&O compounds except GEO yielded significant positive correlations with that of Chl-*a*. For these studied metabolites, strong relativities in each MC congener and MIB and positive correlations between ION and CYC concentrations were demonstrated, respectively. The concentrations of GEO had no relationship with that of ION and CYC, but correlated negatively with the rest metabolites.

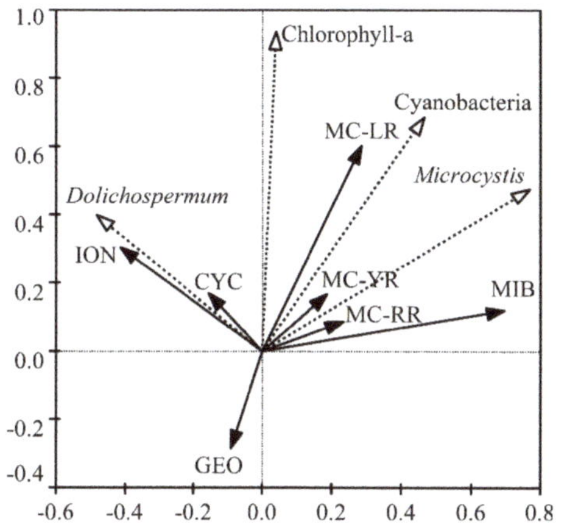

Figure 1. Relationships analyzed by redundancy analysis among cyanobacterial density and concentrations of microcystins and taste-and-odor compounds in the eastern drinking water source of Lake Chaohu (Intake-A, -B and -C). GEO: geosmin; MIB: 2-methyl isoborneol; CYC: β-cyclocitral; and ION: β-ionone.

2.3. The Removal of Chlorophyll-a, MCs and T&O Compounds via Treatment Processes in the DWTP

2.3.1. Coagulation/Flocculation (C/F+)

The concentrations of cyanobacteria and the associated metabolites in raw water sampled from the DWTP showed no significant differences from that sampled at Intake-A in the studied EDWS ($p > 0.05$). As the first step to remove algae and metabolites, the Coagulation/Flocculation effects must be assessed to ensure minimal impacts on the subsequent processes and improve total removal efficiencies of DWTP. In this study, the elimination efficiency of chlorophyll-*a* from raw water to the C/F+ (+, with 9 mg/L potassium permanganate ($KMnO_4$) and 10 mg/L powdered activated carbon (PAC) in summer and autumn with contact time of 30 min) process ranged from 24.6 to 80.9% (45.6% average) (Figure 2). The removal rates of intracellular MCs and T&O compounds ranged from 9.7 to 89.5% (43.8% average) (Figures 3 and 4). The maximum removal rates of chlorophyll-*a* and the cell-bound metabolites by the C/F+ process were both observed in summer. The elimination rates of the dissolved MCs and T&O compounds were lower than 45.7%, and the concentrations of some dissolved metabolites even increased after the C/F+ step. The concentrations of EMC-LR, EMC-RR and *d*-ION increased by 55.2%, 42.8% and 45.3% (on average) in winter, respectively, while those of *d*-CYC and *d*-ION increased by 175.1% and 210.7% in summer, respectively.

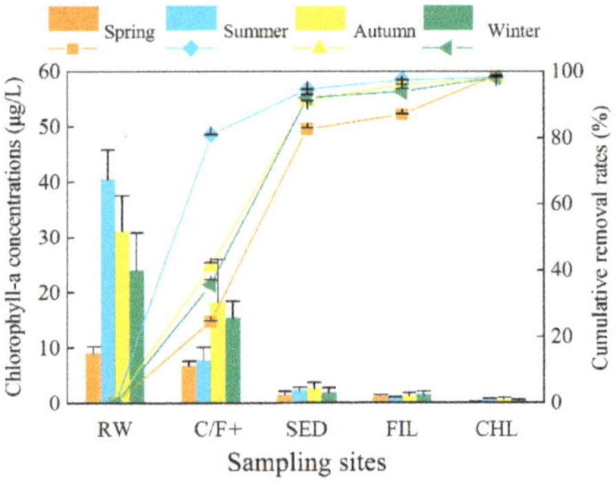

Figure 2. The concentrations of chlorophyll-*a* (left, column) and cumulative removal rates (right, line + scatter) in different seasons of the water treatment processes including raw water (RW), coagulation-flocculation (C/F+, with potassium permanganate and powdered activated carbon in summer and autumn), sedimentation (SED), filtration (FIL) and chlorination (CHL) in the drinking water treatment plant associated with Intake-A. Error bars indicate the standard deviation of data in three months of each season.

2.3.2. Sedimentation (SED)

The sedimentation step in DWTP is used to settle the particles including cyanobacterial flocs in suspension into the sediment and out of the fluid. More than 90% of the chlorophyll-*a* were further removed by sedimentation except in the spring (82.7%), which made up 13.8–58.8% of total removal efficiency in the studied DWTP. Meanwhile, 50.4–97.8% of the three IMC congeners, *p*-CYC and *p*-ION were further removed by the sedimentation step. However, the concentrations increased after sedimentation for *p*-GEO in the spring and autumn and *p*-MIB in the spring. For dissolved metabolites, the sedimentation efficiencies were less than 30.2% compared to the concentrations in the C/F+ effluent. Furthermore, the concentrations of EMC-LR, EMC-RR, *d*-GEO and *d*-CYC increased by 55.7%, 13.0%, 33.1% and 201.7% compared to those in the C/F+ effluent in winter, respectively. In the spring, summer and autumn, the dissolved metabolites in 38.1% of the studied samples increased in a range of 1.6–9.9% compared to those metabolites in the effluent of C/F+ step.

2.3.3. Sand Filtration (FIL)

Sand filtration is primarily used to remove the excess flocs after sedimentation. In this step, the elimination of chlorophyll-*a* represented less than 5% of the total removal rate of the whole processes in the DWTP. In filtered water, the average concentration of chlorophyll-*a* was 1.2 ± 0.2 μg/L. Compared to the concentrations in the sedimentation effluent, 5.7–63.6% (32.9% on average) of the cell-bound metabolites were reduced further by filtration. With an average removal rate of 14.0%, the concentrations of extracellular metabolites increased in 39.2% of the samples after filtration compared to that of in the sedimentation effluent, where summertime rates for EMC-YR and *d*-MIB increased by 84.5% and 28.2%, respectively.

2.3.4. Chlorination (CHL)

Post-treatment chlorination provides the last defense against cyanobacterial cells, MCs, and T&O compounds reaching consumers in water treated by the DWTP. In this study, the chlorophyll-*a* removal rates by chlorination were between 30.1% and 90.2% compared to the concentrations in the filtration effluent. After chlorination, the maximum concentration of each intracellular MC and T&O compound was 0.03 μg/L (IMC-LR in the summer) and 1.9 ng/L (*p*-CYC in the spring), respectively. For extracellular MCs, the chlorination process further reduced the concentrations from 15.7 to 74.2%

compared to those in the filtration effluent. However, the concentration of each T&O compound increased from 31.0 to 266.7% after chlorination except for *d*-GEO in the spring and autumn and *d*-MIB in the summer and autumn, respectively.

In the finished water, the average concentrations of chlorophyll-*a* for each season were between 0.1 ± 0.0 μg/L and 0.7 ± 0.1 μg/L. The maximum concentrations of MC-LR, MC-RR and MC-YR in the finished water were 0.4 ± 0.3 μg/L (in the summer), 0.7 ± 0.3 μg/L (in the autumn) and 0.9 ± 0.3 μg/L (in the autumn), respectively. The concentrations of T&O compounds varied within a wider range compared to those of MCs in the drinking water. The concentrations of GEO above its OTC of 4 ng/L were detected in the summer (8.4 ± 1.6 ng/L) and winter (8.7 ± 3.5 ng/L). The CYC exceeded its OTC throughout the whole year with concentrations ranging from 19.5 ± 2.4 ng/L to 67.2 ± 13.8 ng/L. In the summer, the concentrations of MIB and ION were 52.5 ± 30.5 ng/L and 12.2 ± 3.3 ng/L, respectively. Both readings exceeded their corresponding OTC standard.

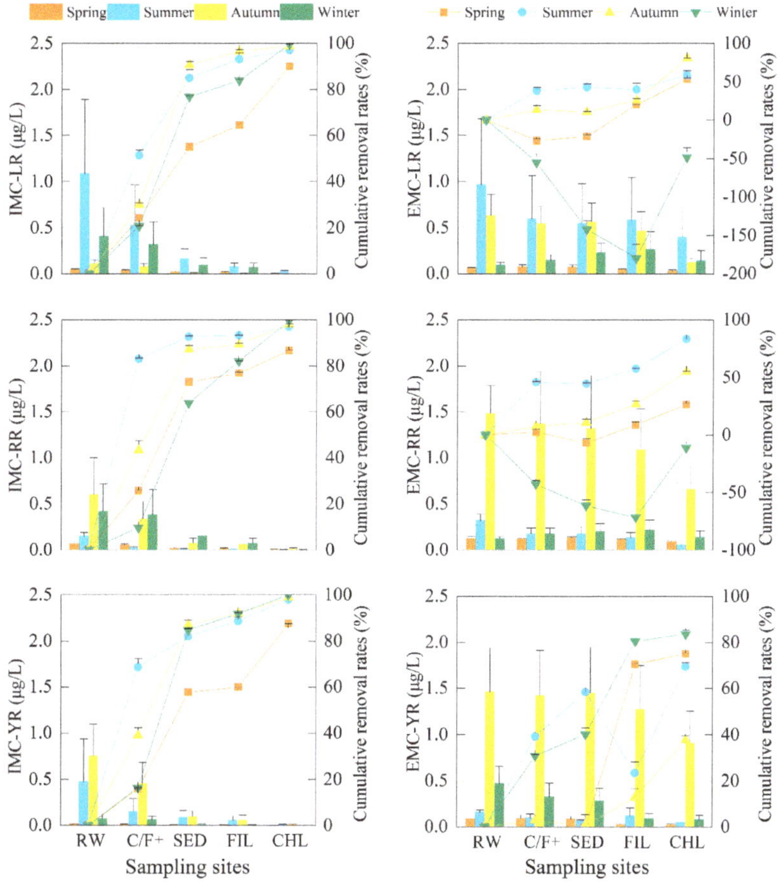

Figure 3. The concentrations of extra- (E) and intra-cellular (I) microcystins (MC-LR, MC-RR and MC-YR) (column) and cumulative removal rates (line + scatter) in different seasons of the water treatment processes including raw water (RW), coagulation-flocculation (C/F+, with potassium permanganate and powdered activated carbon in summer and autumn), sedimentation (SED), filtration (FIL) and chlorination (CHL) in the drinking water treatment plant associated with Intake-A. Error bars indicate the standard deviation of data in three months of each season.

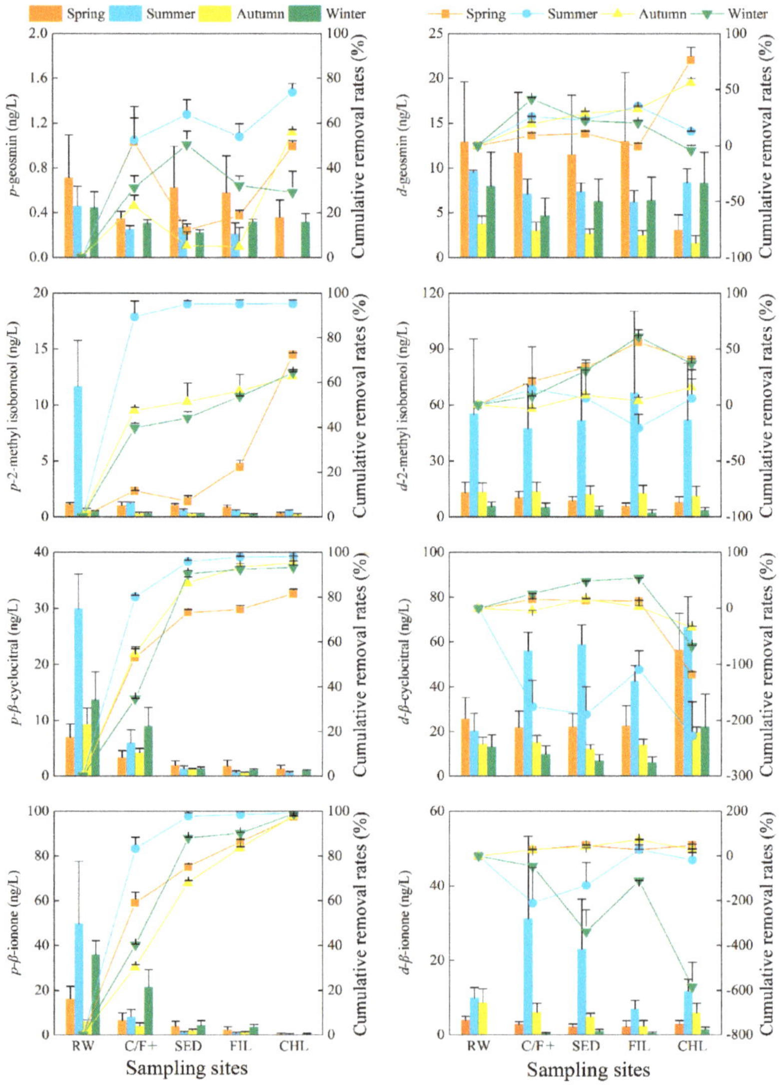

Figure 4. The concentrations of dissolved (*d*-) and particle-bound (*p*-) taste-and-odor compounds (column) and cumulative removal rates (line + scatter) in different seasons of the water treatment processes including raw water (RW), coagulation-flocculation (C/F+, with potassium permanganate and powdered activated carbon in summer and autumn), sedimentation (SED), filtration (FIL) and chlorination (CHL) in the drinking water treatment plant associated with Intake-A. Error bars indicate the standard deviation of data in three months of each season.

2.3.5. Total Removal Rates of Chlorophyll-*a* and the Metabolites

The seasonal average of removal efficiencies of chlorophyll-*a* and the studied metabolites are shown in Figure 5. On average, more than 98% of chlorophyll-*a* was removed from water. The total removal rates of the MC-LR, MC-RR and MC-YR varied from 47.9 to 90.9% with an average of 74.5%. For T&O compounds, the average removal rates during the whole year for GEO and MIB were 36.0%

and 30.2%, respectively. For CYC, the removal efficiencies of 16.3% and 14.8% were detected in the autumn and winter, respectively. However, the concentrations of CYC were increased by 75.7% and 33.7% in the finished water compared to the concentrations in the raw water in the spring and summer, respectively. The ION removal rates were between 58.4% and 94.9% with an average of 80.1% during the whole year.

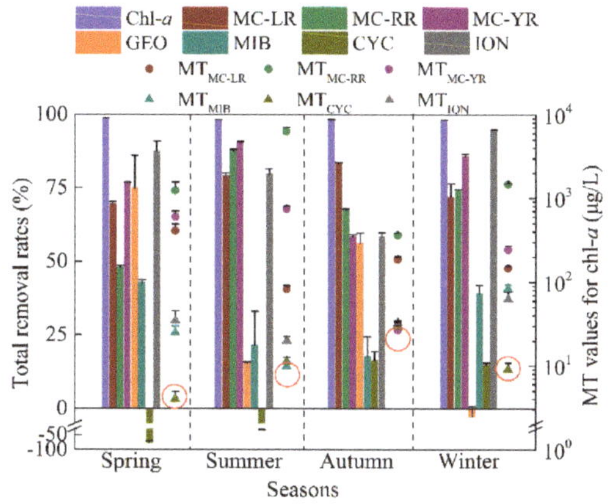

Figure 5. The removal rates of chlorophyll-*a* and the microcystins and taste-and-odor compounds in the drinking water treatment plant associated with Intake-A (left) and the maximum tolerable (MT) values (right) based on MC-LR, MC-RR, MC-YR, 2-methyl isoborneol (MIB), β-cyclocitral (CYC) and β-ionone (ION); the establishment of chlorophyll-*a* (Chl-*a*) limits at intake (CLIs) in the four seasons in the eastern drinking water source of Lake Chaohu (red circle). Error bars indicate the standard deviation of total removal rates in different months of each season and the MT values for Chl-*a* of three sampling sites.

3. Discussion

3.1. The Co-Occurrence of MCs and T&O Compounds Relating to Different Dominant Cyanobacterial Species in the EDWS

In the EDWS of Lake Chaohu, the co-occurrence of MCs and T&O compounds caused adverse effects on the safety of water source. Previously, different conclusions were drawn involving the relationships between MCs and T&O compounds. For particulate T&O compounds, significant correlations between *p*-CYC and *p*-ION with MCs were reported in Lake Taihu [12]. However, no correlation between toxins and T&O compounds concentrations was found in the survey involving 23 Midwestern United States lakes [10]. When two cyanobacterial species, *Microcystis* and *Dolichospermum*, dominated in different seasons, dynamic changes relating to these metabolites became vastly more complicated compared to the water source with the sole dominant species. In the present study, strong relativities in each MC congener and MIB were found and positive correlations between ION and CYC were demonstrated (Figure 1). The GEO had no relationship with ION and CYC, but correlated negatively with the remaining metabolites. The mutable relativities among the three MCs congeners and four T&O compounds could be explained by the community dynamics of synthesizing cyanobacterial strains, different metabolic pathways, and the different physicochemical properties such as volatility and degradability.

The DWTPs usually reinforce treatment technology operation during cyanobacterial blooming at warmer temperatures instead of operating within a dedicated system year round from the perspective of cost savings [16]. However, in the drinking water sources associated with more than one dominant cyanobacterial species such as those found in this study, the concentrations of MCs and T&O

compounds still maintained at high levels in spring and winter and even exceeded the guidelines during the whole year. The differences in the characteristics and seasonal variations of the two classified compounds (MCs and T&O) indicated the high risk involved in producing safe drinking water, which has led to difficult and complex management for the drinking water source and the DWTP.

3.2. The Fate of Chlorophyll-a, MCs and T&O Compounds in the Associated DWTP

In the studied DWTP, the removal efficiencies of the cyanobacteria and associated metabolites varied with seasons and depended on several factors including physicochemical characteristics of metabolites, community dynamics of dominating cyanobacterial species and the mode of operation in the DWTP.

As the primary step to remove pollutants, the C/F+ process gained more effective removal in summer and autumn for chlorophyll-*a* and the cell-bound metabolites compared to that in spring and winter. According to the coagulation mechanism of charge neutralization [13], spherical aggregates of *Microcystis*, which are dominate in summer and autumn, are easier to remove compared to the filamentous *Dolichospermum*, which is dominate in spring and winter. Fan et al. [15] found that the use of $KMnO_4$ in summer and autumn could improve algae coagulation and the removal of cell-bound metabolites by reducing cell stability and inducing cell decay. Ghernaout et al. [13] found the C/F+ step to be inefficient for removal of soluble organics in water, which was further demonstrated by the low removal rates of the dissolved cyanobacterial metabolites in the present study. Even more, increased concentrations of dissolved metabolites after the C/F+ process were measured in half of the samples compared to their concentrations in the raw water. In the winter, increasing concentrations of EMC-LR, EMC-RR and *d*-ION were observed with their cell-bound forms dominating the total concentrations. Therefore, cells lysis and metabolite release occurred due to shear stress in the C/F step aiming to quickly and uniformly disperse coagulants and form flocs [13,21]. However, only the concentrations of *d*-CYC and *d*-ION increased dramatically during the summer, although the intracellular concentrations of MCs and T&O compounds both had high concentrations. This indicated that the release rates of intracellular CYC and ION from damaged cells were faster than that of degradation in the C/F process or by chemicals such as $KMnO_4$ dosed in the reaction tank. The inconsistent performance of MCs with T&O might be because of the differences in the two classes of metabolites in molecular structure, physicochemical characteristics, and release and degradation rates such as the more volatile components of the odorous compounds.

Most suspended particles and flocs with algal cells formed in coagulation can be removed by sedimentation through gravity precipitation. The elimination efficiencies of the chlorophyll-*a* and the cell-bound metabolites except for *p*-GEO were more than 81% during the treatment train and most of them were removed by the C/F+ and sedimentation steps, which were in accordance with the data (62.0–98.9%) summarized by Zamyadi et al. [18]. However, a rising level in dissolved fraction after sedimentation was observed for both MCs and T&O compounds during the whole year especially in winter with no further treatment used. The sludge in the sedimentation tank became a potential source of organic matter as a storage deposit of flocs with treated cyanobacterial cells after the C/F process [22]. Moreover, GEO and MIB could also be produced by benthic microorganisms growing in the sludge bed other than phytoplankton [23]. Therefore, the sludge with cyanobacteria cells from the sedimentation process should be removed rapidly and safely disposed of to prevent additional hazards caused by cyanobacterial cell lysis.

Cell breakthroughs were observed after filtration in the studied DWTP, with an average chlorophyll-*a* concentration of around 1.2 μg/L in the filtered water. In addition, an increase of extracellular metabolites was observed, such as EMC-YR and *d*-MIB in summer. The cyanobacteria accumulated on the filters and biofilm fragments or other impurities attached to filters might cause subsequent organics to be released into the treatment plant water [21]. Hence, it is necessary to increase backwash time and frequency to improve the filtration efficiency in the removal of cyanobacteria and the associated metabolites.

The chlorination process is generally regarded as a step to reduce organic matter, which can breed harmful bacteria once released into the water distribution network. Incomplete removal of cyanobacterial metabolites by the preceding treatment processes requires more chlorine to guarantee enough residual chlorine in the water distribution network. Serious safety problems could occur if disinfection byproducts (DBPs) are formed in drinking water. Although the chlorination step was effective to remove three MCs congeners, the concentrations of T&O compounds increased after chlorination, especially for *d*-CYC and *d*-ION. This phenomenon contradicts the results found by Zhang et al. [24] that CYC and ION could be removed by chlorination if the process followed a pseudo first-order and a pseudo second-order kinetics mechanism, respectively. However, their previous kinetics experiments using pure odorant chemicals diluted in ultrapure water were not disrupted by other competitive organic compounds such as that found in natural (raw) water. Höckelmann and Jüttner [25] found that CYC forms by an oxidative cleavage reaction of β-carotene and is catalyzed by β-carotene oxygenase bound on *cyanobacterial* cells under aerobic conditions; thus, the results in this study might be because chlorination induced the production of CYC and ION via conversion reactions of carotenoids, which coincided with the increased CYC observed in the oxidation of β-carotene with permanganate oxidation [26]. Further investigation is needed to demonstrate the reaction mechanism to reveal the difference between the laboratory scale chlorination experiments and the operational conditions of raw water suffering from off-flavor problem.

3.3. The Establishment of Chlorophyll-a (Chl-a) Limits at Intake (CLIs) for DWTP and Reservoir Management

The conventional DWTP was effective for removing cyanobacterial cells but not for the dissolved cyanobacterial metabolites. Together with the release of metabolites by cyanobacterial cells in the C/F process and accumulated in sludge and filtering material, the dissolved organic matter could not be removed effectively before chlorine dosing disinfection generated DBPs [27]. Therefore, controlling cyanobacteria in a water source or before entering the DWTP is important to guarantee the safety of drinking water.

In this study, the concentrations of metabolites including three MC categories (MC-LR, MC-RR and MC-YR) and three T&O compounds (MIB, ION and CYC) significantly correlated with chlorophyll-*a* (Figure 1) and were used to calculate the CLIs for the DWTP associated with the EDWS. Correspondingly, the GEO was not chosen in the CLI system since no positive correlation was detected between GEO and algae. According to our previous paper, the guideline (G_{met}) of the MC-LR, MC-RR and MC-YR was 1.0 μg/L, 1.0 μg/L and 10.0 μg/L, respectively [19]. The OTCs of MIB, ION and CYC were used as their guideline (G_{met}), respectively. The MT values for the six chosen metabolites in each season were calculated according to their removal rates in the studies as noted by Equations (2)–(5). The CLI for each season was the minimum value of calculated MTs. Finally, the CLIs were 4.0 ± 0.9, 9.8 ± 1.4, 26.9 ± 3.9 and 9.1 ± 1.7 μg/L according to the MT value of CYC, MIB, MC-YR and CYC in the spring, summer, autumn and winter, respectively (Figure 5). These values were between the threshold of 1 and 50 μg/L chlorophyll-*a* for the Alert Level 1 and 2 [2]. The relatively low CLIs obtained in the present study were not only determined according to MC-YR in autumn but also according to T&O compounds in other seasons. It is worth noting that the CLIs might be unusually low for the DWTP which has been in operation for more than 30 years. The sludge and filters in the sedimentation and filtration step have accumulated a large amount of cyanobacteria, microorganisms and other organic contaminants which resulted in increasing concentrations of dissolved microcystins and T&O in the effluent. Consequently, low removal rates are obtained and then low CLIs are required based on the condition of the DWTP.

Compared to the cyanobacterial density or biovolume, chlorophyll-*a* concentrations were easier to detect and more suitable to be used as the indicated index especially for the water sources suffering with different algal species and many variants of cyanotoxins [1,28]. This method to establish chlorophyll-*a* thresholds in different seasons is not only based on MCs but also on T&O compounds to guide the management of the water source and the DWTP and to guarantee the safety of drinking water involved in multiple contaminants. Because the water sources are specific with different water quality standards and cyanobacterial species and because DWTPs each have unique treatment technologies, we recommend that water companies establish their own standards through routine monitoring of water source quality using the proposed evaluation system. Considering the possibilities of forming DBPs after chlorination with the remnant organic matter in the preceding processes, the DBPs and their precursors should also be taken into account for the reservoir and DWTP management.

4. Conclusions

In the EDWS of Lake Chaohu, three MC congeners and four T&O compounds co-occurred in different seasonal dynamics with two cyanobacterial dominant species, *Microcystis* and *Dolichospermum*. The MCs exceeded the WHO [4] recommended guideline value in summer and autumn and the T&O compounds were above their OTCs throughout the whole year except for MIB in winter, which complicated the management of DWTP and posed great risk on drinking water production. The cell-bound metabolites were removed mainly by the C/F+ step with the elevated elimination observed in summer and autumn in the associated full-scale DWTP with water treatment techniques including C/F (with potassium permanganate and powdered activated carbon in summer and autumn), sedimentation, filtration and final chlorination. However, an increasing concentration of dissolved metabolites were observed along with the treatment techniques, especially after C/F+ and chlorination, resulting in some T&O compounds exceeding their corresponding OTCs in the finished water. Due to the co-occurrence of the two classes of cyanobacterial metabolites with different characteristics, CLIs synthesis during their seasonal variations and their elimination rates were proposed to help managers make proper strategies in source water and DWTPs to guarantee the safety of drinking water.

5. Materials and Method

5.1. Chemicals, Standards and Materials

Microcystin standards (MC-LR, MC-RR and MC-YR) with purities ≥95% were purchased from Alexis Biochemicals (Lausen, Switzerland). The stock solution of three MC congeners was prepared in 1 mL of methanol solution at a concentration of 100 mg/L for each variant, which was kept at −20 °C for later use. The C18 solid phase extraction (SPE) cartridges (500 mg, 6 mL) used in MCs analysis were obtained from Anpel Company (Shanghai, China). The standard compounds GEO, MIB and 2-isobutyl-3-methoxypyrazine (IBMP, as the internal standard) were purchased from Sigma-Aldrich Chemical Co. (St. Louis, MO, USA) with the concentration of 100 mg/L in methanol, while CYC and ION with purities ≥97% were purchased from Adamas Reagent Co., Ltd. (Basel, Switzerland). The stock solution of 1 mg/L with four target T&O compounds prepared in methanol was stored in the dark at 4 °C. Sodium chloride (NaCl, Sigma, St. Louis, MO, USA) was added to the samples to enhance the T&O compounds extraction from water.

Methanol, trifluoracetic acid, acetic acid, acetonitrile and acetone were of HPLC grade from Tedia Company (Fairfield, OH, USA). Water used throughout the work was from a Milli-Q water purification system (Millipore, Billerica, MA, USA). All other reagents used were of analytical grade or better.

5.2. Study Site

Lake Chaohu, the fifth largest shallow lake in China, has four distinct seasons influenced by East Asia monsoons with multiple uses such as water supply, commercial fishery and sightseeing. Eastern Lake Chaohu is the sole drinking water source for around 4.23 million people in the city of Chaohu. To evaluate the impact of cyanobacteria and their metabolites on the safety of EDWS, we took three main drinking water intakes at EDWS as sampling sites: Intake-A (117°50′15.13″ E, 31°35′36.34″ N), Intake-B (117°50′54.32″ E, 31°35′24.82″ N) and Intake-C (117°50′58.40″ E, 31°35′32.30″ N) (Figure 6a). The seasonal variations of physicochemical parameters in this drinking water source during the sampling period are summarized in Table 1.

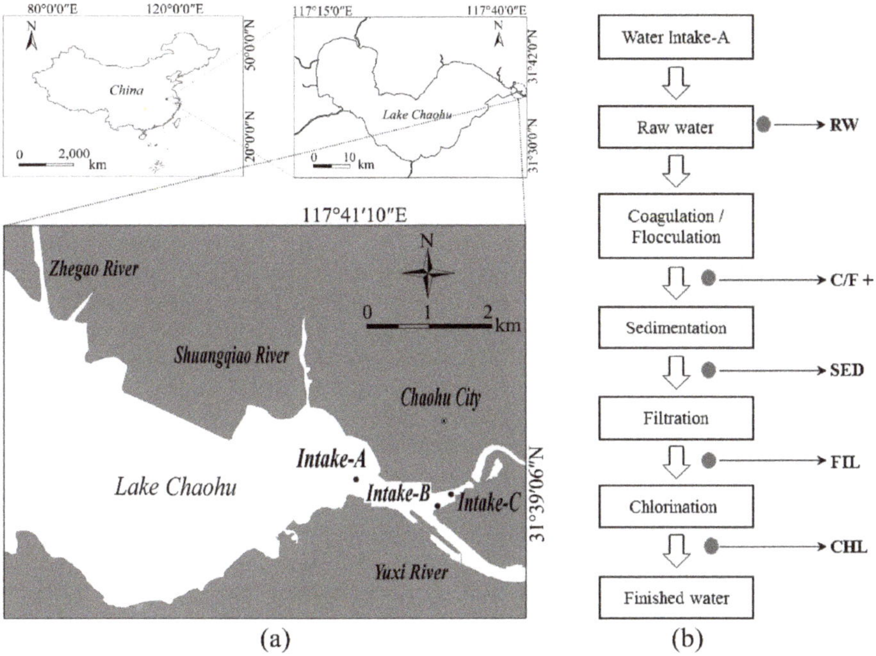

Figure 6. Sampling sites: (**a**) in the eastern drinking water source of Lake Chaohu; and (**b**) in the drinking water treatment plant associated with Intake-A.

The full-scale DWTP associated with Intake-A was chosen to evaluate the elimination of cyanobacteria and their metabolites in treatment processes with the routine sequences of coagulation-flocculation (C/F), sedimentation, filtration and chlorination. The production rate of the DWTP was 40,000 m^3/d. Water from Lake Chaohu was transported through a water pipe at a distance of 2 km. In general, the dosing of coagulant polyaluminum ferric chloride and flocculation aid dimethyl diallyl ammonium chloride were 12 mg/L and 2 mg/L, respectively. In summer and autumn, 9 mg/L of $KMnO_4$ and 10 mg/L of PAC were dosed into a reaction tank with coagulants and flocculants.

5.3. Sampling, Sample Preparation and Phytoplankton Analysis

Water samples were collected monthly from August 2011 to July 2012 at the sampling stations. Integrated water samples were taken from the EDWS of Lake Chaohu (Intake-A, -B and -C), using a 2.5 L sampler. These samples were taken from the entire water column at 0.5 m intervals. Simultaneously, sampling in the DWTP was done throughout the treatment processes and included raw water (site RW), the outflows of the process of C/F (site C/F), sedimentation (site SED), filtration (site FIL) and chlorination (site CHL) (Figure 6b).

The samples for phytoplankton analysis (1 L) were fixed in situ with 1 to 1.5% Lugol's iodine solution. The samples for the chlorophyll-*a* analysis, MC analyses and T&O compounds (without headspace) analysis were stored in 1 L high density polyethylene bottles and sealed with screw caps, respectively. These samples were placed in the dark on ice and transported back to the lab for analysis. Sodium thiosulfate was added into the MC and T&O compounds analysis samples of finished water in the field to prevent further oxidation with free chlorine [29].

Phytoplankton samples were analyzed according to taxonomic keys based on cell structure and dimension [30] using a phase contrast microscope (Nikon, TS100F, Tokyo, Japan). Chlorophyll-*a* concentrations were analyzed by a spectrophotometric method according to Jeffrey and Humphrey [31].

5.4. MCs Extraction and Analysis

Three MCs variants including MC-LR, -RR and -YR with extracellular forms (EMC-LR, EMC-RR and EMC-YR) and intracellular forms (IMC-LR, IMC-RR and IMC-YR) were extracted and condensed from water samples using the method modified according to Barco et al. [32]. Simply, the glass microfiber filter (GF/C, Whatman, Whatman Inc., Clifton, NJ, USA) was used to separate extra- and intra-cellular MCs. The samples of filtered water were concentrated on the SPE cartridges. After freeze-thawed three times, the filters with intra-cellular MCs were extracted using 5% acetic acid once, followed by 75% methanol twice with ultrasonic treatment prior to pre-concentration over the SPE cartridges. Before high performance liquid chromatography (HPLC) analysis, the extracts were stored at −20 °C. The identify and quantity of three MC congeners were determined by HPLC (Agilent 1200, Agilent Technologies, Santa Clara, CA, USA) with quaternary pump (G1311A), autosampler (G1329A), thermostated column compartment (G1316A), diode-array detector (G1315D), and an Agilent Eclipse XDB-C 18 column (4.6 × 150 mm i.d.; 5 μm particle size). Gradient elution of water/0.05% trifluoracetic acid (A) and acetonitrile (B) was used by varying the volume percentage of B from 30 to 40% over 15 min, and held constant for an additional 5 min. Injection volume was 20 μL, and chromatograms were analyzed and integrated at 238 nm. Microcystins in the samples were compared with MC-LR, MC-RR and MC-YR standards based on peak areas and retention times. After concentrating the samples, the detection limits for the three MC congeners in intra- and extra-cellular forms were at the 0.01 μg/L level.

5.5. T&O Compounds Analysis

Four T&O compounds including dissolved (*d*-GEO, *d*-MIB, *d*-CYC and *d*-ION) and particulate T&O compounds (*p*-GEO, *p*-MIB, *p*-CYC and *p*-ION) were analyzed using solid-phase microextraction followed by GC-MS which was modified according to the method of Watson et al. [33]. One liter or more water samples were filtered immediately through a Whatman GF/C glass fiber filter and divided into a dissolved and a particle-bound fraction. The filtrate (80 mL) with the dissolved fraction was transferred into 125 mL vials immediately. The filter residue with the particulate fraction was rinsed into a 125 mL vial with 80 mL of Milli-Q water, and then the cells were disrupted using ultrasound in ice bath. The sample with particle-bound T&O compounds were salted out with 25 g NaCl to improve recovery, stirred vigorously, and extracted at 65 °C for 30 min using a 2 cm long CAR/DVB/PDMS fiber (Supelco). GC-separation was done using HP 6890/5973 GC-MS equipped with a DB-5MS column. The GC was programmed from 60 °C (constant temperature for 2 min) to 200 °C (8 °C/min) with a 2 min hold, and finally to 260 °C (15 °C/min). For the selected ion monitoring mode, m/z 107 and 95 for MIB, m/z 112 and 126 for GEO, m/z 137 and 152 for CYC and m/z 177 and 91 for ION were used. Quantities of each analyte were compared to the corresponding standard curve. The detection limits were 0.2, 0.2, 0.5 and 0.4 ng/L for GEO, MIB, CYC and ION, respectively.

5.6. Statistical Analysis

According to the climate feature of Lake Chaohu, the sampling time was divided into four seasons: spring (March–May), summer (June–August), autumn (September–November) and winter (December–February). In the EDWS of Lake Chaohu, the data listed in Table 1 were expressed with the average value and the range of three sampling sites (Intake-A, -B and -C) for each parameter in the corresponding season. Using the multivariate data analysis software CANOCO 4.5 (Microcomputer Power, Ithaca, NY, USA) [34], redundancy analysis (RDA) was performed to determine the relationship among cyanobacteria, three MC variants and four T&O compounds. The two-tailed student's t test was performed to compare the differences between data from Intake-A in the drinking water source and data from site RW in the DWTP, wherein no differences were accepted as significant at $p > 0.05$. During the removal processes, the values of the parameters in each season were presented using average data and standard deviation.

5.7. Calculation of Removal Rates and CLIs

The removal rates of cyanobacteria and the associated metabolites in the DWTP were calculated using average concentration of targeted pollutants in each season as:

$$\eta_i = (c_{i0} - c_i)/c_{i0} \times 100\% \tag{1}$$

where η is the removal rate of all corresponding compounds designated by its subscript i, which represents chlorophyll-*a*, cyanobacteria and each metabolite. c_{i0} and c_i are the average concentration of chlorophyll-*a*, cyanobacteria and each metabolite measured in each season in the effluent from the preceding step c_{i0} and from this treatment step c_i. Total removal rates were obtained when c_0 and c are the concentrations in the raw water and the finished water, respectively. The values with no detections or below limit of quantitation were set equal to zero for the calculations.

To ensure the safety of drinking water, the CLIs were calculated using Equations (2)–(5):

$$CLI = Min\ (MT_1,\ MT_2,\ MT_3,\ \ldots\ldots) \tag{2}$$

$$MT = G_{Chl-a}/(1 - \eta_{met}) \tag{3}$$

$$G_{Chl-a} = G_{met}/c'_{met} \tag{4}$$

$$c'_{met} = c_{met}/c_{Chl-a} \tag{5}$$

where CLI represents the chlorophyll-*a* limits at intake; MT is the maximum tolerable value for chlorophyll-*a* in the drinking water source, which was modified according to Schmidt, Bornmann, Imhof, Mankiewicz and Izydorczyk [20]; and η_{met} is the total removal rates of the targeted metabolites in the DWTP. G_{Chl-a} is the chlorophyll-*a* concentration equivalent to the guideline standard of the studied metabolites [19]. G_{met} is the guideline standards or OTCs of the studied metabolites' concentration in the drinking water. c'_{met} is the cell quota of metabolites, e.g., in pg per cell; c_{met} is the metabolite concentration in water; and c_{chl-a} is the chlorophyll-*a* concentration in water.

Acknowledgments: This work was supported by grants from National Natural Science Foundation of China (41471075 and 41171366), Major Science and Technology Program for Water Pollution Control and Treatment of China (2017ZX07603-005), and Science Foundation of Nanjing Institute of Geography and Limnology, CAS (NIGLAS2012135013). The authors thank Carla Roberts for the improvement of the written English.

Author Contributions: Lixia Shang and Muhua Feng conceived and designed the experiments; Lixia Shang, Feifei Liu and Xiangen Xu performed the experiments; Fan Ke and Xiangen Xu analyzed the data; Feifei Liu and Wenchao Li contributed reagents/materials/analysis tools; and Lixia Shang, Muhua Feng and Wenchao Li wrote the paper.

Conflicts of Interest: The authors declare no conflict of interest. The founding sponsors had no role in the design of the study; in the collection, analyses, or interpretation of data; in the writing of the manuscript, and in the decision to publish the results.

References

1. BouhaddaFda, R.; Nelieu, S.; Nasri, H.; Delarue, G.; Bouaicha, N. High diversity of microcystins in a *Microcystis* bloom from an Algerian Lake. *Environ. Pollut.* **2016**, *216*, 836–844. [CrossRef] [PubMed]
2. Chorus, I.; Bartram, J. *Toxic Cyanobacteria in Water: A Guide to Their Public Health Consequences, Monitoring and Management*, 1st ed.; E & FN Spon: London, UK, 1999; pp. 183–210.
3. Watson, S.B. Aquatic taste and odor: A primary signal of drinking-water integrity. *J. Toxicol. Environ. Health Pt. A* **2004**, *67*, 1779–1795. [CrossRef] [PubMed]

4. World Health Organization (WHO). *Guidelines for Drinking Water Quality, Incorporating First Addendum*; WHO: Geneva, Switzerland, 2006.
5. El Herry, S.; Fathalli, A.; Rejeb, A.J.-B.; Bouaïcha, N. Seasonal occurrence and toxicity of *Microcystis* spp. and *Oscillatoria tenuis* in the Lebna Dam, Tunisia. *Water Res.* **2008**, *42*, 1263–1273. [CrossRef] [PubMed]
6. Hotto, A.M.; Satchwell, M.F.; Berry, D.L.; Gobler, C.J.; Boyer, G.L. Spatial and temporal diversity of microcystins and microcystin-producing genotypes in Oneida Lake, NY. *Harmful Algae* **2008**, *7*, 671–681. [CrossRef]
7. Sinang, S.C.; Reichwaldt, E.S.; Ghadouani, A. Spatial and temporal variability in the relationship between cyanobacterial biomass and microcystins. *Environ. Monit. Assess.* **2013**, *185*, 6379–6395. [CrossRef] [PubMed]
8. Szlag, D.C.; Sinclair, J.L.; Southwell, B.; Westrick, J.A. Cyanobacteria and cyanotoxins occurrence and removal from five high-risk conventional treatment drinking water plants. *Toxins* **2015**, *7*, 2198–2220. [CrossRef] [PubMed]
9. Zhang, X.; Chen, C.; Ding, J.; Hou, A.; Li, Y.; Niu, Z.; Su, X.; Xu, Y.; Laws, E.A. The 2007 water crisis in Wuxi, China: Analysis of the origin. *J. Hazard. Mater.* **2010**, *182*, 130–135. [CrossRef] [PubMed]
10. Graham, J.L.; Loftin, K.A.; Meyer, M.T.; Ziegler, A.C. Cyanotoxin mixtures and taste-and-odor compounds in cyanobacterial blooms from the midwestern united states. *Environ. Sci. Technol.* **2010**, *44*, 7361–7368. [CrossRef] [PubMed]
11. Ozawa, K.; Fujioka, H.; Muranaka, M.; Yokoyama, A.; Katagami, Y.; Homma, T.; Ishikawa, K.; Tsujimura, S.; Kumagai, M.; Watanabe, M.F. Spatial distribution and temporal variation of *Microcystis* species composition and microcystin concentration in Lake Biwa. *Environ. Toxicol.* **2005**, *20*, 270–276. [CrossRef] [PubMed]
12. Chen, J.; Xie, P.; Ma, Z.; Niu, Y.; Tao, M.; Deng, X.; Wang, Q. A systematic study on spatial and seasonal patterns of eight taste and odor compounds with relation to various biotic and abiotic parameters in gonghu bay of Lake Taihu, China. *Sci. Total Environ.* **2010**, *409*, 314–325. [CrossRef] [PubMed]
13. Ghernaout, B.; Ghernaout, D.; Saiba, A. Algae and cyanotoxins removal by coagulation/flocculation: A review. *Desalin. Water Treat.* **2010**, *20*, 133–143. [CrossRef]
14. Zamyadi, A.; Dorner, S.; Sauvé, S.; Ellis, D.; Bolduc, A.; Bastien, C.; Prévost, M. Species-dependence of cyanobacteria removal efficiency by different drinking water treatment processes. *Water Res.* **2013**, *47*, 2689–2700. [CrossRef] [PubMed]
15. Fan, J.; Hobson, P.; Ho, L.; Daly, R.; Brookes, J. The effects of various control and water treatment processes on the membrane integrity and toxin fate of cyanobacteria. *J. Hazard. Mater.* **2014**, *264*, 313–322. [CrossRef] [PubMed]
16. Srinivasan, R.; Sorial, G.A. Treatment of taste and odor causing compounds 2-methyl isoborneol and geosmin in drinking water: A critical review. *J. Environ. Sci.* **2011**, *23*, 1–13. [CrossRef]
17. Jurczak, T.; Tarczynska, M.; Izydorczyk, K.; Mankiewicz, J.; Zalewski, M.; Meriluoto, J. Elimination of microcystins by water treatment processes-examples from sulejow reservoir, Poland. *Water Res.* **2005**, *39*, 2394–2406. [CrossRef] [PubMed]
18. Zamyadi, A.; MacLeod, S.L.; Fan, Y.; McQuaid, N.; Dorner, S.; Sauve, S.; Prevost, M. Toxic cyanobacterial breakthrough and accumulation in a drinking water plant: A monitoring and treatment challenge. *Water Res.* **2012**, *46*, 1511–1523. [CrossRef] [PubMed]
19. Shang, L.; Feng, M.; Liu, F.; Xu, X.; Ke, F.; Chen, X.; Li, W. The establishment of preliminary safety threshold values for cyanobacteria based on periodic variations in different microcystin congeners in Lake Chaohu, China. *Environ. Sci. Process. Impact.* **2015**, 728–739. [CrossRef] [PubMed]
20. Schmidt, W.; Bornmann, K.; Imhof, L.; Mankiewicz, J.; Izydorczyk, K. Assessing drinking water treatment systems for safety against cyanotoxin breakthrough using maximum tolerable values. *Environ. Toxicol.* **2008**, *23*, 337–345. [CrossRef] [PubMed]
21. Pietsch, J.; Bornmann, K.; Schmidt, W. Relevance of intra-and extracellular cyanotoxins for drinking water treatment. *Acta Hydrochim. Hydrobiol.* **2002**, *30*, 7–15. [CrossRef]
22. Maghsoudi, E.; Prévost, M.; Vo Duy, S.; Sauvé, S.; Dorner, S. Adsorption characteristics of multiple microcystins and cylindrospermopsin on sediment: Implications for toxin monitoring and drinking water treatment. *Toxicon* **2015**, *103*, 48–54. [CrossRef] [PubMed]
23. Pestana, C.J.; Reeve, P.J.; Sawade, E.; Voldoire, C.F.; Newton, K.; Praptiwi, R.; Collingnon, L.; Dreyfus, J.; Hobson, P.; Gaget, V.; et al. Fate of cyanobacteria in drinking water treatment plant lagoon supernatant and sludge. *Sci. Total Environ.* **2016**, *565*, 1192–1200. [CrossRef] [PubMed]

24. Zhang, K.; Gao, N.; Deng, Y.; Zhang, T.; Li, C. Aqueous chlorination of algal odorants: Reaction kinetics and formation of disinfection by-products. *Sep. Purif. Technol.* **2012**, *92*, 93–99. [CrossRef]
25. Höckelmann, C.; Jüttner, F. Off-flavours in water: Hydroxyketones and β-ionone derivatives as new odour compounds of freshwater cyanobacteria. *Flavour Fragr. J.* **2005**, *20*, 387–394. [CrossRef]
26. Zhang, K.; Gao, N.; Yen, H.; Chiu, Y.; Lin, T. Degradation and formation of wood odorant β-cyclocitral during permanganate oxidation. *J. Hazard. Mater.* **2011**, *194*, 362–368. [CrossRef] [PubMed]
27. Li, C.M.; Wang, D.H.; Xu, X.; Wang, Z.J. Formation of known and unknown disinfection by-products from natural organic matter fractions during chlorination, chloramination, and ozonation. *Sci. Total Environ.* **2017**, *587*, 177–184. [CrossRef] [PubMed]
28. Cantoral Uriza, E.; Asencio, A.; Aboal, M. Are we underestimating benthic cyanotoxins? Extensive sampling results from Spain. *Toxins* **2017**, *9*, 385. [CrossRef] [PubMed]
29. Lin, T.; Liu, C.; Yang, F.; Hung, H. Effect of residual chlorine on the analysis of geosmin, 2-MIB and MTBE in drinking water using the SPME technique. *Water Res.* **2003**, *37*, 21–26. [CrossRef]
30. Hu, H.; Wei, Y. *The Freshwater Algae of China: Systematics, Taxonomy and Ecology*; Science Press: Beijing, China, 2006.
31. Jeffrey, S.W.; Humphrey, G.F. New spectrophotometric equations for determining chlorophylls a, b, c_1 and c_2 in higher plants, algae and natural phytoplankton. *Biochem. Physiol. Pflanz.* **1975**, *167*, 191–194. [CrossRef]
32. Barco, M.; Lawton, L.A.; Rivera, J.; Caixach, J. Optimization of intracellular microcystin extraction for their subsequent analysis by high-performance liquid chromatography. *J. Chromatogr. A* **2005**, *1074*, 23–30. [CrossRef] [PubMed]
33. Watson, S.B.; Brownlee, B.; Satchwill, T.; Hargesheimer, E.E. Quantitative analysis of trace levels of geosmin and MIB in source and drinking water using headspace SPME. *Water Res.* **2000**, *34*, 2818–2828. [CrossRef]
34. Braak, C.T.; Šmilauer, P. CANOCO reference manual and CanoDraw for Windows user's guide: Software for canonical community ordination (version 4.5). In *Section on Permutation Methods*; Microcomputer Power: Ithaca, NY, USA, 2002.

toxins

MDPI

Review

A Systematic Literature Review for Evidence of *Aphanizomenon flos-aquae* Toxigenicity in Recreational Waters and Toxicity of Dietary Supplements: 2000–2017

Amber Lyon-Colbert [1], Shelley Su [1] and Curtis Cude [2,*]

[1] School of Biological and Population Health Science, Oregon State University, Corvallis, OR 97331, USA; lyoncola@oregonstate.edu (A.L.-C.); shelley.su@oregonstate.edu (S.S.)
[2] Oregon Health Authority, Public Health Division, Portland, OR 97232, USA
* Correspondence: curtis.g.cude@dhsoha.state.or.us; Tel.: +1-971-673-0975

Received: 15 May 2018; Accepted: 15 June 2018; Published: 21 June 2018

Abstract: Previous studies of recreational waters and blue-green algae supplements (BGAS) demonstrated co-occurrence of *Aphanizomenon flos-aquae* (AFA) and cyanotoxins, presenting exposure risk. The authors conducted a systematic literature review using a GRADE PRISMA-p 27-item checklist to assess the evidence for toxigenicity of AFA in both fresh waters and BGAS. Studies have shown AFA can produce significant levels of cylindrospermopsin and saxitoxin in fresh waters. Toxicity studies evaluating AFA-based BGAS found some products carried the *mcyE* gene and tested positive for microcystins at levels ≤ 1 μg microcystin (MC)-LR equivalents/g dry weight. Further analysis discovered BGAS samples had cyanotoxins levels exceeding tolerable daily intake values. There is evidence that *Aphanizomenon* spp. are toxin producers and AFA has toxigenic genes such as *mcyE* that could lead to the production of MC under the right environmental conditions. Regardless of this ability, AFA commonly co-occur with known MC producers, which may contaminate BGAS. Toxin production by cyanobacteria is a health concern for both recreational water users and BGAS consumers. Recommendations include: limit harvesting of AFA to months when toxicity is lowest, include AFA in cell counts during visible blooms, and properly identify cyanobacteria species using 16S rRNA methods when toxicity levels are higher than advisory levels.

Keywords: *Aphanizomenon flos-aquae*; blue-green algae supplements; cyanotoxins; microcystin; cylindrospermopsin; saxitoxin

Key Contribution: This manuscript informs the debate over toxigenicity of *Aphanizomenon flos-aquae* as it relates to exposure in recreational fresh waters and consumption of relevant blue-green algae supplements.

1. Introduction

Cyanobacteria, also known as blue-green algae, are photosynthetic bacteria that occur in many fresh and salt water environments around the world. Some cyanobacteria species are toxigenic; they have the potential to produce toxins that can harm people, pets and wildlife. *Aphanizomenon flos-aquae* (AFA) is naturally present in fresh-water sources and has been harvested as dietary blue-green algae supplements (BGAS) in the U.S. since the 1980s [1,2]. AFA is one of the most common species of cyanobacteria collected from the natural environment. In freshwater environments, several cyanobacteria species including *Aphanizomenon* spp. form the most common and noxious type of harmful cyanobacteria blooms (CyanoHABs), which have potentially dire consequences for environmental and human health [3,4]. The potential danger that exposure to cyanotoxins presents is widely known [2,5], and has been estimated to cause between 50,000 to 500,000 human intoxications per year from consumption of finfish and shellfish [6]. While the liver is the primary target organ of microcystins (MCs), other organs can be affected as well [7]. Previous studies have shown effects on the heart, nervous system, kidneys, and GI tract [8]. Very few earlier studies addressed the variability of toxin content in BGAS, although this knowledge would have been important for risk assessments. Long-term consumption of BGAS containing harmful cyanotoxins is cause for public health concerns as they are widely available, labeled as safe products and promoted as beneficial for health. In addition, a plethora of fatalities and severe illness/injury have been recorded worldwide in pets, livestock, birds, wildlife and fish [9–12]. Most cyanotoxin poisonings have occurred when animals drink cyanobacterial-laden freshwater, but other aquatic animals, including fish and shellfish, are also affected [13]. These effects include diarrheal illness, acute liver damage, and even more serious and potentially fatal neurotoxic outcomes.

Cyanotoxin guidelines were established by the World Health Organization (WHO) in 1999 [5]. Work is ongoing to determine appropriate lethal dose for 50 percent of the test population (LD_{50}) and tolerable daily intake (TDI), or thresholds for daily reference doses (RfD). Under the Clean Water Act (CWA) 304(a) the Environmental Protection Agency (EPA) recommended a threshold for RfD concentrations of cyanotoxins to protect human health while swimming or participating in other recreational activities in and on the water [14]. States can adopt the EPA criteria into their water quality standards and use these same threshold values as the basis of swimming advisories for public notification purposes at beaches [14]. States may also use the proposed EPA RfD values when adopting new or revised water quality standards (WQS). If adopted as WQS and approved by the EPA under CWA 303(c), the WQS could be used for all CWA purposes. The EPA has also noted that currently, "available data are insufficient to develop quantitative recreational values for cyanobacterial cell density related to health outcomes" [14]. The epidemiologic associations between cell density and specific health outcomes described in the literature are not consistent [14]. In addition, epidemiologic studies addressing species and strains of cyanobacteria and cell densities associated with significant health effects do not provide sufficient information to determine a consistent association between cyanobacteria cell counts and adverse health outcomes [14].

Because of incidents attributed to toxic CyanoHABs, the WHO and the EPA recommended that risk assessment plans and safety levels include cyanobacteria as an environmental parameter to be monitored and assessed. Swimmers can involuntarily swallow water while swimming, and harm from ingestion of recreational water is comparable to that of drinking water with the same toxin content [5]. For recreational water users with whole-body contact, a swimmer can expect to ingest 100–200 mL of water in one session, with sailboard riders and water skiers ingesting more depending upon age and length of exposure [5]. AFA is reported to be capable of producing anatoxin-a (ANTX), cylindrospermopsin (CYN), microcystins (MCs), and the paralytic shellfish poison toxins (PSP toxins), saxitoxins (STXs), though there have been conflicting reports and varying advisory levels among scientists and health departments [5,13–15]. In addition, laboratory studies with pure strains of cyanobacteria have found that environmental factors can induce changes in toxicity or toxin concentration [5]. The factors that control the growth and toxin content of individual strains are unknown, but the regulation of cyanotoxin production is a critical area for further study and understanding [5]. Most existing studies say that cyanobacteria produce the most toxins under favorable environmental conditions for growth, such as temperature, light, and pH levels [4,5]. Although different cyanobacteria species have differing light and temperature requirements, all cyanobacteria strains produce the most toxins when grown under optimal light conditions [5]. Traditionally we have seen seasonal patterns of CyanoHABs with the majority reported during the warmer summer months [14]. This may lead to recommendations for seasonal harvesting of AFA when MC concentration would be lowest, and for increased testing and warnings for recreational use and seafood consumption during times of high MC levels.

The aim of this systematic review is to determine the strength of evidence for the toxigenicity of *Aphanizomenon flos-aquae* in order to evaluate the risks posed to recreational water users and dietary BGAS consumers.

2. Results

2.1. Toxigenicity

Some species of cyanobacteria are considered toxigenic, meaning they carry the genes responsible for producing various toxins that are classified by mode of action into hepatotoxins (MC, CYN), neurotoxins (ANTX, STX), and skin irritants [16]. Several factors determine toxin production by cyanobacteria, including trophic, genetic, hydrological, environmental, and seasonal patterns [3,14]. A level of 100,000 cyanobacterial cells/mL (which is equivalent to approximately 50 mg chlorophyll-a/L if cyanobacteria dominate) is a guideline value for a moderate health alert in recreational waters [5]. *Aphanizomenon* spp. are known to produce a variety of cyanotoxins, including ANTX, CYN and STX [17–19]. *Aphanizomenon* spp. produce toxins in eutrophic conditions; studies suggest that nitrogen limitation may increase the extracellular release of toxin in CYN-producing cultures. The release of toxins through the cell membrane has been linked to NorM MATE proteins encoded by genes *cyrK* (for CYN toxins) [15,19,20] and *sxtM* (for STXs) [20,21]. Extracellular toxins are directly bioavailable and in direct contact with aquatic organisms and water users during bloom development and even after bloom dissipation. Some of these cyanotoxins have slow natural photo- and biodegradation (particularly CYN) [20] and need more monitoring for ecological and health risk assessments in waters commonly affected by *Aphanizomenon* spp.

2.1.1. Anatoxin-a

ANTX is a naturally occurring organophosphate. Toxicity of ANTX has been well documented since the 1960s [22]. It is an acute neurotoxin that has been shown to be toxigenic to both animals and humans. Uncertainty exists about the effects of ingestion of ANTX at low levels over extended periods of time [23]. The LD_{50} of ANTX is 200–250 μg/kg body weight in rats [24,25]. ANTX binds to nicotinic acetylcholine receptors where it mimics the natural ligand, acetylcholine [26], binding at 20 times higher affinity than acetylcholine. Normally, binding of acetylcholine to its receptor in neuronal membranes induces a conformation change in the receptor which results in the opening of an ion channel in the membrane, allowing calcium and/or sodium ions to move across the lipid bilayer [22]. This "depolarization" of the membrane allows for muscle contraction. Normally acetylcholinesterase quickly degrades acetylcholine, halting the activity. ANTX produces the same effect when it binds to the acetylcholine receptor, except that the binding is irreversible. After continuous stimulation of the neuromuscular junction a desensitization or "block" may follow, resulting in death due to respiratory paralysis [22]. Symptoms of ANTX ingestion include muscle tremors (involuntary muscle contractions or fasciculations) that can often be seen under the skin, followed by decreased movement, rigidity, exaggerated breathing, cyanosis, convulsions, respiratory paralysis and death [26]. Depending on the dose and species exposed, death can occur anywhere between a few minutes to 3 h. In some species ANTX exposure can cause significant effects on blood pressure, heart rate, and acidosis [27]. There are no human health standards for ANTX. The highest no-observed-adverse-effect level in mice (NOAEL) is 2.5 mg/kg-day [2]. While ANTX does not have a NOAEL for humans, researchers have developed a guideline value for ANTX in drinking water of 1 μg/L based on mice studies that would offer an adequate margin of safety for humans [23].

2.1.2. Cylindrospermopsin

CYN is a naturally occurring liver toxin that is found typically in warm tropical waters but, has a presence worldwide. CYN is found in cyanobacteria such as *Cylindrospermopsis raciborskii* detected in Australia, Hungary [28], and the U.S. [29]; *Umezakia natans* in Japan [30]; *Anabaena bergii* in Israel [31];

Raphidopsis curvata in China [32]; and *Aphanizomenon ovalisporium* in Australia and Israel [31,33]. CYN is a potent inhibitor of protein synthesis [34]. In addition, studies have shown that this toxin is extremely stable. While CYN rapidly degrades in sunlight, it is resistant to degradation by temperature or pH [2]. P-450 metabolism is thought to be important in the toxicity of CYN. Studies by Runnegar et al. indicate that the toxicity of CYN may be due to the production of a CYN metabolite since toxicity in the presence of P-450 inhibitors is greatly reduced [35]. CYN has also recently been reported to be a suspect carcinogen, as it has been shown to be both mutagenic and cytotoxic [36]. The LD_{50} (oral) in mice was determined to be approximately 4–6 mg/kg [22]. The EPA also evaluated the health effects of CYN and derived an RfD in its 2015 'Health Effects Support Document for the Cyanobacterial Toxin Cylindrospermopsin' [2]. The kidneys and liver appear to be the primary target organs for CYN toxicity. A critical study for the derivation of the CYN RfD was conducted by Humpage & Falconer (2003) based on drinking water exposure in mice [37]. The critical effect of cyanotoxins noted by the EPA was kidney damage, including increased kidney weight and decreased urinary protein [2].

2.1.3. Microcystins

MCs are a class of cyclic peptides, typically containing seven amino acids. Those found in freshwater are predominantly produced by species of *Microcystis, Planktothrix (Oscillatoria), Anabaenopsis, Cylindrospermopsis and Aphanizomenon* [38]. This group of natural toxins include MCs and CYN [38]. MCs are hepatotoxins and once ingested, travel through the body to the liver, where they are stored. MC-LR is the most common MC and is an extremely acute toxin. The LD_{50} by the intraperitoneal route is approximately 25–150 µg/kg of body weight in mice; the oral LD_{50} (by gavage) is 5000 µg/kg of body weight in mice, and higher in rats [39]. With an estimated 100 MC congeners identified and named according to variable amino acids positions, complete evaluation of supplements or cyanobacterial material for all congeners is not feasible [40]. The provisional WHO guideline of 1 µg/L MC-LR is used if local health authorities suspect a risk to human health [4,5,16]. Because of repeated MC findings in BGAS, many countries have elected to develop regulatory limits for the amount of MC in these products, including businesses harvesting and selling AFA [41]. The EPA evaluated the health effects of MCs and derived a RfD in the 2015 document "Health Effects Support Document for the Cyanobacterial Toxin Microcystins". The derivation of the MCs RfD was established by Heinze et al. in 1999, based on rat exposure to MC-LR in drinking water [40]. It is well known that MC-LR is the most toxic form of an estimated ~100 known congeners of MC [14,42]; therefore, the EPA established an RfD for MC-LR and used it as a surrogate value for the other ~100 MC congeners.

2.1.4. Saxitoxin

Another toxin produced by AFA is saxitoxin (STX), a paralytic shellfish poison (PSP) found in both marine and freshwater environments. The term saxitoxin can also refer to a class of more than 50 structurally related neurotoxins (known collectively as "saxitoxins" or STXs). Humans and other animals are commonly poisoned by STXs after they are taken up by filter-feeding bivalves and crustaceans, which ingest and concentrate the toxin within their tissues and organs [4]. STX-related poisonings are well known in Oregon, Washington, California, Alaska and many of the New England states [5,14]. Worldwide STXs have been documented in Australia, New Zealand, South Africa, China, Thailand, Japan and coastal areas of Western Europe. In recent years there appears to be an increase in STX intoxications. However, it is unknown whether this is a true increase in frequency, influenced by warming climatic patterns or simply a reflection of increased awareness or improved methods of detection. STXs exert their toxic effects by binding with high affinity to voltage-gated sodium channels. This binding blocks the passage of sodium ions across biological membranes resulting in a "blockade" of nerve signal transmission. STX is also capable of interfering with potassium and calcium-mediated ion channels. In the case of potassium channels, the STX effect is slightly different. Instead of blocking the potassium channel outright, it binds at the channel site, requiring stronger membrane depolarization to open the channel [43]. This results in an overall reduction of potassium conductance across membranes. STX acts on calcium channels in a manner similar to its action on sodium channels, although its ability to block calcium channels is not as efficient [43]. STX is fatal to humans even in very low doses. STX is also capable of entering the human body via a cut or other open wound, and the predicted human LD_{50} based on previous mice studies, via this route of exposure has been estimated to be 50 μg/kg body weight [5,14]. Currently the acute RfD, used in place of the NOAEL, is 0.5 μg STX equivalent/kg-day [44]. Humans develop a variety of symptoms, ranging from slight tingling or numbness of the face and extremities, to complete respiratory paralysis. Neurological symptoms present themselves shortly after ingestion [45].

2.2. *Fresh Water*

AFA is naturally present in freshwater sources [14]. CyanoHABs occur during optimal environmental conditions for cyanobacterial growth, including warm water (temperatures between 50–86°F), phosphorus concentrations greater than 30 μg/L, and high nitrogen content [4]. These conditions are often found during water pollution events, such as runoff from agricultural operations, or golf course or residential chemical pollution. During CyanoHABs many species of cyanobacteria produce toxins, which can be lethal or extremely harmful to both humans and animals, even at very low concentrations [46]. Cyanotoxins are toxic substances released by certain species of both marine and freshwater cyanobacteria during CyanoHABs, which are characterized by rapid, unchecked growth of cyanobacteria promoted by the presence or sudden addition of nutrients (especially nitrogen and phosphorus) into waters [14].

Throughout the literature AFA has demonstrated the ability to produce ATX, CYN, MC, and STXs under certain conditions and in certain geographical areas. Overall the *Aphanizomenon* spp. show evidence for moderate to high toxin production particularly with MCs, CYN (11–41% of total toxins, up to 58–63% under certain conditions), ATX (7–47% of total toxins), and STXs (7–35% of total toxins) [15,17]. While it is generally accepted that the *Aphanizomenon* spp. carry toxigenic genes and produce ATX, CYN, MCs and STXs in varying degrees, uncertainty exists over whether AFA has the genes for toxigenicity and if it does, whether it produces toxins in differing environmental conditions. In 2015, genome analysis by Šulcius et al. revealed the presence of non-coding sequences belonging to ANTX gene cluster that indicates AFA may have contained genes responsible for the production of cyanotoxins [47]. While the non-coding genes may have been present, the two AFA strains (2012/KM1/D3 and 2012/KM1/C4) that were isolated from bloom samples obtained from the Curonian Lagoon did not produce any cyanotoxins [47]. Although high concentrations of MC toxins were found in an oligo-mesotrophic lake in the Baltic Lake District, Germany, researchers discovered that the MCs were not produced by the dominant cyanobacteria present (*Dolichospermum circinale* and AFA) but by small numbers of *Microcystis cf. aeruginosa* and *Planktothrix rubescens* that co-habited the blooms sampled [48]. AFA was tested with HEPF/HEPR primers, but without any amplification for the *mcyE* gene responsible for MC toxin production [48]. Since AFA was responsible for 80% of the bloom's total biomass [48], contamination with *Microcystis aeruginosa* is likely. This cyanobacterial co-occurrence also supports findings by Palus et al. in 2007, where cyanobacterial blooms were found to be most intense between August and September, producing the highest concentrations of MC toxins seen in the study [46]. *Microcystis aeruginosa* and AFA were the most dominant cyanobacteria present within the blooms during periods of high MC concentration [46], making determination of the cyanobacteria responsible for the MC production more difficult.

More direct evidence for AFA toxigenicity was found in the following studies. In 2007 Fastner et al. detected measurable amounts of CYN toxin in 21 German lakes with high concentrations of *Aphanizomenon* spp. [28,49]. CYN was detected in 19 of the 21 lakes at concentrations ranging from 0.002–0.484 μg/L (phytoplankton + suspended particles) and 0.08–11.75 μg/L dissolved in water [28]. The maximum CYN measured in a total sample of water was 12.1 μg/L [28]. *Aphanizomenon gracile* is highly correlated with CYN concentrations in the lakes and was suspected to be the major producer of the detected CYN toxin [49], and may have led to potential contamination. In 2006

Preussel et al. identified three isolates of AFA from two German lakes, which were found to produce large amounts of CYN [17]. This was considered the first report of CYN in AFA strains [17]. CYN-synthesis of the strains was shown both by liquid chromatography-tandem mass spectrometry (LC-MS/MS) analysis and detection of PCR products of gene fragments [17]. The strains contained CYN in the range of 2.3–6.6 mg/g of cellular dry weight [17]. In 2007 Faster et al. also confirmed that AFA was a CYN-producing species, which often inhabits water bodies in temperate regions such as Germany and Portugal [17,28,50]. A risk assessment is recommended to confirm findings in geographic locations, suggesting that toxigenicity may vary by region [28]. *Aphanizomenon var klebahnii* was identified in 2009 in the Czech Republic as a potential producer of CYN toxins [51]. ANTX was also listed as a toxin produced by AFA, found in flamingos in Lakes Bogoria and Nakuru in Kenya. [1,52]. The birds were observed staggering and convulsing prior to death, and post-mortem studies found ANTX levels in the liver capable of causing death at 1.06–5.82 μg/L fresh weight [53]. In addition to ANTX, MC-LR was observed from 0.21 to 0.93 μg/L fresh weight within the bird livers post-mortem [53].

In addition to CYN and ANTX, there was evidence of STX when AFA was the dominant species in the Montargil Reservoir, Portugal [54]. To confirm the production of neurotoxins, a strain of AFA was isolated and established in culture and confirmed by high performance liquid chromatography with post column fluorescence derivatization HPLC-FLD [54]. A study in 1986 similarly found that AFA (NH-5) may produce neo-saxitoxin and saxitoxin, which was confirmed with thin-layer chromatography and HPLC [53]. A 2001 study by Ferreira et al. also detected STX toxins in cultures of AFA, isolated from the Crestuma-Lever Reservoir, using high performance liquid chromatography (HPLC) [53]. This finding was supported by claims from Liu et al. in 2006 that AFA has been reported in several countries to produce STXs and associated toxic effects [8]. Acute toxicity testing was performed by mouse bioassay using extracts from the lyophilized material and clear symptoms of STX intoxication were observed [8]. High performance liquid chromatography with post column fluorescence derivatization (HPLC-FLD) and liquid chromatographic mass spectrometry technique (LC-MS) analysis of extracts from cultured material demonstrated that STXs were produced by AFA blooms in China [8]. This was the first study reporting chemically and toxicologically confirmed STXs related to AFA in China [8]. Gkelis et al. identified AFA and *C. raciborskii* as potential producers of STX, while confirming *A. gracile* did, in fact, produce STX [55]. It was also noted that the STX gene cluster may vary biogeographically [55].

2.3. Dietary Supplements

Dietary supplements are largely self-regulated, although there are some safeguards in place following the passage of the 1994 Dietary Supplement Health and Education Act (DSHEA) by the United States Congress and its implementation by the Food and Drug Administration (FDA) [9]. In the United States, the FDA has determined that AFA is a dietary supplement; therefore, it is not subject to regulation as a drug, provided that the health benefits claimed by the manufacturer do not include the cure or treatment of a specific disease such as depression or cancer [41].

The harvesting of cyanobacteria for production as dietary supplements has recently come under scrutiny, as the production of these BGAS suffer from less strict quality controls than other food products or pharmaceuticals. In addition, BGAS are marketed internationally and sold widely over the counter and via the Internet. Although insufficient evidence exists, BGAS are reported to have beneficial health effects including weight loss, increasing alertness and energy, and mood elevation for people suffering from depression [2]. In children, they have been used as an alternative, natural therapy to treat attention deficit hyperactivity disorders (ADHD) [2]. Although AFA-based BGAS have generally been found to be non-toxic, the methods of cultivation in natural waters with inadequate quality controls allows for contamination by other toxigenic species. AFA is primarily harvested from Upper Klamath Lake in Oregon, USA where harvesting for supplement use started in the 1980's. Other cyanobacteria species have been found to contaminate the products, since the AFA is harvested from the natural environment. As a potential consequence of contamination, the presence of microcystins (MCs) and alkaloid cyanotoxins have been found in BGAS [56]. Studies have found BGAS contaminated with MCs ranging from 1 μg/g up to 35 μg/g [57]. It is uncertain whether AFA or contamination by other species of cyanobacteria is the source of the MC toxins.

For nearly 30 years, AFA has been harvested for blue-green algae supplements and constitutes a significant portion of the health food supplement industry in North America [41]. AFA is primarily harvested from cyanobacteria collected from Upper Klamath Lake in Oregon to produce BGAS [10,58]. In southern Oregon, growth of *Microcystis aeruginosa* is a regular occurrence together in blooms with AFA [10]. Because *M. aeruginosa* regularly coexists with AFA it can be collected inadvertently during the harvesting process, resulting in MC contamination of BGAS [10]. In 1996, in response to bloom advisories and BGAS consumer concern, the Oregon Health Authority (OHA, then known as Oregon Health Division) and the Oregon Department of Agriculture (ODA) established a regulatory limit of 1 μg/g for MCs in BGAS and tested BGAS products for the presence of MCs [10]. MCs were detected in 85 of 87 samples tested, with 63 samples (72%) containing concentrations greater than the regulatory limit of 1 μg/g. HPLC and ELISA toxin detection methods identified MC-LR as the predominant congener within the BGAS [10].

Saker et al. (2007) supports the findings that many AFA-based BGAS may be contaminated with MC-producing strains of *Microcystis* spp., and recommends that genetic testing be done to identify the organism responsible for MC production [59]. Previous studies of AFA samples have relied on the ELISA assay, which does not give information on the various chemical forms of MC [59]. Although the health benefits of BGAS are promoted, there are a growing number of studies showing the presence of toxins, some of which (for example MCs) are known to adversely affect human health [59]. In 2012, Heussner et al. analyzed 13 commercially available BGAS in Germany for the presence of cyanotoxins. All samples were analyzed and confirmed by PCR for the presence of the mcyE gene, a part of the MC and nodularin gene cluster [11]. Of all products tested, ten consisted of AFA, five of Spirulina platensis, and three of Chlorella pyrenoidosa [59]. Spirulina spp. are generally regarded as safe within the BGAS industry as of this time [11]. Only AFA BGAS products tested positive for MCs as well as the presence of the mcyE gene [11]. In 2015, a confirmed case of MC poisoning in a dog was documented following the consumption of a BGAS containing organic AFA daily for just 3 weeks [9]. While several brands of supplements of AFA have been found in the past to contain ANTX and its congeners [60,61], neither ANTX nor CYN were found in any of the supplements in the 2012 study. All products containing AFA tested positive for MCs at levels $\leq$ 1 μg MC-LR equivalents/g dry weight [11]. In 2017, Roy-Lachapell found that out of the 18 BGAS products analyzed, 8 contained cyanotoxins at levels exceeding the tolerable daily intake values [56]. Toxic strains of AFA may be found occasionally in some supplements, and have been reported to produce STX and ANTX toxins [56]. Some strains of AFA are known to produce STX and beta-methylamino-L-alanine (BMAA) [56]. Dietary supplements containing AFA had total MC toxin contents of between 0.8 and 8.2 μg/g [56].

3. Discussion

Cyanotoxin advisories and guideline values vary from state to state, and not all states have advisory and guideline values. To date, fifteen states have included recreational guideline values for cyanotoxins as part of their response protocol. The EPA currently lists Oregon's options for recreational water guidance/action level as public health advisory over >100,000 cells/mL of "toxigenic species", or >40,000 cells/mL of *Microcystis* or *Planktothrix*, and toxin testing for MC >10 μg/L, ANTX >20 μg/L, CYN >6 μg/L, and STX >100 μg/L [14]. Currently, for the purpose of issuing public health advisories, Oregon Health Authority excludes AFA from calculation of combined cell counts of toxigenic species. Other states include all *Aphanizomenon* spp. in their list of potentially toxic cyanobacteria, e.g., toxigenic taxa include *Anabaena, Microcystis, Planktothrix, Nostoc, Coelosphaerium, Anabaenopsis, Aphanizomenon, Gloeotrichia, Woronichinia, Oscillatoria, and Lyngbya* [4].

There is evidence to support *Aphanizomenon* spp. as a known toxin producer specifically for cyanotoxins: CYN (11–41%, up to 58–63% under certain conditions), ANTX (7–47%), and STXs

(7–35%) [15]. In 2017, researchers Chernova et al. found evidence through PCR and LC-MS/MS of ANTX in Sestroretskij Razliv Lake, Russia, produced specifically by AFA [62]. Environmental conditions as well as bloom composition should be monitored closely due to the potential for changes in toxicity under increases in sunlight and temperature. Mariani et al. discovered that AFA and *Aphanocapsa* spp. dominated total CyanoHABs during periods of maximum MC concentration in Sardina, Italy [3]. There is still disagreement on whether AFA has the ability to produce cyanotoxins and this may be dependent upon environmental conditions and location. In 2012, a molecular study was carried out using 16S rRNA sequencing and results concluded that AFA did not show amplification for the toxin producing *mcyE* gene, and did not produce ANTX or STX [48]. Gkelis et al. also supported this theory of biogeographic toxicity stating that the *Aphanizomenon* spp. STX gene cluster may be biogeographically differentiated by county [55]. In Ireland AFA is commonly associated with blooms that have high MC concentration, but may be attributed to other contaminating species such as *M. aeruginosa* or *A. gracile* [63]. Researchers recommended further studies using molecular detection methods to determine whether AFA is a MC producer [63].

The first report of CYN in AFA was reported by Preussel et al. in Germany, 2006 through LC-MS/MS analysis and detection of PCR products of gene fragments [17]. In 2007, Fastner et al. also reported AFA as a CYN producer using microscopy and mass spectrometry identification methods [28]. CYN production was also reported in France and the Czech Republic where AFA was found to be a potential producer of CYN due to AFA dominated CyanoHABs where CYN was detected [51,64]. This was not confirmed with molecular detection methods and would be an area for future research. Ferreira et al. (2001) was the first report of STX toxins GTX4, GTX1, GTX3 [53]. This was followed in 2006 by Liu et al. who first reported STXs among AFA isolates discovered in China, through HPLC-FLD and LC-MS detection methods [8]. Among dietary supplements, the toxin-producing *mcyA* gene was detected in all AFA dietary supplements tested, which could be attributed to contamination by other toxic cyanobacteria such as *Microcystis* spp. [59]. Similarly, in 2012 Heussner et al. found all AFA products in their study tested positive for MCs and the *mcyE* gene [11].

The evidence for production of ANTX, CYN, MC, and STX, by AFA is of increasing concern, given that AFA occurs with high frequency in freshwater ecosystems throughout the U.S., and is regularly and increasingly used as a dietary supplement. Not only is AFA abundant in CyanoHAB events, but it appears to survive well in poor growth conditions as well, even at low temperatures [5]. AFA production of CYN and ANTX still remains uncertain [15]. Apart from AFA, the remaining seven *Aphanizomenon* spp. of this genus have only been described morphologically [15]. No phylogenetic data are available which can support their assignment to the genus *Aphanizomenon* and distinguish them as separate species [15]. AFA may have seasonality and/or biogeographical requirements for toxin production which should be established in order to confirm safety of both recreational water use and dietary supplements.

One finding of this review is the large amount of uncertainty that exists over the potential for AFA to carry toxin genes and produce toxins. Several earlier studies have misidentified *Aphanizomenon* spp. for AFA and therefore their findings cannot be confirmed or used as evidence to support the presence of toxin production by AFA [32,65]. Without proper species identification, it is impossible to tell whether the cyanobacteria are AFA, or another species of *Aphanizomenon* such as *A. gracile*, which is a known toxin producer [15,66]. In addition, the data are currently insufficient to develop solid quantitative recreational values for cyanobacterial cell densities related to negative health outcomes [67]. Previous studies by Heinze et al. 1999 [40] and Humpage and Falconer in 2003 [37], have served as references for WHO, EPA and the Oregon Health Authority [58] to develop RfD for MCs and CYN toxins. No current NOAELs exist for ANTX or STX toxins, therefore TDI and guideline values cannot be properly assessed at this time. Because exposures can be acute (e.g., drinking water), sub-acute (e.g., recreational swimming exposure), and chronic (e.g., dietary supplement use), a single guideline value for any cyanotoxins is not appropriate [5]. Rather, it is recommended that a series of guideline values associated with incremental severity and probability of health effects should be defined at all three levels (acute, sub-acute, and chronic) [5].

4. Conclusions

Cyanobacteria blooms are commonly a mixture of toxic and non-toxic genotypes, and associated toxin concentrations can be highly variable both spatially and temporally [52,68]. Additionally, there is evidence that AFA carries toxin genes and produces significant levels of CYN and STX [17,28,50,51,69]. These toxins can potentially harm people, pets, livestock, aquatic animals, birds and other wildlife. In past analysis of AFA supplements, researchers have found the presence of a multitude of toxins including ANTX, STX, CYN, BMAA and MC [11]. MCs have been identified as the most common contaminants of BGAS [9,11]. Uncertainty remains regarding whether AFA was properly identified in earlier studies and was even reclassified to unidentified species of *Aphanizomenon* in two studies [32,65]. AFA-based BGAS may be contaminated with other more toxigenic species of *Aphanizomenon* or *M. aeruginosa* [10,41,59]. Overall, all studies looking more broadly at *Aphanizomenon* spp. have shown a moderate to high production of toxins.

The presence of toxins in recreational waters and BGAS production environments remains a public health concern when CyanoHABs are present, with elevated levels of MCs being found and increasing reports of illness. Accumulated evidence from the peer-reviewed literature suggests that AFA carries toxigenic genes and has the potential to produce toxins under the right environmental conditions. Toxin production of cyanobacteria is highly variable, both within and between blooms and given the timing or seasonality of the bloom [52]. Evidence for production of ANTX, CYN, MC and STX by AFA is of immediate concern, given that AFA occurs with high frequency in freshwater ecosystems throughout the U.S. and is regularly and increasingly used as a dietary supplement. This issue should be evaluated regardless of the species of *Aphanizomenon* due to the potential for cross-contamination among species, and lack of regulation and testing among dietary supplements. Further investigation is warranted to improve our understanding of the effects that AFA dietary supplements have on consumers [57]. Surveys isolating individual colonies of *Microcystis* spp. from Upper Klamath Lake should be followed by laboratory cultivation and subsequent toxicological analyses to provide information on the toxicity of cyanobacterial contaminants [59].

The following recommendations are based on the most current global studies (2000–2017) to inform public health, regulatory and natural resource management agencies regarding AFA toxicity to make the most informed decisions.

4.1. Recommendation to Post Educational Signs as a Precautionary Measure at First Sight of Visible Scum

These signs will help to educate and alert the public of the potential health risks associated with recreational water use during CyanoHABs. This step is particularly important for recreational users with small children or dogs. It is recommended that signs be posted at first visual confirmation of scum, while cell counts and/or toxicity analyses are underway. This action would reduce the risk of exposure to cyanotoxins.

4.2. Recommendation to Limit Harvesting of AFA to Months When Toxicity Is Lowest

The majority of studies indicate that cyanobacteria produce most toxins under environmental conditions, which are most favorable for their growth (warmer summer months). When temperature and light requirements are optimal for toxin production, AFA may produce more MCs than during times with less than optimal temperature and light [5]. This may lead to recommendations for seasonal harvesting of AFA when MC concentration would be lowest, and for increased testing and warnings for BGAS production and recreational use during times of high MC levels. This action would reduce global risk of exposure to humans and domestic animals from BGAS contamination with cyanotoxins.

4.3. Recommendation to Include AFA in Cell Counts during Visible Blooms

It has been demonstrated that AFA has the potential to produce toxins, and should be included in the cell counts of toxigenic cyanobacteria for public health advisories related to CyanoHABs. Ending exclusion of AFA from CyanoHAB advisory criteria would lead to increased frequency and duration of public health advisories and would reduce the risk of exposure to cyanotoxins. This may be evaluated and revised after additional evidence.

4.4. Recommendation to Reduce Health Advisory Guideline Value for Cyanotoxin Levels

The EPA's cyanotoxin guidelines were established under scientific rigor based on existing knowledge from published peer-reviewed scientific evidence on the adverse human health effects of toxins, criteria methodologies, and recreation-specific exposure parameters, reviewed by the Health and Ecological Criteria Division, Office of Science and Technology Office of Water, U.S. [14]. The health advisory guideline value for MCs should be reduced to match EPA's proposed guidelines of 4 μg/ L [67]. Adoption of the EPA's values as recreational use and swimming advisory values at the state level should be adopted under CWA 303(c) as ambient water quality standards. Addition of cyanotoxin water quality standards would eventually improve water quality and reduce exposure risks.

4.5. Recommendations for Proper Species Identification Using 16S rRNA Methods When Toxicity Levels Are Higher than Advisory Levels

Prior to 2007, several research studies had misidentified another toxin-producing cyanobacteria species for AFA. Previous attempts to differentiate toxigenic from non-toxic strains of the same species using microscopic methods have failed. Therefore, it is not recommended to identify species of cyanobacteria using only microscopic methods. During CyanoHABs events when cyanotoxin levels are above health advisory guideline values, advanced molecular methods such as 16S rRNA PCR should be used to determine specific strains of toxigenic cyanobacteria. This information would help to focus analysis for cyanotoxins and inform future evaluation of cyanobacteria toxigenicity.

4.6. Recommendations for Laboratory-Based Research to Confirm the Ability of AFA to Produce Toxins under Differing Environmental Conditions

Further research would provide an evidence base to improve recommendations for recreational and BGAS guidance values. It should be noted whether there is a visible algal bloom (include picture) at the time the sample is taken in addition to recording other environmental factors such as temperature (ambient and water), pH, and whether it is raining/snowing. Protocols for sample collection, preparation and storage should be evaluated for improvement. Both microscopic evaluation and ELISA testing should be done regardless of cell count. Cell count and ELISA results should be recorded and compared across time to note any predictive environmental factors for increased toxicity and to definitively prove whether AFA is a toxin producer. Molecular techniques should be employed to detect toxicity genes. This work will help to either confirm or refute the presence of toxin producing genes and better understand the risks associated with AFA in recreational waters and dietary supplements.

5. Materials and Methods

A systematic review of the peer-reviewed literature was conducted per the preferred reporting items for systematic review and meta-analysis protocol (PRISMA-P, Figure 1). The PRISMA-P is a 27-item checklist developed to strengthen the methodological quality, assessment of bias and quality of systematic reviews. No existing review protocols exist for this systematic review [70].

5.1. Search Strategy

A systematic search was conducted using the following electronic databases: NCBI's PubMed, Web of Science, Cochrane CEN-TRAL. The following combinations of search terms were used: Toxicity and *Aphanizomenon flos-aquae*, cyanobacteria, harmful algal blooms, *Aphanizomenon* and toxicity in fresh waters, lakes, and dietary supplements. Two independent researchers screened titles and abstracts to identify eligible studies for inclusion within this systematic review. A literature search in NCBI's PubMed database for '*Aphanizomenon flos aquae*' returned 145 articles. Search term 'Toxicity and *Aphanizomenon flos aquae*' returned 50 articles. Ten more articles were identified using Science Direct through Oregon State University's library; these mentioned AFA as being found in CyanoHABs but did not specifically describe AFA toxigenicity. Three additional review articles were identified through Cochrane. Articles which did not include measurement of toxin levels, or that did not specifically evaluate whether AFA was the producer of measured toxins, were excluded. In addition, systematic reviews were excluded in the analysis, but were considered for background information and article identification. Articles that included *Aphanizomenon* spp. were independently evaluated for relatedness. Titles with keywords "*Aphanizomenon, Aphanizomenon flos aquae*, toxicity including ATNX, CYN, MC, STX or PSP toxins, toxigenic, dietary supplements, and/or fresh-water" were included in the review. All relevant peer-reviewed articles up to 1 August 2017 were considered for inclusion.

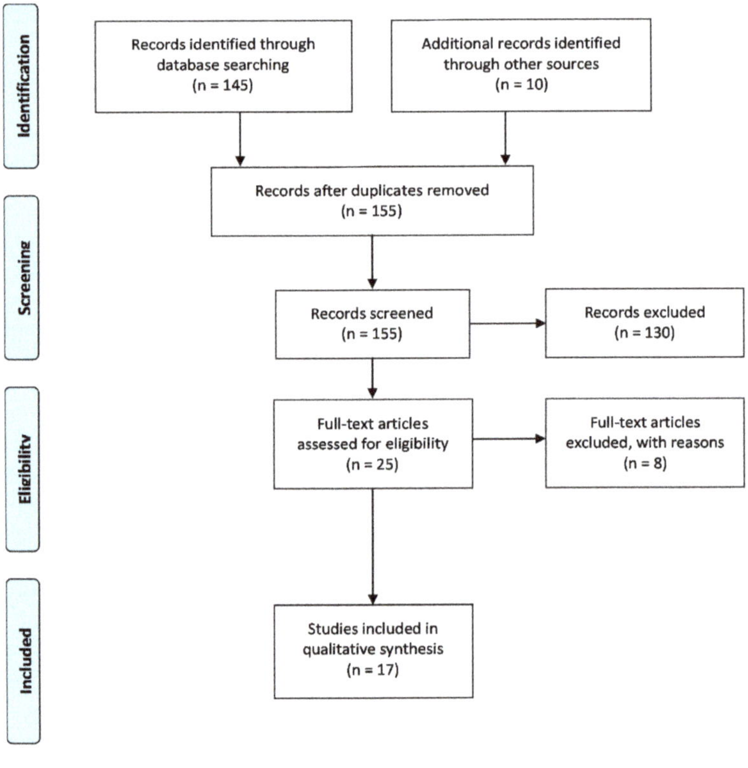

Figure 1. Article selection (PRISMA-P) flow diagram.

5.2. Inclusion Criteria

Included studies were all peer-reviewed research articles measuring toxin levels and toxigenicity of AFA either in recreational waters, dietary supplements, or measuring toxicity and/or illness in either human cases or animal studies. Peer-reviewed articles and published books or reports with an ISBN number were included. All seventeen articles for inclusion listed in Table 1 referenced directly measured toxin levels associated with *Aphanizomenon* spp. dominated blooms or specifically measured genes and toxin production by AFA.

Table 1. Characteristics of studies included in the systematic review for AFA toxigenicity (2000–2017).

Lead Author (Year)	Location	Findings	Method of Detection	Conclusions
Roy-Lachapelle (2017)	Canada	Out of 18 products tested, 8 contained cyanotoxins at levels exceeding WHO's TDI. Supplements containing AFA had MC concentrations between 0.8 and 8.2 μg/g. Low amounts of BMAA (neurotoxin) were also found.	Lemieux and Adda oxidation. Chemical derivatization, laser diode thermal desorption, and liquid chromatography.	Some dietary products could be harmful upon long-term consumption due to the presence of cyanotoxins. A critical need exists for better monitoring for all BGAS, and guidelines for maximum intake.
Chernova (2017)	Russia	*Aphanizomenon flos-aquae* found to produce ANTX in Sestroretskij Razliv Lake	PCR and LC-MS/MS and Restriction Fragment Length Polymorphism (RFLP) analysis	Both dominant species *Aphanizomenon flos-aquae* and *Dolichospermum planctonicum* are ANTX producers.
Cires (2016)	Global	*Aphanizomenon* spp. known toxin producer specifically: CYN (11–41% of total toxins, up to 58–63% under certain conditions), ANTX (7–47% of total toxins), and SXTs (7–35% of total toxins)	Literature Review.	Although *Aphanizomenon* spp. are known toxin producers, the toxigenicity of AFA is still uncertain.
Mariani (2015)	Sardinia	AFA and *Aphanocapsa* spp. dominated total cyanobacteria.	ELISA, Mass spectrometer.	Species composition during periods of maximum MC concentration differed from typical in other Mediterranean sites.
Sulcius (2015)	Lithuania/Russia	Concentrations of cyanotoxins in scum materials increased from ~30~300 fold compared to bloom samples. AFA comprised ~19% of total cyanobacteria biomass. The most common toxin-producing cyanobacteria from Curonian Lagoon belong to the genera of *Aphanizomenon* spp., *Microcystis*, and *Planktothrix*.	Microscopic, and chemical. Extraction of saxitoxins with 4 mM ammonium formate buffer and acetonitrile 2:3 ratio, Mass spectrometer, information dependent acquisition mode, and multiple reaction monitoring.	Larger concentrations of cyanotoxins were found in scum compared to blooms.
Dadheech (2014)	Germany	Although AFA dominated total phytoplankton at >80% contribution to total biomass, AFA did not show amplification for the *mcyE* gene, or STX and ANTX production.	Molecular analysis:16S rRNA sequencing, BLAST identification.	Differences seen in dominant taxon of field sample from *Dolichospermum circinale* in 2011 to AFA in 2012, with reduction in total MC content seen from 27.32 μg/L to 4.25 μg/L.
Gkelis (2014)	Greece	*C. raciborskii* and AFA are *potential* STX producers. *A. gracile* confirmed STX producer. MC: 3.9–108 μg/L, CYNs: 0.3–2.8 μg/L, and STXs: 0.4–1.2 μg/L *Aphanizomenon* spp. STX gene cluster may be biogeographically differentiated by county.	Microscopy, molecular, and immunologic methods: ELISA.	AFA was not found to be the dominant species in blooms, or a producer of toxins. Co-occurrence of more than one cyanotoxins in sites used for drinking water, agriculture, or recreation represent potential health risks.
Heussner (2012)	Germany	All AFA products tested positive for MCs and the *mcyE* gene. The contamination levels of the MC-positive samples were ≤1 μg MC-LR equivalents per g dry weight.	Colorimetric protein phosphatase inhibition assay (cPPIA), Adda-ELISA, Cell Culture, Liquid chromatography tandem mass spectrometry (LC-MS/MS), DNA extraction and PCR.	Recommendation for prohibition of marketing and sale of AFA-based dietary supplements in order to prevent acute and chronic exposure to MCs.
Mooney (2011)	Ireland	AFA, *Gomaphosphaeria* spp. and *Microcystis aeruginosa* were the most dominant cyanobacterial species associated with high MC concentration. AFA was dominant in 1/14 sites with lake area of 382 km^2, and MC concentration of 1652 ng/μg Chl*a*	Synoptic survey of 14 sites, used high performance liquid chromatography-tandem mass spectrometry (HPLC-MS/MS).	Further studies are recommended to use molecular detection methods to determine whether AFA is a MC producer. It is unknown which species produced the toxins, only recorded dominant cyanobacteria and total toxin per area.

Table 1. *Cont.*

Lead Author (Year)	Location	Findings	Method of Detection	Conclusions
Blahova (2009)	France	CYN was found at 3 localities with *Aphanizomenon* spp. sub-dominated water blooms. Concentrations determined by ELISA (0.4–4 μg/L) were systematically higher than concentrations determined by LC/MS (0.01–0.3 μg/L).	ELISA and LC/MS.	AFA is a potential producer of CYN.
Brient (2009)	Czech Republic	*AFA var klebahnii* found to be a potential producer of CYN. Intracellular concentrations of CYN ranged between 1.55 and 1.95 μg/L.	LC-MS/MS.	AFA is a potential producer of CYN.
Palus (2007)	Poland	AFA dominated blooms August–October. The concentration of MC in water did not exceed 1 μg/L, Cyanobacteria co-occurrence found with *E. coli*.	Protein phosphatase inhibition assay (PPIA), ELISA and HPLC.	Phytoplankton biomass and genotoxicity of CyanoHABs should be assessed to avoid public health issues.
Fastner (2007)	Germany	Concentrations reached up to 73.2 μg CYN/g dry weight. Study confirmed AFA is a CYN-producing species frequently inhabiting water bodies in temperate climatic regions.	Microscopy, Mass-spectrometer.	*Aphanizomenon* spp. may be an important CYN toxin producer in Germany waters. A world hazard analysis and risk assessment is recommended to confirm by geographic regions.
Saker (2007)	Australia and Canada	*mcyA* gene was detected in all 12 AFA dietary supplements, suggesting contamination by *Microcystis* spp.	Multiplex PCR.	Dietary supplements containing AFA are more at risk for contamination by *Microcystis* spp., and should be monitored. Laboratory and toxicological analysis of Upper Klamath Lake cyanobacteria would provide useful information to inform and protect human health.
Preussel (2006)	Germany	Toxin CYN detected in the range of 2.3–6.6 mg/g of cellular dry weight.	LC-MS/MS analysis and detection of PCR products of gene fragments.	First report of CYN in AFA strains.
Ferreira (2001)	Portugal	Presence of PSP toxins: GTX4, GTX1, GTX3, and Cs toxin present either in cells of AFA or in other toxic isolates.	High performance liquid chromatography (HPLC) using 2 isocratic elution systems.	AFA known STX producer, but *A. circinalis* is also found to be potential STX producer. More work is needed to understand the toxicological profiles of cyanobacteria.
Liu (2006)	China	STXs produced by AFA bloom. Significant glutathione-S-transferase (GST) and lactate dehydrogenase (LDH) increases, together with decrease of the glutathione (GSH) level, were measured.	High performance liquid chromatography with post-column fluorescence derivatization (HPLC-FLD) and liquid chromatographic mass spectrometry technique (LC-MS).	The results indicate a potential role of STXs intoxicating and metabolizing in test animals.

5.3. Exclusion Criteria

Articles were excluded if no English abstract was available. Review articles, student theses, newspapers or magazine reports, commentaries, correspondences, and letters were excluded from this review. Additionally, articles were excluded if there was no evaluation of toxin levels or direct human health outcomes due to exposure to AFA in either recreational waters or dietary supplements. Although two articles were found to be relevant to the search terms, it was later discovered there was a re-classification of the cyanobacteria from AFA to *Aphanizomenon* spp. These are considered for background and discussion but were excluded from the systematic review.

5.4. Study Quality Assessment

For each included study, the quality of evidence was assessed using the GRADE (Grading of Recommendations: Assessment, Development and Evaluation) method. The GRADE method rates the quality of evidence presented in research studies for evaluation. Evidence is assessed based on five factors: risk of bias, imprecisions, inconsistency, indirectness of evidence and publication bias. The quality of evidence is rated on the following scale: "High Quality", "Moderate Quality", "Low Quality" and "Very Low Quality." Randomized control trials are considered "High Quality" evidence but, can be downgraded if there is a serious risk of bias in the study. Additionally, observation studies begin as "Low Quality" evidence and can be upgraded due to large effect size (Relative Risk or Odds Ratio > 2), presence of a dose response or the presence of confounders against bias. The methodology is widely used and accepted by numerous organizations including the WHO. Articles were assessed, and tables were produced using the online GRADEpro toolkit, summary of findings table.

Funding: This journal article was supported by Cooperative Agreement Number 1 NUE1EH001347-01, funded by the Centers for Disease Control and Prevention. Its contents are solely the responsibility of the authors and do not necessarily represent the official views of Oregon State University, the Oregon Health Authority, the Centers for Disease Control and Prevention or the Department of Health and Human Services.

Acknowledgments: The authors want to thank Rebecca Hillwig, David Farrer, and Thomas Jeanne and three anonymous reviewers for their thorough review and commentary on drafts of this article.

Conflicts of Interest: The authors declare no conflict of interest. The funding sponsors had no role in the design of the study; in the collection, analyses, or interpretation of data; in the writing of the manuscript, and in the decision to publish the results.

References

1. Funari, E.; Testai, E. Human health risk assessment related to cyanotoxins exposure. *Crit. Rev. Toxicol.* **2008**, *38*, 97–125. [CrossRef] [PubMed]
2. Environmental Protection Agency. Health Effects Support Document for the Cyanobacterial Toxin Cylindrospermopsin. 2015. Available online: https://www.epa.gov/sites/production/files/2017-06/documents/cylindrospermopsin-support-report-2015.pdf (accessed on 10 November 2017).
3. Mariani, M.A.; Padedda, B.M.; Kaštovský, J.; Buscarinu, P.; Sechi, N.; Virdis, T.; Lugliè, A. Effects of trophic status on microcystin production and the dominance of cyanobacteria in the phytoplankton assemblage of Mediterranean reservoirs. *Sci. Rep.* **2015**, *5*, 17964. [CrossRef] [PubMed]
4. Environmental Protection Agency. Contaminant Candidate List (CCL) and Regulatory Determination. 2016. Available online: https://www.epa.gov/ccl/contaminant-candidate-list-4-ccl-4-0 (accessed on 10 November 2017).
5. World Health Organization (WHO). Toxic Cyanobacteria in Water: A Guide to Their Public Health Consequences, Monitoring and Management 1999. Available online: http://www.who.int/water_sanitation_health/resourcesquality/toxcyanchap3.pdf (accessed on 1 October 2017).
6. Wang, D.Z. Neurotoxins from Marine Dinoflagellates: A Brief Review. *Mar. Drugs* **2008**, *6*, 349–371. [CrossRef] [PubMed]

7. Del Chierico, F.; Vernocchi, P.; Bonizzi, L.; Carsetti, R.; Castellazzi, A.M.; Dallapiccola, B.; de Vos, W.; Guerzoni, M.E.; Manco, M.; Marseglia, G.L.; et al. Early-life gut microbiota under physiological and pathological conditions: The central role of combined meta-omics-based approaches. *J. Proteom.* **2012**, *75*, 4580–4587. [CrossRef] [PubMed]
8. Liu, Y.; Chen, W.; Li, D.; Shen, Y.; Li, G.; Liu, Y. First report of aphantoxins in China—Waterblooms of toxigenic *Aphanizomenon flos-aquae* in Lake Dianchi. *Ecotoxicol. Environ. Saf.* **2006**, *65*, 84–92. [CrossRef] [PubMed]
9. Bautista, A.C.; Moore, C.E.; Lin, Y.; Cline, M.G.; Benitah, N.; Puschner, B. Hepatopathy following consumption of a commercially available blue-green algae dietary supplement in a dog. *BMC Vet. Res.* **2015**, *11*, 136. [CrossRef] [PubMed]
10. Gilroy, D.J.; Kauffman, K.W.; Hall, R.A.; Huang, X.; Chu, F.S. Assessing potential health risks from microcystin toxins in blue-green algae dietary supplements. *Environ. Health Perspect.* **2000**, *108*, 435–439. [CrossRef] [PubMed]
11. Heussner, A.H.; Mazija, L.; Fastner, J.; Dietrich, D.R. Toxin content and cytotoxicity of algal dietary supplements. *Toxicol. Appl. Pharmacol.* **2012**, *265*, 263–271. [CrossRef] [PubMed]
12. Stewart, I.; Seawright, A.A.; Shaw, G.R. Cyanobacterial poisoning in livestock, wild mammals and birds—An overview. *Adv. Exp. Med. Biol.* **2008**, *619*, 613–637. [PubMed]
13. Watson, S.B.; Miller, C.; Arhonditsis, G.; Boyer, G.L.; Carmichael, W.; Charlton, M.N.; Confesor, R.; Depew, D.C.; Höök, T.O.; Ludsin, S.A.; et al. The re-eutrophication of Lake Erie: Harmful algal blooms and hypoxia. *Harmful Algae* **2016**, *56*, 44–66. [CrossRef] [PubMed]
14. Environmental Protection Agency. Cyanobacteria/Cyanotoxins, Nutrient Policy Data. 2017. Available online: https://www.epa.gov/nutrient-policy-data/cyanobacteriacyanotoxin (accessed on 1 October 2017).
15. Cires, S.; Ballot, A. A review of the phylogeny, ecology and toxin production of bloom-forming *Aphanizomenon* spp. and related species within the *Nostocales* (*cyanobacteria*). *Harmful Algae* **2016**, *54*, 21–43. [PubMed]
16. WHO. Cyanobacterial Toxins: Microcystin-LR in Drinking-Water Background Document for Development of WHO Guidelines for Drinking-Water Quality. 2003. Available online: http://www.who.int/water_sanitation_health/dwq/chemicals/cyanobactoxins.pdf (accessed on 10 November 2017).
17. Preussel, K.; Stüken, A.; Wiedner, C.; Chorus, I.; Fastner, J. First report on cylindrospermopsin producing *Aphanizomenon flos-aquae (Cyanobacteria)* isolated from two German lakes. *Toxicon* **2006**, *47*, 156–162. [CrossRef] [PubMed]
18. Ballot, A.; Fastner, J.; Wiedner, C. Paralytic shellfish poisoning toxin-producing cyanobacterium *Aphanizomenon gracile* in northeast Germany. *Appl. Environ. Microbiol.* **2010**, *76*, 1173–1180. [CrossRef] [PubMed]
19. Stuken, A.; Jakobsen, K.S. The cylindrospermopsin gene cluster of *Aphanizomenon* spp. strain 10E6: Organization and recombination. *Microbiology* **2010**, *156*, 2438–2451. [PubMed]
20. Ballot, A.; Cerasino, L.; Hostyeva, V.; Cirés, S. Variability in the *sxt* Gene Clusters of PSP Toxin Producing *Aphanizomenon gracile* Strains from Norway, Spain, Germany and North America. *PLoS ONE* **2016**, *11*, e0167552. [CrossRef] [PubMed]
21. Soto-Liebe, K.; López-Cortés, X.A.; Fuentes-Valdes, J.J.; Stucken, K.; Gonzalez-Nilo, F.; Vásquez, M. In silico analysis of putative paralytic shellfish poisoning toxins export proteins in cyanobacteria. *PLoS ONE* **2013**, *8*, e55664. [CrossRef]
22. Botana, L.M. Marine toxins and the cytoskeleton. *FEBS J.* **2008**, *275*, 6059. [CrossRef] [PubMed]
23. Fawell, J.K.; Mitchell, R.E.; Hill, R.E.; Everett, D.J. The toxicity of cyanobacterial toxins in the mouse: II anatoxin-a. *Hum. Exp. Toxicol.* **1999**, *18*, 168–173. [CrossRef] [PubMed]
24. Cook, W.O.; Iwamoto, G.A.; Schaeffer, D.J.; Carmichael, W.W.; Beasley, V.R. Pathophysiologic effects of anatoxin-a(s) in anaesthetized rats: The influence of atropine and artificial respiration. *Pharmacol. Toxicol.* **1990**, *67*, 151–155. [CrossRef] [PubMed]
25. Sivonen, K.; Skulberg, O.M.; Namikoshi, M.; Evans, W.R.; Carmichael, W.W.; Rinehart, K.L. Two methyl ester derivatives of microcystins, cyclic heptapeptide hepatotoxins, isolated from *Anabaena flos-aquae* strain CYA 83/1. *Toxicon* **1992**, *30*, 1465–1471. [CrossRef]
26. Araoz, R.; Molgo, J.; de Marsac, N.T. Neurotoxic cyanobacterial toxins. *Toxicon* **2010**, *56*, 813–828. [CrossRef] [PubMed]

27. Carmichael, W.W.; Drapeau, C.; Anderson, D.M. Harvesting of *Aphanizomenon flos-aquae Ralfs ex Born. & Flah. var. flos-aquae (Cyanobacteria)* from Klamath Lake for human dietary use. *J. Appl. Phycol.* **2000**, *12*, 585–595.
28. Fastner, J.; Rücker, J.; Stüken, A.; Preußel, K.; Nixdorf, B.; Chorus, I.; Köhler, A.; Wiedner, C. Occurrence of the cyanobacterial toxin cylindrospermopsin in northeast Germany. *Environ. Toxicol.* **2007**, *22*, 26–32. [CrossRef] [PubMed]
29. Antunes, J.T.; Leao, P.N.; Vasconcelos, V.M. *Cylindrospermopsis raciborskii:* Review of the distribution, phylogeography, and ecophysiology of a global invasive species. *Front. Microbiol.* **2015**, *6*, 473. [CrossRef] [PubMed]
30. Harada, K.I.; Ohtani, I.; Iwamoto, K.; Suzuki, M.; Watanabe, M.F.; Watanabe, M.; Terao, K. Isolation of cylindrospermopsin from a cyanobacterium *Umezakia natans* and its screening method. *Toxicon* **1994**, *32*, 73–84. [CrossRef]
31. Ballot, A.; Ramm, J.; Rundberget, T.; Kaplan-Levy, R.N.; Hadas, O.; Sukenik, A.; Wiedner, C. Occurrence of non-cylindrospermopsin-producing *Aphanizomenon ovalisporum* and *Anabaena bergii* in Lake Kinneret (Israel). *J. Plankton Res.* **2011**, *33*, 1736–1746. [CrossRef]
32. Li, R.; Carmichael, W.W.; Brittain, S.; Eaglesham, G.K.; Shaw, G.R.; Liu, Y.; Watanabe, M.M. First report of the cyanotoxins cylindrospermopsin and deoxycylindrospermopsin from *Raphidiopsis curvata (cyanobacteria)*. *J. Phycol.* **2001**, *37*, 1121–1126. [CrossRef]
33. Shaw, G.R.; Sukenik, A.; Livne, A.; Chiswell, R.K.; Smith, M.J.; Seawright, A.A.; Norris, R.L.; Eaglesham, G.K.; Moore, M.R. Blooms of the cylindrospermopsin containing cyanobacterium, *Aphanizomenon ovalisporum (Forti)*, in newly constructed lakes, Queensland, Australia. *Environ. Toxicol.* **1999**, *14*, 167–177. [CrossRef]
34. Froscio, S.M.; Humpage, A.R.; Burcham, P.C.; Falconer, I.R. Cylindrospermopsin-induced protein synthesis inhibition and its dissociation from acute toxicity in mouse hepatocytes. *Environ. Toxicol.* **2003**, *18*, 243–251. [CrossRef] [PubMed]
35. Runnegar, M.T.; Xie, C.; Snider, B.B.; Wallace, G.A.; Weinreb, S.M.; Kuhlenkamp, J. In Vitro Hepatotoxicity of the Cyanobacterial Alkaloid Cylindrospermopsin and Related Synthetic Analogues. *Toxicol. Sci.* **2002**, *67*, 81–87. [CrossRef] [PubMed]
36. Bazin, E.; Huet, S.; Jarry, G.; Hégarat, L.L.; Munday, J.S.; Humpage, A.R.; Fessard, V. Cytotoxic and genotoxic effects of cylindrospermopsin in mice treated by gavage or intraperitoneal injection. *Environ. Toxicol.* **2012**, *27*, 277–284. [CrossRef] [PubMed]
37. Humpage, A.R.; Falconer, I.R. Oral toxicity of the cyanobacterial toxin cylindrospermopsin in male Swiss albino mice: Determination of no observed adverse effect level for deriving a drinking water guideline value. *Environ. Toxicol.* **2003**, *18*, 94–103. [CrossRef] [PubMed]
38. Meneely, J.P.; Elliott, C.T. Microcystins: Measuring human exposure and the impact on human health. *Biomarkers* **2013**, *18*, 639–649. [CrossRef] [PubMed]
39. Fawell, J.K.; Mitchell, R.E.; Everett, D.J.; Hill, R.E. The toxicity of cyanobacterial toxins in the mouse: I microcystin-LR. *Hum. Exp. Toxicol.* **1999**, *18*, 162–167. [CrossRef] [PubMed]
40. Heinze, R. Toxicity of the cyanobacterial toxin microcystin-LR to rats after 28 days intake with the drinking water. *Environ. Toxicol.* **1999**, *14*, 57–60. [CrossRef]
41. Schaeffer, D.J.; Malpas, P.B.; Barton, L.L. Risk assessment of microcystin in dietary *Aphanizomenon flos-aquae*. *Ecotoxicol. Environ. Saf.* **1999**, *44*, 73–80. [CrossRef] [PubMed]
42. Campos, A.; Vasconcelos, V. Molecular mechanisms of microcystin toxicity in animal cells. *Int. J. Mol. Sci.* **2010**, *11*, 268–287. [CrossRef] [PubMed]
43. Cusick, K.D.; Sayler, G.S. An overview on the marine neurotoxin, saxitoxin: Genetics, molecular targets, methods of detection and ecological functions. *Mar. Drugs* **2013**, *11*, 991–1018. [CrossRef] [PubMed]
44. European Food Safety Authority (EFSA). *Scientific Opinion: Marine Biotoxins in Shellfish-Saxitoxin Group;* EFSA: Parma, Italy, 2009; pp. 1–76.
45. Drobac, D.; Tokodi, N.; Simeunović, J.; Baltić, V.; Stanić, D.; Svirčev, Z. Human exposure to cyanotoxins and their effects on health. *Arh Hig Rada Toksikol* **2013**, *64*, 305–316. [CrossRef] [PubMed]
46. Palus, J.; Dziubałtowska, E.; Stańczyk, M.; Lewińska, D.; Mankiewicz-Boczek, J.; Izydorczyk, K.; Bonisławska, A.; Jurczak, T.; Zalewski, M.; Wąsowicz, W. Biomonitoring of cyanobacterial blooms in Polish water reservoir and the cytotoxicity and genotoxicity of selected cyanobacterial extracts. *Int. J. Occup. Med. Environ. Health* **2007**, *20*, 48–65. [CrossRef] [PubMed]

47. Sulcius, S.; Pilkaitytė, R.; Mazur-Marzec, H.; Kasperovičienė, J.; Ezhova, E.; Błaszczyk, A.; Paškauskas, R. Increased risk of exposure to microcystins in the scum of the filamentous cyanobacterium *Aphanizomenon flos-aquae* accumulated on the western shoreline of the Curonian Lagoon. *Mar. Pollut. Bull.* **2015**, *99*, 264–270. [CrossRef] [PubMed]
48. Dadheech, P.K.; Selmeczy, G.B.; Vasas, G.; Padisák, J.; Arp, W.; Tapolczai, K.; Casper, P.; Krienitz, L. Presence of potential toxin-producing cyanobacteria in an oligo-mesotrophic lake in Baltic Lake District, Germany: An ecological, genetic and toxicological survey. *Toxins* **2014**, *6*, 2912–2931. [CrossRef] [PubMed]
49. Butler, N.; Carlisle, J.; Linville, R. *Toxicological Summary and Suggested Action Levels to Reduce Potential Adverse Health Effects of Six Cyanotoxins*; Office of Environmental Health Hazard Assessment, California Environmental Protection Agency: Sacramento, CA, USA, 2012; pp. 1–119.
50. Ferreira, F.M.; Soler, J.M.F.; Fidalgo, M.L.; Fernández-Vila, P. PSP toxins from *Aphanizomenon flos-aquae (cyanobacteria)* collected in the Crestuma-Lever reservoir (Douro river, northern Portugal). *Toxicon* **2001**, *39*, 757–761. [CrossRef]
51. Brient, L.; Lengronne, M.; Bormans, M.; Fastner, J. First occurrence of cylindrospermopsin in freshwater in France. *Environ. Toxicol.* **2009**, *24*, 415–420. [CrossRef] [PubMed]
52. Van Apeldoorn, M.E.; Van Egmond, H.P.; Speijers, G.J.; Bakker, G.J. Toxins of cyanobacteria. *Mol. Nutr. Food Res.* **2007**, *51*, 7–60. [CrossRef] [PubMed]
53. Mahmood, N.A.; Carmichael, W.W. Paralytic shellfish poisons produced by the freshwater cyanobacterium *Aphanizomenon flos-aquae* NH-5. *Toxicon* **1986**, *24*, 175–186. [CrossRef]
54. Pereira, P.; Onodera, H.; Andrinolo, D.; Franca, S.; Araújo, F.; Lagos, N.; Oshima, Y. Paralytic shellfish toxins in the freshwater cyanobacterium *Aphanizomenon flos-aquae,* isolated from Montargil reservoir, Portugal. *Toxicon* **2000**, *38*, 1689–1702. [CrossRef]
55. Gkelis, S.; Zaoutsos, N. Cyanotoxin occurrence and potentially toxin producing cyanobacteria in freshwaters of Greece: A multi-disciplinary approach. *Toxicon* **2014**, *78*, 1–9. [CrossRef] [PubMed]
56. Roy-Lachapelle, A.; Solliec, M.; Bouchard, M.F.; Sauvé, S. Detection of Cyanotoxins in Algae Dietary Supplements. *Toxins* **2017**, *9*, 76. [CrossRef] [PubMed]
57. Dietrich, D.; Hoeger, S. Guidance values for microcystins in water and cyanobacterial supplement products (blue-green algal supplements): A reasonable or misguided approach? *Toxicol. Appl. Pharmacol.* **2005**, *203*, 273–289. [CrossRef] [PubMed]
58. Farrer, D.; Counter, M.; Hillwig, R.; Cude, C. Health-based cyanotoxin guideline values allow for cyanotoxin-based monitoring and efficient public health response to cyanobacterial blooms. *Toxins* **2015**, *7*, 457–477. [CrossRef] [PubMed]
59. Saker, M.L.; Welker, M.; Vasconcelos, V.M. Multiplex PCR for the detection of toxigenic cyanobacteria in dietary supplements produced for human consumption. *Appl. Microbiol. Biotechnol.* **2007**, *73*, 1136–1142. [CrossRef] [PubMed]
60. Draisci, R.; Ferretti, E.; Palleschi, L.; Marchiafava, C. Identification of anatoxins in blue-green algae food supplements using liquid chromatography-tandem mass spectrometry. *Food Addit. Contam.* **2001**, *18*, 525–531. [CrossRef] [PubMed]
61. Rellan, S.; Osswald, J.; Saker, M.; Gago-Martinez, A.; Vasconcelos, V. First detection of anatoxin-a in human and animal dietary supplements containing cyanobacteria. *Food Chem. Toxicol.* **2009**, *47*, 2189–2195. [CrossRef] [PubMed]
62. Chernova, E.; Sidelev, S.; Russkikh, I.; Voyakina, E.; Babanazarova, O.; Romanov, R.; Kotovshchikov, A.; Mazur-Marzec, H. *Dolichospermum* and *Aphanizomenon* as neurotoxins producers in some Russian freshwaters. *Toxicon* **2017**, *130*, 47–55. [CrossRef] [PubMed]
63. Mooney, K.M.; Hamilton, J.T.; Floyd, S.D.; Foy, R.H.; Elliott, C.T. Initial studies on the occurrence of cyanobacteria and microcystins in Irish lakes. *Environ. Toxicol.* **2011**, *26*, 566–570. [CrossRef] [PubMed]
64. Blahova, L.; Oravec, M.; Maršálek, B.; Šejnohova, L.; Šimek, Z.; Bláha, L. The first occurrence of the cyanobacterial alkaloid toxin cylindrospermopsin in the Czech Republic as determined by immunochemical and LC/MS methods. *Toxicon* **2009**, *53*, 519–524. [CrossRef] [PubMed]
65. Li, R.; Carmichael, W.W.; Pereira, P. Morphological and 16S rRNA Gene Evidence for Reclassification of the Paralytic Shellfish Toxin Producing *Aphanizomenon Flos-Aquae* LMECYA 31 as *Aphanizomenon Issatschenkoi (Cyanophyceae)*. *J. Phycol.* **2003**, *39*, 814–818. [CrossRef]

66. Rucker, J.; Stüken, A.; Nixdorf, B.; Fastner, J.; Chorus, I.; Wiedner, C. Concentrations of particulate and dissolved cylindrospermopsin in 21 *Aphanizomenon*-dominated temperate lakes. *Toxicon* **2007**, *50*, 800–809. [CrossRef] [PubMed]
67. EPA. Guidelines and Recommendations-Nutrient Policy Data, 2017. 2017. Available online: https://19january2017snapshot.epa.gov/nutrient-policy-data/guidelines-and-recommendations_.html (accessed on 1 October 2017).
68. Catherine, Q.; Susanna, W.; Isidora, E.S.; Mark, H.; Aurelie, V.; Jean-François, H. A review of current knowledge on toxic benthic freshwater cyanobacteria—Ecology, toxin production and risk management. *Water Res.* **2013**, *47*, 5464–5479. [CrossRef] [PubMed]
69. Zhang, D.; Hu, C.; Wang, G.; Li, D.; Li, G.; Liu, Y. Zebrafish neurotoxicity from aphantoxins—Cyanobacterial paralytic shellfish poisons (PSPs) from *Aphanizomenon flos-aquae* DC-1. *Environ. Toxicol.* **2013**, *28*, 239–254. [CrossRef] [PubMed]
70. Moher, D.; Liberati, A.; Tetzlaff, J.; Altman, D.G.; Prisma Group. Preferred reporting items for systematic reviews and meta-analyses: The PRISMA statement. *Int. J. Surg.* **2010**, *8*, 336–341. [PubMed]

toxins

MDPI

Review

In Vitro Toxicological Assessment of Cylindrospermopsin: A Review

Silvia Pichardo, Ana M. Cameán * and Angeles Jos

Area of Toxicology, Faculty of Pharmacy, University of Sevilla, C/Profesor García González 2, 41012 Sevilla, Spain; spichardo@us.es (S.P.); angelesjos@us.es (A.J.)
* Correspondence: camean@us.es; Tel.: +34-954-556-762

Academic Editor: Vítor Vasconcelos
Received: 6 November 2017; Accepted: 13 December 2017; Published: 16 December 2017

Abstract: Cylindrospermopsin (CYN) is a cyanobacterial toxin that is gaining importance, owing to its increasing expansion worldwide and the increased frequency of its blooms. CYN mainly targets the liver, but also involves other organs. Various mechanisms have been associated with its toxicity, such as protein synthesis inhibition, oxidative stress, etc. However, its toxic effects are not yet fully elucidated and additional data for hazard characterization purposes are required. In this regard, *in vitro* methods can play an important role, owing to their advantages in comparison to *in vivo* trials. The aim of this work was to compile and evaluate the *in vitro* data dealing with CYN available in the scientific literature, focusing on its toxicokinetics and its main toxicity mechanisms. This analysis would be useful to identify research needs and data gaps in order to complete knowledge about the toxicity profile of CYN. For example, it has been shown that research on various aspects, such as new emerging toxicity effects, the toxicity of analogs, or the potential interaction of CYN with other cyanotoxins, among others, is still very scarce. New *in vitro* studies are therefore welcome.

Keywords: cylindrospermopsin; *in vitro*; cytotoxicity; oxidative stress; genotoxicity

1. Introduction

The cyanobacterial toxin Cylindrospermopsin (CYN) is a tricyclic alkaloid that consists of a tricyclic guanidine moiety combined with hydroxymethyluracil [1]. Owing to its zwitterionic nature, CYN is a highly water-soluble compound [2]. Currently, five analogs of CYN are known, namely CYN, 7-epi-CYN, 7-deoxy-CYN, and the two recently characterized congeners, 7-deoxydesulfo-CYN and 7-deoxydesulfo-12-acetyl-CYN [3,4].

It was first reported in 1979 after a hepatoenteritis outbreak occurred in Palm Island, northern Queensland, Australia [5], owing to a *Cylindrospermopsis raciborskii* bloom in the local drinking water supply. Nowadays, the variety of identified CYN-producing cyanobacteria species has increased considerably (i.e., *Umezakia natans*, *Aphanizomenon ovalisporum*, *Raphidiopsis curvata*, *Anabaena bergii*, *Aphanizomenon flos-aquae*, *Anabaena lapponica*, etc.). *Aphanizomenon gracile* and *A. flos-aquae* are the most important CYN producers in Europe [6]. Moreover, for most of the known CYN-producing species, both CYN-producing and nonproducing strains have been observed [4].

The occurrence of CYN and/or CYN-producing species has been reported worldwide, in Germany, Saudi Arabia, Australia, China, Israel, Spain, United States of America (USA), Italy, Finland, Poland, Portugal, France, etc. [4,7]. The ever-expanding distribution of CYN producers into temperate zones is heightening concern that this toxin will represent serious human, as well as environmental, health risks across many countries [8]. Among the reasons for the increase in extension and frequency of their blooms, i.e., cyanobacterial growth at high densities, are anthropogenic activities and climate changes.

With regard to environmental concentrations of CYN, it is usually found in the range of 1–10 μg/L [9], the highest values reported being 589 μg/L in an aquaculture pond in Queensland [10] and 800 μg/L in a farm dam in Australia [11,12].

CYN can adversely affect both humans and the environment. Human exposure to CYN may occur by different pathways. Dermal contact with CYN may occur during showering or bathing, or during recreational activities such as wading, swimming, boating, or water skiing. Also, by ingesting toxin-contaminated water during recreational activities or by the ingestion of food or water contaminated with the toxin. In fact, it has been demonstrated that cyanobacterial toxins (including CYN) are able to accumulate in edible plants [13,14], fish [15], crustaceans [10], etc., an aspect that has been reviewed by Gutiérrez-Praena et al. [16]. To protect consumers from the adverse effects of CYN, a provisional Tolerable Daily Intake (TDI) of 0.03 μg/kg body weight (b.w.) has been established [17]. Moreover, these authors also proposed a guideline value of 1 μg/L for CYN in drinking water.

CYN appears to be a molecule with a wide range of toxic effects. The toxin primarily targets the liver, but it is also a general cytotoxin that attacks the eye, spleen, kidney, lungs, thymus, heart, etc. [18]. The lack of a specific target for CYN hinders further efforts to understand its potent toxicity and to define acceptable thresholds of exposure [4]. In this line, the European Authority of Food Safety (EFSA) considers that there are also data gaps regarding the characterization of the toxicological profile of cyanotoxins other than microcystins [19].

If we focus on the toxicological evaluation that is required for hazard characterization purposes, *in vitro* methods play an important role. The use of *in vitro* model systems in toxicity testing has many advantages, including a decrease in animal numbers, a reduced cost of animal maintenance and care, a small quantity of chemicals needed for testing, shortening of the time needed, and an increase in throughput for evaluating multiple chemicals and their metabolites [20]. *In vitro* systems can also be used to study chemical metabolism, evaluate toxicity mechanisms, measure enzyme kinetics, and examine dose-response relationships [20,21]. Thus, the aim of this work was to compile and evaluate the *in vitro* data dealing with CYN that are available in the scientific literature, focusing on its toxicokinetics and its main toxicity mechanisms. This analysis would be useful to identify research needs and data gaps in order to complete knowledge about the toxicity profile of CYN.

2. Basal Cytotoxicity Assays and Morphological Studies

Tables 1 and 2 show the various CYN *in vitro* studies that are dealing with these two basic toxicological features, respectively [9,22–42]. *In vivo* studies in mice suggest that liver is a major target organ; in fact, CYN has traditionally been classified as a hepatotoxin [43]. Consequently, most of the first *in vitro* studies performed to investigate the cytotoxicity of this cyanotoxin used primary rodent hepatocytes [44]. Primary rat hepatocytes exposed for 18 h to 3.3–5 μM of CYN isolated from *C. raciborskii* cultures resulted in significant cell death [22]. Subsequently, the same authors studied the toxicity of natural and synthetic CYN and its analogs in rat hepatocytes in order to investigate the role of various chemical groups [24]. They showed that the sulfate group and the orientation of the hydroxyl group at C-7 were not relevant in CYN biological activity. Recently, the toxicity of four CYN analogs, which are differing in the length of tether guanidine and uracil groups, and the presence or absence of a hydroxyl group, was studied. Preliminary findings revealed that the −OH group at C-7 of the toxin was responsible of toxic effects induced on human neutrophils [9]. In addition, Neumann et al. [26] compared the toxicity of CYN and its analog deoxycylindrospermopsin (deoxy-CYN), showing similar effects on cell viability and proliferation in different cell lines.

Table 1. *In vitro* cytotoxicity studies performed with Cylindrospermopsin (CYN).

Toxin/Cyanobacteria	Experimental Model	Assays Performed	Exposure Conditions Concentration Ranges	Main Results	Reference
Purified extract from *Cylindrospermopsis raciborskii*	Primary rats hepatocytes	Lactate dehydrogenase (LDH) activity	0.5–5 μM for 0–18 h	After 18 h of incubation with 3.3 and 5 μM, significant cell death (40% and 67%, respectively) was found. No measurable cell lysis within the first 12 h of exposure to CYN, although slight signs of rounding were observed.	[22]
Commercial CYN pure standard	Primary rat hepatocytes and KB cells	3-(4,5-dimethylthiazol-2-yl)-2,5 diphenyltetrazolium bromide (MTT) assay	0–10,000 ng/mL for 0, 24, 48, 72 h	Toxic effects were observed at after 72 h, being the LC50 40 ng/mL in the case of exposure to rat hepatocytes while in KB cells it was 200 ng/mL.	[23]
Purified extract from *Cylindrospermopsis raciborskii*. Synthetic CYN and analogues: racemic CYN (RAC-CYN), CYN-DIOL, AB-MODEL, Epi-cylindrospermopsin (EPI-CYN), and EPI-DIOL	Primary rats hepatocytes	LDH release	CYN: 0.16–10 μM RAC-CYN: 1.25–20 μM CYN-DIOL: 0.16–5 μM EPI-CYN: 0.075–12.5 μM EPI-DIOL: 0.1–50 μM 19 h	When hepatocytes were exposed to 20 μM of RAC-CYN cell death increased from 14% to 23%, while in the case of CYN-DIOL cell death enhanced up to 38%. Similar results were observed in exposures to 10 μM CYN, 12.5 μM EPI-CYN and 50 μM EPI-DIOL with increases of 33.4%, 38.5% and 35%, respectively.	[24]
Purified extract from *Cylindrospermopsis raciborskii*	Primary rat hepatocytes; Caco-2 and HepG2 cells	MTT assay	24 h to primary rat hepatocytes 48 h to permanent cells lines	None of the 7 isolates of *C. raciborskii* contained CYN; however, they were all toxic. The methanolic extracts were generally more toxic than the aqueous extracts.	[25]
Commercial CYN pure standard	Primary mouse hepatocytes	LDH release	0.05–25 μM for 18, 21, 24 h	Time- and concentration- dependent increases in LDH leakage was observed after exposure to CYN. The EC50 at 18 h was 0.47 μM. The concentration response was very steep, with concentrations of 1 μM and above producing greater than 75% LDH leakage within 18 h whereas concentrations below 0.1 μM had no effect.	[26]
Commercial CYN pure standard	HDF, HepG2, and Caco-2 cells	MTS assay and LDH leakage	0.1–5 μg/mL CYN for 24, 48, 72 h	Although it was not possible to calculate the IC50 for the MTS assay due to lack of data for higher concentrations, a time-dependent effect was observed in all three cell types. However, no effect was observed in the LDH assay in rHepG2 and Caco-2 cells, but HDF cells reached 30% of the lysed controls at concentrations above 1 μg/mL CYN (2.4 μM) after 72 h.	[27]
Purified extract from *Cylindrospermopsis raciborskii*	CHO-K1 cells	Annexin V- fluorescein isothiocyanate/propidium ionide (FITC) apoptosis detection kit	0.05–2 μg/mL for 3, 16, 21 h	CYN increases the frequency of necrotic cells in a dose and time-dependent manner, but very slight impact on apoptosis was observed. In addition, when cells are metabolic activated the susceptibility to necrotic cell death increases, whereas it has no impact on apoptosis.	[28]
Purified extract containing CYN and deoxyCYN	HepG2, BE-2, Caco-2, MNA, HDF	Trypan blue exclusion test (TBET) and MTS assays	0.1–5 μg/mL for 24, 48, 72 h	Both CYN and deoxyCYN exerted toxic effects to all exposed cells in a concentration and dependent way, being deoxyCYN slightly less cytotoxic than CYN.	[29]

Table 1. *Cont.*

Toxin/Cyanobacteria	Experimental Model	Assays Performed	Exposure Conditions Concentration Ranges	Main Results	Reference
Commercial CYN pure standard	Primary human granulosa cells	MTT assay	0–1 μg/mL for 2, 4, 6, 24, 48, 72 h	No effect was recorded in cells exposed up to 1 μg/mL in short 2–6 h exposures. However, cell viability decreased in a concentration-dependent way at longer exposures (24–72 h).	[30]
Commercial CYN pure standard	C3A, HepG2, NCI-87, HCT-8, HuTu-80, Caco-2, and Vero cells	MTT assay and LDH leakage	0.4–66 μM for 1, 2, 4, 6, 24 h	The 24 h IC50 for CYN cytotoxicity was set at 1.5 μM for hepatic cell lines (C3A and HepG2 cells), while for colonic cells (Caco-2) the IC50 was 6.5 μM. Similar onset was found in hepatic cells (C3A) in long-term exposures up to 7 days. No recovery of the toxicity caused by CYN was evidenced in C3A cells after exposure for 1–6 h.	[31]
Commercial CYN pure standard	Vero-GFP cells	MTS (3-(4,5-dimethylthiazol-2-yl)-5-(3-carboxymethoxyphenyl)-2-(4-sulfophenyl)-2H-tetrazolium) assay	0.1–100 μM for 4, 24 h	The IC50 found for CYN after 24 h was 5.9 μM. The use of other protein inhibitors indicated that the toxicity exerted by CYN was not only related to protein synthesis mechanism but other effects may contribute to the toxicity observed.	[32]
Commercial CYN pure standard	CHO cells	Annexin V-FITC assay	0.1–10 μM for 12, 18, 24, 48 h	CYN cause apoptosis at low concentrations (1–2 μM) and over short exposure periods (12 h). Necrosis was observed at higher concentrations (5–10 μM) and following longer exposure periods (24 or 48 h).	[33]
Commercial CYN pure standard	PLHC-1	Protein content (PC), neutral red uptake (NRU) and MTS assay	0.3–40 μg/mL for 24, 48 h	Cytotoxic effects were observed in all the endpoints assayed in a time and concentration-dependent manner. Regarding the EC50 values, the most sensitive endpoint was PC for 24 h of exposure, with an EC50 of 8 μg/mL, and MTS assay for 48 h with an EC50 of 2.2 μg/mL.	[34]
Purified extract containing CYN	Primary *Prochilodus lineatus* hepatocytes	NRU	0.1–10 μg/L for 72 h	Cell viability decreased 8% in hepatocytes exposed to 0.1 and 1 μg/L. However, at the highest concentration assayed (10 μg/L) no significant change was observed in comparison to the control.	[35]
Commercial CYN pure standard	Caco-2	PC, NRU and MTS assays	0.3–40 μg/mL for 24, 48 h	The most significant endpoint was MTS assay. This endpoint revealed significant cytotoxicity in Caco-2 cells exposed to all concentrations assayed except for the lowest concentration after 24 h. The EC50 were 2.5 μg/mL for 24 h and 0.6 μg/mL for 48 h.	[36]
Commercial CYN pure standard	HUVEC	PC, NRU and MTS assays	0.3–40 μg/mL for 24, 48 h	The higher cytotoxic effects were observed in NRU. Very low rates of cell viability were reported at 40 μg/mL, being 20% and 3% after 24 and 48 h, respectively. Similarly, low EC50 were found, 1.5 μg/mL for 24 h and 0.8 for 48 h.	[37]

Table 1. *Cont.*

Toxin/Cyanobacteria	Experimental Model	Assays Performed	Exposure Conditions Concentration Ranges	Main Results	Reference
Commercial CYN pure standard	Primary rat hepatocytes	Alarm blue assay	10–360 nM for 24, 48 h	CYN reduced cell viability in hepatocytes exposed to 90, 180 and 360 nM CYN. The two higher concentrations (180 and 360 nM) decreased cell viability around 50% after 48 and 24 h, respectively.	[38]
Commercial CYN pure standard	Caco-2 and Clone 9 cells	Alarm blue assay	0.1–10 μM for 8, 10, 12, 24, 48, 72 h	No cytotoxicity was observed for Caco-2 cells exposed to CYN up to 72 h. However, a time and concentration-dependent decrease in viability of Clone 9 cells exposed to CYN in comparison to the controls.	[39]
Commercial CYN pure standard	Primary human T-lymphocytes	FAM caspase activity kit	1 μg/mL for 6, 24, 48 h	The viability of human T-lymphocytes decreased in a concentration and time dependent way. Significant decreases were observed in exposure to 1 μg/mL, with the highest alterations observed after 24 h of exposure.	[40]
Purified extract containing CYN	HepG2	NRU and MTT assay	0.001–100 μg/L for 4, 12, 24, 48 h	CYN was not toxic to HepG2 cells after 48 h of exposure, except for the higher concentration (100 μg/L) with a decrease of 11%. At concentrations bellow 10 μg/L cell viability increased.	[41]
Synthetic CYN analogues (11a, 11b, 11c and 22), 1 and guanidinoacetate (GAA)	human neutrophils	MTT assay	2.0 μg/mL for 1 h	The general toxicity decreased in the following order: 11c > 11a > 1 > 11b > 22 > GAA. No remarkable toxic effect was observed for the two last compounds (22 and GAA).	[42]

Abbreviations: BE-2 (Caucasian bone-marrow neuroblastoma cell line); Caco-2 (human colorectal adenocarcinoma cell line); CHO (Chinese hamster ovary cell line); CHO (a subclone from the parental CHO cell line); cylindrospermopsin (CYN); C3A (human hepatocellular carcinoma); effective mean concentration (EC50); guanidinoacetate (GAA); HCT-8 (human ileal adenocarcinoma); HDF (human dermal fibroblast cell line); HepG2 (human liver hepatocellular carcinoma cell line); HuTu-80 (human duodenal adenocarcinoma); HUVEC (human vascular endothelium cell line); IC50 (inhibitory mean concentration); KB (human cervix carcinoma); MNA (mouse neuroblastoma cell line); MTS (3-(4,5-dimethylthiazol-2-yl)-5-(3-carboxymethoxyphenyl)-2-(4-sulfophenyl)-2H-tetrazolium); MTT (3-(4,5-dimethylthiazol-2-yl)-2,5 diphenyltetrazolium bromide); NCI-N87 (human gastric carcinoma); PLHC-1 (*Poeciliopsis lucida* hepatocellular carcinoma cell line); SHE (Syrian hamster embryo cell line); Vero (African green monkey kidney).

Table 2. *In vitro* morphological studies dealing with CYN.

Toxin/Cyanobacteria	Experimental Model	Microscopy Used	Exposure Conditions Concentration Ranges	Main Results	Reference
Purified extract containing CYN and deoxyCYN	HepG2, BE-2 and MNA cells	Light microscope	0.5–5 μg/mL for 24, 48 h	The BE-2 and MNA cells underwent shrinkage and cell rounding at 2.5 and 5 μg/mL, respectively. These findings indicated apoptosis process.	[29]
Commercial CYN pure standard	SHE cells	Light microscope	1×10^{-7}–1×10^{-3} ng/mL for 7 days	CYN induced morphological cell transformation after 7 days of treatment with CYN (1×10^{-7}–1×10^{-2} ng/mL), with nuclear enlargement. The morphologically transformed phenotype also showed loss of contact inhibition and density-dependent inhibition of cells.	[42]
Commercial CYN pure standard	Caco-2 cells	Light and electron microscopes	0.625, 2.5 μg/mL for 24, 48 h	The most remarkable ultrastructural changes were lipid degeneration, mitochondrial damage and nucleolar segregation with altered nuclei.	[36]
Commercial CYN pure standard	HUVEC cells	Light and electron microscopes	0.3–40 μg/mL for 24, 48 h	The main findings observed were nucleolar segregation with altered nuclei, degenerated Golgi apparatus, increases in the presence of granules and apoptosis.	[37]
Commercial CYN pure standard	Clone 9 cells	Light microscope	5 μM for 24, 48 h	After 24 h of treatment with CYN no discernible effect was observed, although after 48 h signs of damage and detachment of cells were reported.	[39]

Abbreviations: BE-2 (Caucasian bone-marrow neuroblastoma cell line); Caco-2 (human colorectal adenocarcinoma cell line); Clone 9 (*Rattus norvegicus* epithelial liver cell line); cylindrospermopsin (CYN); deoxycylindrospermopsin (deoxyCYN); HepG2 (human liver hepatocellular carcinoma cell line); HUVEC (human vascular endothelium cell line); MNA (mouse neuroblastoma cell line); SHE (Syrian hamster embryo cell line).

Given that metabolism plays an important role in the toxicity of CYN [24,28,45], the influence of metabolic inhibitors has also been studied. Co-exposure of CYN and CYP inhibitors on primary mouse hepatocytes for 18 h demonstrated the effectiveness of ketoconazole and SKF525A in decreasing CYN toxicity, while furafylline and omeprazole showed a moderate protective effect [26]. Similarly, various isoforms of CYP have been induced by ethanol, rifampicin, and phenobarbital, in order to assess the influence of each CYP on CYN biotransformation and further toxicity [41]. The authors showed that CYP-induced HepG2 cells were more sensitive to CYN exposure with regard to the decrease of cell viability after 24 h of exposure to 10 µg/L CYN.

In addition, the toxicity of CYN on primary hepatocytes and permanent cell lines has been compared. After 72 h of exposure to pure standard CYN, the viability of KB cells decreased from 200 ng/mL, whereas the effective concentration was around 10-fold lower (25 ng/mL) in rat hepatocytes [23]. The higher sensitivity of isolated hepatocytes to CYN in comparison to permanent cell lines will be discussed further. However, CYN causes severe toxicity in a wide range of human cell lines from various target organs, such as liver, kidney, and intestine [23,27,31,32]. After rodent hepatocytes, the most used permanent cell lines have been Caco-2 cells (from human intestinal carcinoma) and HepG2 cells (derived from human hepatoma). The hepatic-derived cells have proved to be more sensitive than intestinal ones, while colon-derived cells are even less sensitive than the others [31]. This finding may be related to the limited CYN uptake in colon cells that is reported by several authors [39,46].

Most of the experiments that have demonstrated toxic effects have used 24 and 48 h of exposure. In fact, Young et al. [30] showed no cytotoxic effects in human granulosa cells that were exposed for 6 h to 1 µg/mL, although at longer exposure times (24–72 h) cell viability decreased. In contrast, short-term exposure of hepatic cells (C3A) to CYN (1–6 h) was shown to induce cytotoxicity at 24 h despite a washout and recovery incubation, demonstrating the apparently irreversible nature of CYN toxicity [32]. Apart from the time-dependent cytotoxicity, the toxic effects of CYN also increase with concentration [26,28,30,34,36–40]. Surprisingly, CYN decreased cell viability in fish hepatocytes that were exposed to the lowest concentrations (0.1 and 1 µg/L), but no significant effect was recorded in the exposure to the highest concentration assayed (10 µg/L) [41]. This is the only work performed on primary fish hepatocytes, so the unexpected behavior cannot be compared with similar experiments. The only report available carried out on fish cells did not use isolated hepatocytes and instead used the permanent fish cell line, PLHC-1, a hepatocellular carcinoma of the cyprinid fish *Poeciliopsis lucida* [34]. However, this study revealed time- and concentration-dependent cytotoxic effects, but at higher concentrations than in the above-mentioned fish hepatocytes (0.3–40 µg/mL). When considering that most of the studies have been performed on mammalian cells, more research is needed using *in vitro* experimental models of aquatic origin, because they may easily be exposed to CYN.

Cell death was also determined using the Annexin V kit in rat hepatocytes that were exposed to CYN [38]. After 6 h of exposure, cells were stained with Annexin V, indicating that apoptosis is rapidly induced by CYN. Only at the highest concentrations assayed (180 and 360 nM CYN) did the cells suffer loss of membrane integrity after 48 h, demonstrated by propidium iodide staining. In order to determine whether necrosis was also induced by CYN, the LDH released to the medium was determined in hepatocyte cultures treated with the toxin, showing positive results after 72 h of treatment. However, CHO-K1 cells that were exposed to CYN for 3 and 16 h did not result in a statistically significant enhancement of the frequency of early apoptotic cells [28]. Only at a longer incubation time (21 h) was a dose-dependent increase in the frequency of early apoptotic cells observed, which was significant at 1 and 2 µg/mL CYN. Similar results were observed in necrotic cells, which increased steadily after 21 h of exposure. Poniedziałek et al. [40] also reported that 1 µg/mL CYN was able to cause both apoptosis and necrosis in human lymphocytes after 6 h of exposure, although at 72 h only necrotic cells were found. In general, cell death may occur by apoptosis or necrosis, depending on physiological conditions, developmental stages, cell type, and nature of the death signal [33]. These authors reported 19% of apoptotic cells and 9% of necrosis after incubation with 10 μM CYN for 18 h. At longer exposure times (24 h, 48 h) in the presence of CYN, apoptosis became a necrotic process, attaining about 75% after 48 h of incubation, with 10 μM CYN.

Although morphological studies are very scarce, they are of great interest because they are more sensitive than cytotoxicity studies and can be used as an early indicator of damage that is induced in cells [36]. In this context, using microscopy, Gutiérrez-Praena et al. [37] observed apoptosis in human endothelium cells (HUVEC), which showed pleomorphic nuclei after being exposed to 0.375 μg/mL CYN. In addition, they also reported nucleolar segregation with altered nuclei, degenerated Golgi apparatus, and increases of granules. These authors found lipid degeneration, mitochondrial damage, and nucleolar segregation with altered nuclei in Caco-2 cells after exposure to 2.5 μg/mL CYN [36]. Also, Maire et al. [42] showed nuclear alteration in Syrian hamster embryo (SHE) cells after seven days of exposure to 10^{-7}–10^{-2} ng/mL CYN. In the case of hepatic cells, no remarkable morphological changes were observed after 24 h incubation of Clone-9 cells with 5 mM CYN. However, after 48 h of exposure to the toxin, the cells showed evident disturbance, with signs of damage and detachment from the substrate [39]. Finally, significant morphological changes were also observed in various cell lines, BE2, MNA, and HepG2 cells, after exposure to 2.5 μg/mL CYN and deoxy-CYN. Moreover, BE2 and MNA cells underwent morphological changes that were indicative of apoptosis, such as cell shrinkage and cell rounding [29].

3. Toxicokinetic Studies

In vitro experiments provide a means of selectively measuring and estimating absorption, distribution, metabolism, and excretion (ADME) parameters [47]. Relevant studies on this topic are compiled in Table 3 [23,32,39,41,46,48]. The results that are obtained in these *in vitro* assays, once they have been properly analyzed, can be extrapolated reliably to the *in vivo* situation [47,49]. However, these data should also be completed with those that are obtained in physiologically-based toxicokinetic (PBTK) models for a better approach in the study of the toxicity of chemicals [50]. Such combinatorial approaches are very promising for the investigation of interspecies and intraspecies differences [51]. However, very few studies have been performed so far to study the ADME of CYN; and, they have focused mainly on absorption and metabolism.

The exact uptake mechanism of CYN has not been fully elucidated yet. The chemical characteristics of CYN, its size (415 Da), and hydrophilic nature indicate that it would be unlikely to cross the lipid bilayer of cell membrane, and therefore would need to be transported across the cell membrane [24,32]. Transport inhibitors have been used in order to clarify the CYN uptake mechanism that is involved [23]. Incubation with bile acids, cholate, and taurocholate, resulted in limited CYN uptake. A protective effect of bile acids was observed only after 48 h, but not at 72 h. These results showed that, although the bile acid transport system may participate in CYN uptake, another mechanism could be involved. In this context, a facilitated transport mechanism and active transport have been studied. Competition experiments excluded the uracil nucleobase transporter system as a potential mechanism for CYN uptake in Vero-GFP cells [32]. In addition, these authors confirmed that the uptake process is not energy-dependent because CYN entry also occurred at 4 °C, and at this temperature the energy-dependent cell processes are minimized. Similarly, no significant changes in CYN uptake were reported at 4 °C in comparison to 37 °C in Caco-2 intestinal cells [46]. However, a significant reduction in CYN transport was observed in the secretory direction when the temperature was decreased. Moreover, the main pathway that is involved in CYN uptake in intestinal cells was the paracellular route. As has been suggested with regard to other cell lines, a minor carrier-mediated transcellular transport has been indicated as a possible CYN uptake mechanism. This transport through the intestinal monolayer may be H^+ and GSH-dependent, and energy and Na^+-independent [46]. However, apart from these insights, intestinal uptake of CYN has not been reported, and it seems clear that intestinal absorption of CYN through Caco-2 cells is very limited. Fernandez et al. [39] reported that the passage of CYN across the intestinal monolayer was about 2.5% after 3 h and up to 20.5% after 24 h. Similarly, the permeability coefficients found in Pichardo et al. correlate well with very low *in vivo* absorption (below 20%) [46].

Table 3. *In vitro* toxicokinetics studies performed with CYN.

Toxin/Cyanobacteria	Experimental Model	Assays Performed	Exposure Conditions Concentration Ranges	Main Results	Reference
Commercial CYN pure standard	Primary rat hepatocytes and KB cells	Incubation with cholate and taurocholate and measurement of CYN uptake across hepatocyte	800 ng/mL for 0, 24, 48, 72 h	There was no protection against the toxicity of CYN at 72 h by both bile acids, although some protection was observed after 48 h. This suggests that bile acid transport may be involve in certain extent in the uptake of the toxin.	[23]
Commercial CYN pure standard	Vero-GFP cells	Monitoring CYN uptake in Vero cells expressing green fluorescent protein (GFP)	0.1–100 μM for 4, 24 h	CYN effects on GFP signal increased 6 fold over 4–24 h incubation indicating slow, progressive uptake of the toxin. However, the mechanism involved was not elucidated.	[32]
Commercial CYN pure standard	Caco-2 cells	Study of intestinal permeability of CYN	1–10 μM for 3, 10, 24 h	The CYN uptake across Caco-2 cells is limited. Only 2.4–2.7% of CYN was detected in the basolateral side after 3 h, increasing slightly up to 16.7–20.5% after 24 h.	[39]
Commercial CYN pure standard	HepG2 cells	Study the influence of cytochrome P450 (CYP) inductors on the cytotoxicity of CYN by means of viability assays	1, 10 μg/L for 4, 12, 24, 48 h	CYPs induction made HepG2 cells more sensitive to CYN toxic effects. Moreover, low concentrations of CYN increased the metabolism in HepG2 cells.	[41]
Commercial CYN pure standard	HepaRG cells and liver tissue fractions	Study of the metabolism of CYN by means of neutral red uptake assay with and without ketaconazol as well as by measuring CYN by LC/MS	0.1–50 μM for 24 h	The use of ketoconazole, a CYP3A4 inhibitor, led to a decreased cytotoxicity of CYN. However, no decrease of CYN was reported after co-incubation with the inhibitor both in HepaRG and liver fractions measured by high resolution mass spectrometry.	[48]
Commercial CYN pure standard	Caco-2 cells	Study of intestinal transport of CYN	0.8 mg/L for 30, 60, 90, 120 min	The paracellular route was pointed out as the most important pathway in CYN absorption. Although a second mechanism was not identified, some insights were reported. This minor carrier-mediated transcellular transport may be independent of energy and Na^+ and dependent of H^+ and GSH.	[46]

Abbreviations: Caco-2 (human colorectal adenocarcinoma cell line); cylindrospemopsin (CYN); HepaRG: (human hepatoma cells); HepG2: (human hepatoma cells); KB (human cervix carcinoma); Vero (African green monkey kidney).

As mentioned earlier, differences in the toxicity exerted by CYN have been observed in a variety of experimental models [52]. These differences could be related to a highly active transport process in primary hepatocytes or other primary cells that may be absent in immortalized cell lines [39]. However, the grade of metabolic competency of each *in vitro* experimental model may also be of concern. In this context, the influence of CYN metabolism is an important key to understanding the toxicity that is exerted by CYN. In fact, it has been proposed that the higher sensitivity of hepatocytes exposed to CYN is due to bioactivation-dependent events [39,45,53]. In this regard, the activity of the cytochrome P450 (CYP450) enzyme system has been shown to be important for the development of CYN toxicity in hepatocyte cultures [22,53]. This finding could be the explanation for the lower toxicity observed in permanent cell lines, such as KB cells [23], HeLa cell types [11], and CHO-K1 cells [28], in comparison to primary rat hepatocytes, suggesting that CYP450 activity is higher in hepatocytes. Moreover, some authors have reported that the metabolic activation of CYN intensified the cytotoxic effect, indicating that S9 fraction-induced metabolism of CYN is important for its cytotoxic activity [22,28,45], and also in genotoxicity effects (see Section 4.3). Despite the use of broad-spectrum CYP inhibitors, the isoforms that are involved have not been identified so far [41]. However, other authors have reported that preinduction of expression of xenobiotic metabolism enzymes, such as CYPs, does not increase the toxicity of CYN [41]. Similarly, no evidence was found for phase I metabolites of CYN when studying metabolic conversion using HepaRG cells and different liver tissue fractions, so this metabolic activation plays only a minor role for CYN toxicity [48]. With regard to the other two toxicokinetic phases, no *in vitro* experiment has been carried out to study the distribution or excretion of CYN as far as we know, and *in vivo* studies are scarce [54].

4. Toxicity Mechanisms

4.1. Protein Synthesis Inhibition

The first evidence that CYN induced an irreversible protein synthesis inhibition *in vivo* in mice was reported by Terao et al. [55]. Moreover, they verified this finding and also found inhibitory effects of the toxin on globin synthesis in a rabbit reticulocyte cell-free system. Subsequently, Froscio et al. [45] also confirmed this effect on primary mouse hepatocytes. These authors stated that protein synthesis inhibition was a sensitive early indicator of cellular responses to CYN. Moreover, the inhibition of CYP450 activity diminished the toxicity of CYN, but not the effects on protein synthesis. This suggests that the parent compound and the possibly formed metabolites could exert toxicity with a different mechanism, also depending on CYN concentrations [7].

4.2. Oxidative Stress

Oxidative stress is one of the toxic mechanisms postulated as being responsible for CYN toxicity. The term oxidative stress has been defined as a serious imbalance between reactive oxygen species (ROS) production and antioxidant defenses [56]. Sies [57] defined it as "a disturbance in the pro-oxidant–antioxidant balance in favour of the former, leading to potential damage". Other authors suggest that oxidative stress may be better defined as a disruption of redox signaling and control [58]. Table 4 [9,22,24,26,34–39,41,59–63] shows the *in vitro* studies available in the scientific literature dealing with oxidative stress induced by CYN.

Glutathione (GSH) is one of the major endogenous antioxidants that is produced by cells, so a deficit of it can play an important role in the potential induction of oxidative damage by xenobiotics. It is well known that CYN inhibits GSH synthesis. This statement derives from the studies performed by Runnegar et al. [22,24,59], who found that CYN caused a significant GSH fall in rat primary hepatocytes, and that potentiating effects were observed when cells were exposed concomitantly to CYN and a GSH inhibitor (propargylglycine) [22]. Moreover, they also addressed whether the fall in GSH was due to decreased GSH synthesis or increased GSH consumption, and they found that the inhibition of GSH synthesis was the predominant mechanism for the CYN-induced fall in GSH [59]. The GSH depletion was related to the cytotoxicity observed.

Table 4. *In vitro* oxidative stress studies dealing with CYN exposure.

Toxin/Cyanobacteria	Experimental Model	Assays Performed	Exposure Conditions Concentration Ranges	Main Results	Reference
Purified extract from *Cylindrospermopsis raciborskii*	Primary rat hepatocytes	GSH	0.8–5 µM CYN for 18 h. Exposure also to CYN + PPG (a GSH synthesis inhibitor)	1.6 µM CYN caused a significant fall (~50%) in cell GSH. At 5 µM GSH cell was only 12.5%. The fall in GSH preceded an increase in LDH release. Reduction of GSH contributes to toxicity. Potentiating effects were found when cells were exposed to CYN and PPG. GSH is most likely to be involved in the detoxification of CYN *in vivo*.	[22]
Purified extract from *Cylindrospermopsis raciborskii*	Primary rat hepatocytes	GSH GSH accumulation GSH efflux GSH synthesis	0, 2.5, 5 µM CYN. Different exposure times depending on the experiment	GSH was depleted significantly after 16 h exposure to 2.5 µM CYN and after 10 h to 5 µM. CYN caused a C-dependent inhibition in GSH accumulation. There was no effect on GSH efflux. GSH synthesis was not altered by 2.5 µM CYN in cell free extracts but the high dilution of the cytosolic content (~500-fold) could avoid the detection of the GSH synthesis inhibition. Authors considered that the inhibition of GSH synthesis is the predominant mechanism for the CYN-induced fall in GSH.	[59]
Purified extract from *Cylindrospermopsis raciborskii* RAC-CYN (synthetic CYN) CYN-DIOL (intermediate from CYN synthesis) EPI-CYN and EPI-DIOL (epimers of CYN at C-7)	Primary rat hepatocytes	GSH	0, 0.16, 0.32, 0.63, 1.25, 2.5 and 5 µM natural CYN and CYN-DIOL 1.25, 2.5, 5, 7.5, 10, 20 µM RAC-CYN	GSH IC_{50} values for CYN and RAC-CYN were 2.38 and 8.99 µM, respectively. When the racemic nature of RAC-CYN and the uncertainty in the original amounts of the synthetic analogues are taken into account, the decrease in GSH by RAC-CYN is almost equivalent to that of natural CYN. CYN-DIOL was as potent as CYN in lowering GSH levels, with the IC_{50} at 2.33 µM. Hepatocytes incubated with 6.25 µM EPI-CYN and EPI-DIOL had cell GSH levels of 39 ± 2.5 and 66 ± 14% of control respectively.	[24]
Purified extract from *Cylindrospermopsis raciborskii*	Primary mouse hepatocytes	GSH LPO by MDA assay	GSH: 0, 1, 5 µM MDA: 5 µM CYN with and without BCNU, an inhibitor of GSSG-Rd	GSH levels were depleted by CYN concentrations of 1 µM and above after a 18-h exposure, and 5 µM produced a significant reduction after 10 h with almost complete depletion after 18 h. However, 5 µM CYN did not elevate levels of lipid peroxidation, as measured by MDA production, and furthermore, inhibition of glutathione reductase by BCNU did not increase MDA production.	[26]
Commercial CYN pure standard	PLCH-1 cells derived from a hepatocellular carcinoma of the topminnow *P. lucida*	ROS GSH GCS activity	0, 2, 4 and 8 mg/mL for 24 h	ROS content increased in a C-dependent way. GSH and GCS activity showed a similar pattern: a significant increase at lowest concentration and a significant reduction at the highest one. The initial increase is considered a try to face the toxic insult. The depletion of GSH may be due to an inhibition of its synthesis.	[34]
Purified extract containing CYN	*Prochilodus lineatus* primary hepatocytes	RONS GST activity G6PDH activity 2GSH/GSSG ratio PCO LPO	0.1, 1.0 or 10 µg/L for 72 h	Cells exposed to the all concentrations of CYN have similar GST and G6PDH activities in comparison to the control group. However, GST activity of the hepatocytes exposed to 10 µg/L was 12% lower than of those exposed to 1 µg/L. G6PDH showed a similar pattern with significant differences between CYN treated cells but not in comparison to the control. No significant alterations were observed for GSH concentration and also for the 2GSH/GSSG ratio. RONS increased 25% in all CYN-exposed groups. PCO did not change. LPO increased in all CYN-exposed groups.	[35]

Table 4. *Cont.*

Toxin/Cyanobacteria	Experimental Model	Assays Performed	Exposure Conditions Concentration Ranges	Main Results	Reference
Commercial CYN pure standard	Human intestinal Caco-2 cell line	ROS GSH GCS activity	0, 0.625, 1.25 and 2.5 μg/mL for 24 h	ROS content was significantly increased only at the concentration of 1.25 mg/mL CYN. GSH and GCS activity were only significantly increased at 2.5 mg/mL. The decrease of ROS at the highest concentration can be related to the higher GSH levels due to its higher synthesis.	[36]
Commercial CYN pure standard	Human vascular endothelium (HUVEC)	ROS GSH GCS activity	0, 0.375, 0.75 and 1.5 μg/mL for 24 h	When HUVEC cells were exposed to 0.375 μg/mL CYN, ROS content was significantly enhanced, while at higher concentrations it decreased to the levels of the control group. GCS activity increased at the highest concentrations (0.75 and 1.5 μg/mL) with enhancements of 2.25 and 3.5-folds, respectively. GSH content underwent concentration-dependent enhancements, with a 3-fold increase at the highest concentration used in comparison with the control group. The recovery of basal ROS content can be related to the concentration-dependent increase in the GSH and the GCS activity observed.	[37]
Commercial CYN pure standard	Human hepatoma cells HepG2	ROS	0.05, 0.1 and 0.5 μg/mL for 5 h	A C-dependent statistically significant increase of ROS was observed in cells treated with 0.05, 0.1 and 0.5 μg/mL CYN already after 30 min of exposure, which steadily increased with incubation time. After 5 h incubation, the fluorescence intensity at the highest dose of CYN was about five times higher than in the control cells.	[60]
Commercial CYN pure standard	Primary rat hepatocytes	ROS Nrf2 transcription factor	0, 90, 180, 360 nM CYN for 24 and 48 h 0, 360 nM CYN with/without 10 or 20 μM resveratrol	CYN induced oxidative stress at all the concentrations tested after 24 and 48 h of incubation. A 3-fold increase in fluorescence was observed in hepatocytes treated with 360 nM CYN for 48 h. Resveratrol partially rescued the cells in a concentration dependent manner after 24 and 48 h of treatment. The increase in cell viability in cultures treated with CYN plus 20 μM resveratrol was about 32% and 7% after 24 and 48 h, respectively, when compared to that of CYN treated cells. A higher level of Nrf2 (transcription factor that regulates the expression of antioxidant enzymes) in toxin treated cells after 48 h was observed.	[38]
Commercial CYN pure standard	Rat hepatic cell line, Clone 9	GSH GCS level	1 μM or 5 μM CYN for 4, 12, 24 and 48 h	Both treatments with CYN (1 and 5 mM) showed a clear and gradual increase of the GSH levels over time, especially at 48 h. No significant changes were observed on GCS level over time in cells exposed to 1 mM. 5 mM CYN, on the contrary, clearly increased levels of GCS time-dependently.	[39]
Commercial CYN pure standard	*Cyprinus carpio* L. leucocyte cell line (CLC)	ROS SOD GSH/GSSG	0, 0.1, 0.5 or 1 μg/mL for 3.5 h	A CYN-induced increase of ROS in exposed CLC cells was observed at each toxin concentration. The results were concentration dependent, with a growing tendency observed until the end of the experiment. In cells exposed to the lowest CYN concentration (0.1 μg/mL) SOD activity was elevated in a statistically significant manner, reaching 179% of the enzyme activity detected in the control cells. At the other tested CYN concentrations SOD activity was also slightly enhanced, however, these increases were not statistically significant. The toxin at each tested concentration increased the total GSH content in the cells, with the concomitant reduction of the GSH/GSSG ratio.	[61]

Table 4. *Cont.*

Toxin/Cyanobacteria	Experimental Model	Assays Performed	Exposure Conditions Concentration Ranges	Main Results	Reference
Purified extract from *Cylindrospermopsis raciborskii*	Human hepatoma cells HepG2	ROS GST activity LPO Superoxide production in mitochondria	0, 0.001, 0.01, 0.1, 1, 10 and 100 μg/L CYN for 48 h with 10% FBS 0, 0.1, 1, 10 μg/L CYN for 12 and 24 h with 2% FBS and/without CYP induction with phenobarbital	No concentration-dependent changes in superoxide production by the mitochondria, ROS and LPO. Actually, LPO decreased. GST activity only increased significantly at 100 μg/L. The 10% FBS could reduced toxicity. ROS increased at both exposure times in an approximate concentration–response pattern, with and without prior CYPs induction. LPO response was very variable; it decreased in non-induced cells exposed to CYN for 12 h and increased in the cells exposed to the highest CYN concentration for 24 h. GST activity only increased after 12 h exposure to 10 μg/L CYN. But on the contrary after 24 h a decreased was observed. CYPs-induction with phenobarbital has led generally to similar results as those observed in non-induced cells for the tested biomarkers.	[41]
Commercial CYN pure standard	Human lymphocytes	ROS SOD activity GPx activity CAT activity LPO	0, 0.01, 0.1 and 1 μg/mL CYN for 0.5–48 h to evaluate ROS production 0, 0.01, 0.1 and 1 μg/mL CYN for 3 and 6 h for the other biomarkers	CYN elevated ROS level in a concentration-dependent manner. The increase was observed within a time as short as 0.5 h of exposure and reached its maximum after 3 and 6 h. SOD level was decreased in a concentration-dependent manner. The greatest depletion (45% respect to the control) was observed after 6 h with 1.0 μg/mL. CAT also decreased after 6 h of exposure to 0.1 and 1 and after 3 h exposure to the highest concentration. GPx activity increased. This was particularly observed after 6 h of exposure. CYN treatments resulted in increased peroxidation of lipids in lymphocytes exposed to 0.1 (after 6 h) and 1 μg/mL (after 3 and 6 h).	[62]
Purified extract from *Cylindrospermopsis raciborskii*	*Hoplias malabaricus* hepatocytes	ROS CAT activity SOD activity GPx activity GST activity G6PDH activity Non-protein thiols GR activity LPO Protein carbonylation	0, 0.1, 1.0, 10, and 100 μg/L for 72 h	The activities of SOD, CAT, GPx, GST and G6PDH were not altered by the exposure to CYN in all groups tested. Non-protein thiols concentration increased 72% only in the cells exposed to the highest CYN concentration. CYN caused a concentration-dependent decrease of GR activity in the cells exposed to >1.0 μg/L. ROS levels increased 40% only in the cells exposed to the highest CYN concentration. No significant damage to lipids (peroxidation), and proteins (carbonylation) was observed.	[63]
CYN Guanidinoacetate (the primary substrate in CYN biosynthesis) 4 CYN synthetic analogs	Human neutrophils	ROS LPO	ROS: 2 μg/mL for 5–60 min LPO: 2 μg/mL for 1 h	All the compounds tested had the ability to temporarily increase the intracellular ROS levels to different extents. LPO levels were significantly increased.	[9]

Abbreviations: BCNU: 1,3-bis(chloroethyl)-l-nitrosourea; CAT: Catalase; CYP: cytochrome P450; G6PDH: Glucose-6-phosphate dehydrogenase; FBS: Fetal Bovine Serum; GCS: Gamma Glutamylcysteine Synthetase; G6PDH: glucose-6-phosphate dehydrogenase; GPx: glutathione peroxidase; GR: Glutathione Reductase; GSH: Glutathione; GST: Glutathione S-transferase; GSSG-Rd: Glutathione disulfide reductase; LPO: Lipid peroxidation; MDA: Malondialdehyde; PCO: Protein carbonylation; PPG: Propargylglycine; RONS: Reactive oxygen/nitrogen species; ROS: Reactive Oxygen Species; SOD: Superoxide dismutase.

The effect of CYN on GSH content is one of the oxidative stress biomarkers that has been most extensively studied in the scientific literature. Apart from Runnegar et al. [22,24,59], there are other authors who also found a depletion (i.e., Humpage et al. [26]), but in other cases, different results have been reported. Thus, Liebel et al. [35] did not find changes in this parameter in *Prochilodus lineatus* primary hepatocytes, while other authors showed a significant increase. Gutiérrez-Praena et al. [34] observed a dual response in the fish PLHC-1 cell line, with a significant increase at the lowest CYN concentration assayed (2 μg/mL) and a significant reduction at the highest one (8 μg/mL). In human HUVEC cells, on the other hand, they found a concentration-dependent increase (from 0.375 to 1.5 μg/mL CYN) [37], and in human intestinal cells, the increase was only evident at the highest concentration tested (2.5 μg/mL) [36]. Other authors who found a significant increase were Fernández et al. [39], in the rat hepatic cell line Clone 9, and Silva et al. [63], in *Hoplias malabaricus* hepatocytes. In any case, it has been suggested that the GSH reduction does not contribute significantly to CYN acute toxicity *in vivo* [64].

Various studies have also investigated the effect of CYN on Gamma Glutamylcysteine Synthetase (GCS) activity, as this is the limiting enzyme in GSH synthesis. Runnegar et al. [59] concluded that CYN inhibits GSH synthesis, and this statement was based on the finding that an excess of a GSH precursor (*N*-acetylcysteine), which supported GSH synthesis in control cells, did not prevent the fall in GSH or toxicity that was induced by CYN. Other authors, however, observed a different response pattern. Thus, Gutiérrez-Praena et al. [34] found a significant increase at the lowest concentration that was assayed and a significant reduction at the highest one (8 μg/mL) in PLHC-1 cells. In human cell lines, on the other hand, only significant increases were observed at 2.5 μg/mL CYN in Caco-2 cells [36] and at 0.75–1.5 μg/mL in HUVEC cells [37]. Fernández et al. [39] also found a time-dependent increase of GCS levels in the rat hepatic cell line (Clone 9) that was exposed to 5 μM CYN.

A GSH depletion could be directly correlated, among other responses, with an increase in ROS levels. In this regard, it is remarkable that, in all of the reports selected, CYN exposure induced an enhancement of ROS content. This may play an important role in other toxic mechanisms, for instance, genotoxicity [26]. Other important oxidative biomarkers, however, such as lipid peroxidation, have scarcely been investigated, and different results have been obtained. Humpage et al. [26] observed no remarkable effects in primary mice hepatocytes, while increases were reported by Poniedziałek et al. [62] in human lymphocytes, and by Liebel et al. [35] in fish primary hepatocytes. These authors also found that CYN produced a variable effect in HepG2 cells [41]. It decreased LPO in cells that were not previously induced by phenobarbital (PHE) exposed for 12 h, and increased it in PHE-induced cells exposed to the highest CYN concentration (10 μg/L). After 24 h of exposure, however, LPO experienced an increase in both cell types only at 10 μg/L CYN.

In the cellular environment, ROS increases are counteracted by enzymatic and non-enzymatic defensive mechanisms. In the first group, the enzymes superoxide dismutase (SOD), catalase (CAT), and glutathione peroxidase (GPx), among others, play an important role. SOD, CAT, GPx, Glutathione S-transferase (GST), and glucose-6-phosphate dehydrogenase (G6PDH) were not altered in *Hoplias malabaricus* hepatocytes that were exposed to up to 100 μg/L CYN for 72 h [63]. Different results were obtained by Poniedziałek et al. (2015) in human lymphocytes. SOD and CAT levels decreased and GPx activity increased. These effects were mainly observed after 6 h of exposure. Liebel et al. [41] also investigated GST activity in a HepG2 cell line and found an increase. Moreover, a higher level of Nrf2 (a transcription factor that regulates the expression of antioxidant enzymes) in toxin-treated rat primary hepatocytes after 48 h was observed by López-Alonso et al. [38].

From a general perspective, the variability of the results that were obtained for a particular biomarker (GSH, ROS, etc.) is high. These differences could be due to the different experimental models (primary cells or cell lines of various origins), CYN concentrations, or exposure periods that were employed. Moreover, given that CYN is also an environmental contaminant, it is noteworthy that few reports are available with fish *in vitro* models [34,35,61,63].

Another point to highlight is that all of these studies have been performed with a CYN pure standard or CYN isolated from *Cylindrospermopsis raciborskii*. The differences between the toxic effects that were induced by pure and extracted CYN have not been systematically studied so far. It is known, however, that in the case of a different cyanobacterial toxin, microcystins, extracts may contain other compounds that can influence the toxicity observed [65]. Therefore, it would be necessary also to test the effect of an extract from a non-CYN-producing culture in order to establish the contribution of these substances to the final response. Also, other oxidative stress biomarkers (protein or DNA oxidation) have not yet been studied, nor has the effect of different chemoprotectants on oxidative stress biomarkers [66]. In this regard, only López-Alonso et al. [38] observed that resveratrol partially reduced the cytotoxicity that was induced by CYN in primary rat hepatocytes. These authors argued that oxidative stress is involved in the cytotoxicity induced by CYN at lower concentrations in primary rat hepatocytes. The explanation was that the low toxin concentrations and long exposure times induced apoptosis, while in the case of necrosis induction the insult to the cells produced by the toxin could be of such a magnitude, and cell death so rapid, that oxidative stress could not be observed.

From all of these results (and also from those obtained *in vivo* and not considered here), it has been shown that CYN induces oxidative stress. It should be of interest to investigate the repercussion that this can have on human and environmental health, as ROS generation may account for the increased risk of cancer development in the aged [67].

4.3. Genotoxicity

Besides being considered as a cytotoxic toxin, CYN has also been described as genotoxic [68]. Several studies imply that it is pro-genotoxic [69], although the genotoxicity of CYN (and/or its metabolites) is still controversial [7]. Genotoxic and even carcinogenic effects of CYN have been reported *in vivo* in mice by several authors [70,71], the liver being the most affected organ. This suspicion is based on the nucleotide structure of CYN, which contains potentially reactive guanidine and sulfate groups [70,72]. The presence of uracil led researchers to suggest a possible interaction with nucleic acids [27].

With regard to *in vitro* studies, various assays have been performed (Table 5) [26–28,35,53,60,63,72–82], mainly in mammalian systems, which demonstrated the pro-genotoxic activity of CYN.

It is important to note that no single genotoxicity test is capable of detecting all relevant genotoxic agents, and therefore various international organizations recommend a test battery of genotoxicity assays to elucidate the genotoxic potential of a biotoxin (such as CYN), xenobiotics, new medical devices, or substances in contact with food, etc. The data compiled in Table 5 indicate that, in general, there are few studies, and only one work that has been performed regarding the mutagenic profile of CYN in bacteria [76], showing negative results. In mammal cells, the alkaline comet assay is the genotoxic assay that is most frequently used to test a pure standard of CYN [74,75,78,82], or CYN isolated and purified from crude extracts obtained from cyanobacterial cultures (usually from *C. raciborskii* [26,73,80]. After applying the comet assay in various metabolic cells, most of the results showed a positive response of CYN, increasing the comet tail length, area, or moment, while cell alterations, but no DNA fragmentations were induced by CYN in metabolism-deficient Chinese hamster ovary K1 (CHO-K1) cells [73]. At molecular level, research has been carried out on changes in the expression of genes that are involved in the response to DNA damage induced by CYN alone on HepG2 cells [60,74,79], and, more recently, by a binary mixture of cyanotoxins CYN and MC-LR [82], indicating the mechanisms involved (oxidative stress, etc.). Further studies in this direction are needed in different experimental models.

Table 5. *In vitro* mutagenicity and genotoxicity studies performed with CYN.

Toxin/Cyanobacteria	Experimental Models	Assays Performed	Exposure Conditions Concentration Ranges	Main Results	Reference
Purified extract from freeze-dried *C. raciborskii* culture	Human lymphoblastoid cell line WIL2-NS	Cytokinesis-block micronucleus (CBMN) assay. Micronuclei (MN) were counted in binucleated cells (BNCs)	1,3,6, and 10 μg CYN/mL, 24 h and 48 h	CYN induced significant increases in the frequency of MN in BNCs exposed to 6 and 10 μg/mL, and a significant increase in centromere (CEN)-positive MN at all concentrations tested. At the higher concentrations, both CEN-positive and CEN-negative MI were induced.	[72]
Purified extract from a *Cylindrospermopsis raciborskii* Australia strain (AWQC CYP-026J)	Chinese hamster ovary K1 (CHO-K1) cells	Comet assay	0.5–1.0 μg CYN/mL, 24 h	No significant induction of DNA strand breaks could be detected after 24 h treatment. However, cell growth was inhibited, as well as cell blebbing and rounding.	[73]
Purified extract from *Cylindrospermopsis raciborskii*	Primary mouse hepatocytes	Comet assay	0.05–0.5 μM CYN. Moreover, cells were preincubated with inhibitors of CYP450: SKF525A, omeprazole	CYN induced increases in comet tail length, area and tail moment at 0.05 μM. The CYP450 inhibitors completely inhibited the genotoxicity of CYN.	[26]
Purified extract from *C. raciborskii* (AWT205)	Human dermal fibroblasts (HDFs), Caco-2, HepG2 and C3A cells	Quantification of mRNA levels for selected p53-regulated genes using qRT-PCR	1, 2.5, or 5 μg/mL CYN for 6 h or 24 h	After 6 h exposure to CYN, concentration-dependent increases in mRNA levels were observed for the p53 target genes *CDKN1A*, *GADD45α*, *BAX* and *MDM2*, indicating an early activation of p53, which remained elevated after 24 h of exposure.	[27]
Purified extract from two cultures of *C. raciborskii* (AWT 205, and CYN-Thai)	CHO-K1 cells	Chromosome aberration (CA) assay	0.05–2 μg CYN/mL were assayed. DNA damage was determined after 3, 16 and 21 h of exposure and the assay was performed with and without metabolic activation (S9)	CYN with and without S9 had no significant influence on the frequency of CA.	[28]
Commercial pure standard CYN (>98% purity)	Caco-2 and HepaRG cells, differentiated and undifferentiated cells	Cytokinesis-block micronucleus (CBMN) assay	0.5–2 μg CYN/mL, and the CYP450 inhibitor ketoconazole (1–5 μM) for 34 h	CYN increased the frequency of binucleated cells in both cell lines, and ketokonazole reduced both the genotoxicity and cytotoxicity induced by CYN	[53]
Purified extract containing CYN	Hepatocytes of the fish *Prochilodus lineatus*	Comet assay	0.1, 1.0, or 10 μg CYN/L for 72 h	No significant effects on DNA strand breaks were found.	[35]
Commercial pure standard CYN	HepG2 cell line (human hepatoma cell line)	Comet assay, and MN, nuclear bud (NBUD), nucleoplasmic bridge (NPB) formation. Changes in the expression of genes involved in the response to DNA damage and in CYN metabolism were investigated using real-time quantitative PCR (qPCR)	0–0.5 μg CYN/mL) for 4, 12, and 24 h	Non cytotoxic concentrations of CYN (0–0.5 μg/mL) induced increased DNA strand breaks after 12 and 24 h of exposure. Increased frequency of MN, NBUDs and NPBs after 24 h exposure in a dose-dependent manner was reported. CYN upregulated the expression of the *CYP1A1*, *CYP1A2* genes, and the expression of the P53 downstream-regulated genes *CDKN1A*, *GADD45α*, and *MDM2*.	[74]

Table 5. *Cont.*

Toxin/Cyanobacteria	Experimental Models	Assays Performed	Exposure Conditions Concentration Ranges	Main Results	Reference
Commercial pure standard CYN	Human peripheral blood lymphocytes (HPBLs)	Comet assay and the cytokinesis-block micronucleus (CBMN) assay. Gene expression of *CYP1A1, CYP1A2, P53, MDM2, GAdd45α, CDKN1A, BAX, BCL-2, GCLC, GPX1, GSR, SOD1* and *CAT*, using the qPCR	The whole blood was treated with CYN (0, 0.05, 0.1 and 0.5 μg/mL) for the comet and CBMN assays. For the mRNA expression the isolated HPBLs were exposed to 0.5 μg/mL of CYN for 4 and 24 h	In HPBLs CYN induced the formation of DNA single strand breaks (comet assay), a time and dose-dependent increase in the frequency of MN and NBUD was observed, and a slight increase in the number of NPB. CYN up-regulated the genes *CYP1A1* and *CYP 1A2*, and the mRNA expression of some DNA damage (*P53, GADD45α, MDM2*), apoptosis responsive genes (*BAX, BCL-2*), and some genes involved in the antioxidant enzymes (*GPX, GSR, GCLC, SOD1*) whereas no changes were detected in *CDKN1A* and *CAT*.	[75]
Commercial pure standard CYN and crude extracts from cyanobacterial blooms. Mixture of commercial pure toxins: CYN, MC-LR and anatoxin-a	*Salmonella typhimuriun* strains (TA 98, TA 100, TA1535, TA 1537) and *Escherichia coli* WP2 uvrA and WP2 (pKM101)	Mutagenicity: Ames test	Pure CYN: 0.312, 0.625, 1.25, 2.5, 5 and 10 μg/mL. The mixture of pure toxins (CYN, MC-LR and anatoxin-a) at 1 μg/mL was also tested but only in two *Salmonella typhimuriun* strains (TA 98, TA 100)	Mutagenicity was detected in four of the ten extracts assayed, mainly against *S. typhimurium* TA100. By contrast, pure CYN was not mutagenic towards all the six bacterial strains up to a concentration of 10 μg/mL. No effects were detected after bacteria exposure to the mixture of purified toxins.	[76]
Commercial CYN Pure standard	Common carp (*Cyprinus carpio*) leukocytes	Alkaline version of comet assay	0.5 μg CYN/mL for 18 h	The cells treated with CYN were affected to a lesser extent in comparison to the damage induced by MC-LR.	[77]
Commercial CYN Pure standard	HepG2 cell line (human hepatoma cell line)	Formation of double strand breaks (DSBs). Analysis of the cell-cycle by flow-cytometry	0–0.5 μg CYN/mL for 24–96 h	CYN induced formation of DSBs after 72 h exposure. The toxin has impacts on the cell cycle, indicating G0/G1 arrest after 24 h and S-phase arrest after longer exposure (72 and 96 h).	[78]
Commercial CYN Pure standard	HepG2 cells	Gene expression was analyzed by qPCR	0.5 μg CYN/mL for 12 and 24 h	CYN increased expression of the immediate early response genes, and strong up-regulation of the growth arrest and DNA damage inducible genes (*GADD45α, GADD45β*), and genes involved in DNA damage repair (*XPC, ERCC4* and others). Up-regulation of metabolic enzyme genes provided evidence for the involvement of phase I and phase II enzymes in the detoxification response and potential activation of CYN.	[79]
Commercial CYN Pure standard	HepG2 cells	Alkaline comet assay and Fpg -enzyme modified assay	0–0.5 μg CYN/mL) for 4, 12 and 24 h	No DNA damage was observed after 4 h exposure to CYN. After 12 and 24 h, CYN (0.25–0.50 μg/mL) induced significant increase of DNA strand breaks, but not oxidative damage. CYN did not induce apoptosis.	[60]

Table 5. *Cont.*

Toxin/Cyanobacteria	Experimental Models	Assays Performed	Exposure Conditions Concentration Ranges	Main Results	Reference
Treated water; crude extract of *C. raciborskii* (CYP-011K); crude extract containing CYN; no toxic extract	HepG2 cells	Comet assay	Cells were exposed to all extracts at concentration of 0.1, 0.5 and 1 μg of dry material/mL, and also to treated water only, for 24, 48 and 72 h	DNA damage was detected only under toxic *C. raciborskii* extract, at the concentration of 1 μg/mL from 24 h of exposure, and at 0.5 μg/mL after 48 and 72 h.	[80]
Commercial CYN Pure standard	HepG2 cells with a plasmid that encodes the fluorescent protein DsRed2 under the control of the *CDKN1A* promoter, (HepG2CD-KN1A-DsRed cells)	The induction of the DsRed fluorescence intensity was determined by spectrofluorimetry, fluorescence microscopy and flow cytometry	Cells were exposed to CYN and the DsRed fluorescence was determined at 24 and 48 h of exposure; the cell viability was determined at 48 h	LOEC [2]: 0.12 μM and RDF [3]: 1.53 μM	[81]
Commercial CYN Pure standard	*Cyprinus carpio* L. leucocyte cell line (CLC)	The cytokinesis-block micronucleus (CBMN) assay. The fluorimetric OxyDNA assay kit was also employed	0.1, 0.5, or 1 μg CYN/mL, for 24 h	CYN increase the number of MN, and oxidative DNA damage was also detected.	[61]
Commercial CYN and MC-LR pure standards, and mixtures MC-LR/CYN	HepG2 cells	Alkaline comet and CBMN assays were performed. The expression of selected genes was analyzed by quantitative time PCR	CYN: 0.01–0.05 μg CYN/mL; MC-LR: 1 μg/mL, and MC-LR/CYN mixtures for 4 h and 24 h	CYN after 24 of exposure induced DNA stand breaks and genomic instability. The MCLR/CYN mixture induced DNA strand breaks after 24 h exposure, but to a lesser extent as CYN alone. The induction of genomic instability and changes in the expression of selected genes induced by the mixture were similar to those induced by CYN alone. CYN alone resulted in changes in the expression of genes involved in the metabolism (*CYP1A1*, *CYP1A2*, *NAT2*), genes involved in immediate-early response/signaling (*FOS*, *JUN*, *TGFB2*), and DNA damage (*MDM2*, *CDJN1A*, *GADD45A*, *ERCC4*), while MC-LR alone down-regulated the expression of *NAT2* and *TGFB2*. The binary mixture exhibit similar results that CYN alone.	[82]
Purifies extract from the strain *C. raciborskii* CYPP011K	*Hoplias malabaricus* hepatocytes	Comet assay	0.1–100 μg/L of CYN for 72 h	No significant DNA damage was observed	[63]

[1] Fpg: Formamidopyrimidine glycosylase enzyme; [2] LOEC: Lowest effective concentration that induced ≥1.5-fold increase in relative DsRed fluorescence, over the solvent-treated control; [3] RDF: Relative DsRed fluorescence induction detected at LOEC.

Simultaneously, with the comet assay, the *in vitro* cytokinesis-block micronucleus assay (CBMN) is being increasingly used in the evaluation of CYN [53,72,74,75,82], rather than the *in vitro* chromosome aberration (CA) assay, which applied only in CHO-K1 cells (Lankoff et al., 2007) [28]. This may be due to some advantages that are offered by the MN test, such as the high number of analyzable cells, simplicity of the technique, possible automation, and the ability to detect aneugens more accurately [83,84].

It is important to note that, to the best of our knowledge, the enzyme-modified comet assay, using endonuclease III (Endo III) and formamidopyrimidine DNA glycosylase (Fpg) to detect oxidation of pyrimidines or purine DNA bases, respectively [85], has rarely been employed to evaluate the role of this mechanism in the genotoxic potential of CYN. Only one study has investigated the oxidation of purine bases (Fpg assay) by CYN [61]. In addition, the mouse lymphoma gene mutation assay (MLA) (OECD 476) [86], which is preferred because it detects the broadest set of genotoxic mechanisms—such as chromosomal, gene, base pair substitutions, and frame-shift mutations [87,88]—that are associated with carcinogenesis activity, has not been performed either, despite the indications of CYN carcinogenicity for humans) [7,69].

In comparison to mammals, the genotoxicity of CYN in fish has been poorly studied, and contradictory results have been found by the alkaline comet assay in various cells from several fish species [35,63,77]. A positive response has been shown by CYN in the CBMN assay on a leucocyte cell line from *Cyprinus carpio* L. [77].

Recently, some studies have evaluated the potential mutagenicity/genotoxicity of CYN in combination with other cyanobacterial toxins, mainly MC-LR [76,82], because in real life, organisms are exposed to mixtures of several biotoxins, rather than to a single compound [89], and following the recommendations that are given by international organizations, such as the European Food Safety Authority (EFSA) [19].

More detailed information about the results and conclusions stated in the genotoxic studies compiled in this review is provided below.

The only evidence from bacterial test systems (Ames test) indicated that CYN pure standard was not mutagenic toward the bacterial strains (*S. typhimurium* and *E. coli*) assayed up to a concentration of 10 μg/mL, a concentration higher than those considered ecotoxicologically relevant [76]. Negative responses were also found for the pure standard solutions of the cyanotoxins MC-LR and anatoxin-a under the conditions assayed. Neither an increase in the number of revertants nor an inhibition of the growth of bacteria was observed, with or without metabolic activation. Similarly, there were negative results after exposure of bacteria to the mixture of pure toxins. By contrast, extracts that were obtained from cyanobacterial bloom-forming cells harvested from environmental waters were evidently mutagenic, mainly against *S. typhimurium* TA100 strain, and only contained CYN in a low concentration (0.89 μg/L). It was concluded that neither CYN nor other cyanotoxins that were tested were responsible per se for the observed mutagenicity of the extracts, and perhaps some other components of cyanobacterial extracts were responsible for the induction of mutations. The authors suggested that, while it can be stated that CYN and MC-LR are not mutagenic for the bacterial strains that are used, there are many reasons for considering these compounds as mutagens for eukaryotic cells. In addition, the metabolic activation enzyme system (S9 fraction) derived from rat livers employed in the Ames test may differ from the metabolism occurring in human cells [76]. Further studies are needed to confirm these preliminary results, especially in the case of CYN, and to elucidate potential synergistic interactions between cyanotoxins.

In mammalian systems, Humpage et al. [72] showed that CYN could induce micronuclei (MN) *in vitro* in human lymphoblastoid WIL2-NS cells, and this effect was mainly linked to an aneugenic effect, and, to a lesser extent, to a clastogenic one. These authors suggested that CYN acts to induce cytogenetic damage using two mechanisms: one at the level of the DNA to induce strand breaks, producing acentric fragments and giving rise to centromere-negative micronuclei; the other at the level of the kinetochore/spindle function, to induce loss of whole chromosomes owing to malsegregation of chromosomes during anaphase, which may possibly be explained by the known effects of CYN on protein synthesis. Metabolism of CYN by WIL2-NS cells is yet to be confirmed, so the involvement of CYN metabolites is not clear [69].

By contrast, when the comet assay was employed in CHO-K1 cells that were exposed to CYN isolated and purified from cultures of *C. raciborskii*, no DNA damage was detected 24 h after treatment, although inhibition of cell growth was reported, and also blebbing and rounding of the cells, linked to cytoskeletal reorganization but not to apoptosis [73]. The authors concluded that CYN did not react directly with DNA, but they pointed to the potential role of its metabolization in the generation of genotoxic products. As no exogenous metabolic activation was used in this study, the lack of DNA damage could be due to the low metabolizing enzyme activity of these cells. This was the first time that the need for further research taking into account the importance of CYN was highlighted.

In this context, to understand the role of CYP450-activated CYN metabolites in the *in vitro* genotoxicity of CYN, Humpage et al. [26] applied the comet assay in primary mouse hepatocytes, both in the presence and in the absence of CYP450 inhibitors, such as omeprazole and SKF525A. The direct assay revealed a statistically significant concentration-dependent increase in comet tail length, area, and moment in cells that were treated for 18 h with CYN (0.05–0.5 μM), and significant DNA fragmentation at a concentration as low as 0.05 μM. The genotoxicity of CYN at subcytotoxic concentrations, below the EC30 where cell death-related DNA digestion should not be detectable [90], suggests that it is a specific and primary effect of CYN. The fact that CYP450 inhibitors, such as omeprazole (100 μM, an inhibitor of CYP 3A4/2C19) and SKF525A (50 μM, a broad-spectrum CYP inhibitor), completely inhibited the genotoxicity that was induced by CYN indicated that CYP450-derived metabolites of the toxin are responsible for its genotoxicity [26].

Other experiments performed to know whether the metabolism could be a prerequisite for CYN-genotoxicity were carried out in CHO-K1 cells, and no chromosome aberrations (CA), with or without metabolic activation (S9 fraction), were detected [28]. The results revealed that CYN was not clastogenic in CHO-K1 cells, irrespective of S9 fraction-induced metabolic activation. However, the toxin significantly decreased the frequencies of mitotic indices and cell proliferation, irrespective of the metabolic activation system. This lack of genotoxicity of CYN confirmed the previous results that were found in the comet assay in the same cell line [73], and showed that, despite the use of metabolic activation (S9 mix, post-mitochondrial supernatant, known to be a potent enzymatic inducer), the frequency of CA was not affected by CYN. Consequently, CYN itself and S9-derived metabolites of the toxin are non-clastogenic under these experimental conditions in CHO-K1 cells [28]. Various factors may be responsible for the discrepancy between the cytogenetic assay results and the comet assay study performed with hepatocytes [26]: (1) the lack of an appropriate metabolic system, because the liver S9 elevates the levels of several CYN metabolizing enzymes, but it does not cover their total spectrum (e.g., CYP1A1 or CYP2E1 are low or inactivated in S9 fraction); (2) diffusion pathways are longer for externally generated active metabolites; (3) some genotoxic metabolites may be formed only within specific target cells; (4) the doses of CYN that were efficient to induce DNA single-strand breaks visible in the comet assay were too low to induce CA; (5) the cytotoxic property of CYN may be a confounding factor in the comet assay, giving false positive results; and, (6) differences in CYN uptake in different cell lines. In conclusion, the metabolic activation of CYN influenced the cytotoxicity of CYN by increasing the susceptibility to necrotic cell death, and the positive comet assay results observed by others could be due to cytotoxicity rather than to genotoxicity. Although CYN did not induce DNA damage and CA in CHO-K1 cells, it affected the microtubular structure in this cell line, which could disrupt spindle or centromere function and may lead to loss of whole chromosomes [33,69].

In contrast, CYN induced MN formation in two human cell lines, hepatocyte (HepaRG) and enterocyte (Caco-2) cell lines, models of CYN target organs [53]. After exposure to 1.25–1.5 μg CYN/mL, significant increases in MN (3-fold above controls) in both differentiated and undifferentiated Caco-2 cells were detected. No increase in MN formation was detected in undifferentiated HepaRG cells, and a positive response at 0.06 μg CYN/mL in differentiated HepaRG cells (1.8-fold) was reported.

This last genotoxic effect was found in a similar dose range in primary hepatocytes by the comet assay, suggesting that the CYN-metabolizing capabilities of differentiated HepaRG may be similar to hepatocytes. There were differences in genotoxicity in the two differentiated cells, the increase in MN frequency being greater in Caco-2 cells. The assay was also performed using the inhibitor of CYP450, ketoconazole, which is widely known to inhibit CYP3A4, a potent inhibitor of CYP1A1, and a moderate inhibitor of CYP2C, CYP1A2, and CYP2D6. Ketoconazole strongly protected undifferentiated Caco-2 cells and reduced cytotoxicity and induction of MN to 50%, in agreement with the findings reported by Humpage et al. [26] with omeprazole. However, the pretreatment with ketoconazole showed no effect on MN-induction by CYN in differentiated HepaRG cells. Therefore, it seems that CYN genotoxicity is mediated through its metabolites, suggesting that this toxin is a progenotoxin, and that minor CYP isoforms may play a role in its metabolic activation [69]. Until now, there was only indirect evidence for the formation of reactive CYN metabolites; consequently, the reduction of CYN toxicity in the presence of CYP inhibitors could be due to alternative pathways [48].

On the other hand, toxicogenomic approaches could elucidate CYN toxicity mechanisms [74,75,79,82]. In human hepatoma HepG2 cells, at non-cytotoxic concentrations, Štraser et al. [74] indicated that CYN induced DNA breakage, and a dose-dependent increase in the frequencies of MN, nuclear buds (NBUD), and nucleoplasmic bridge (NPB) formation, which were associated with upregulation of DNA damage responsive genes CDKN1A, GADD45α, and MDM2. The authors also showed that CYN induced upregulation of some genes presumably involved in CYN metabolism, such as CYP1A1 and CYP1A2. These results are in agreement with previous studies in primary hepatocytes [26], indicating a similar metabolizing capacity of both *in vitro* models; and, with the induction of MN by CYN in the three human cell lines mentioned above, the lymphoblastoid cell line WIL2-NS [72], liver HepaRG cells, and colon-derived Caco-2 cells [53]. It is noteworthy that Štraser et al. [74] provided the first evidence that exposure to CYN induced transcription of CYP1A1 and 1A2 isoforms, supporting the assumption that they are involved in CYN metabolic activation to genotoxic intermediates. Moreover, as CYN induced NBUD and NBP formation, which correlated with increased MN formation, these authors indicated that CYN induced complex genomic alterations, including gene amplification and structural chromosomal rearrangements. The toxicogenomic analysis indicated the upregulation of DNA damage responsive genes, confirming the previous study by Bain et al. [27], who detected upregulation of CDKN1A, GADD45α, MDM2, and BAX in HepG2 cells and in human dermal fibroblasts that are exposed to CYN.

In the same cell line, it was also demonstrated for the first time that CYN caused double-strand breaks (DSBs) and had impacts on the cell cycle, providing evidence that the toxin is a directly acting genotoxin [60]. Similarly, in human peripheral blood lymphocytes (HPBLs), Žegura et al. [75] found, after exposure to pure standard CYN, that the toxin induced the formation of DNA single-strand breaks (comet assay) and a time- and dose-dependent increase in the frequency of MN and NBUD, and only a slight increase in NPB, confirming the previous results that were reported for HepG2 cells. The effects of CYN on mRNA expression of selected genes was again similar to the effects found in HepG2 cells: the genes involved in CYN metabolism (CYP1A1 and CYP1A2) were upregulated, indicating that they are involved in CYN metabolic activation, although other CYP isoforms might also be implicated. In addition, CYN induced significant upregulation of the P53 gene, as well as its downstream regulated genes (MDM2 and GADD45α), apoptosis genes (BCL-2 and BAX), and, for the first time, some stress responsive genes (GPX1, SOD1, GSR, GCLC). Subsequently, these authors confirmed the time-dependent upregulation of the growth arrest and DNA-damage inducible genes (GADD45α and GADD45β), and the genes involved in DNA damage repair (XPC, ERCC4, and others), indicating cell-cycle arrest and induction of nucleotide excision and double-strand break repair [79]. In relation to detoxification response, evidence for the involvement of phase I and phase II enzymes was also demonstrated. After longer exposure (24 h), CYN could induce the possible depletion of glutathione and minor oxidative stress, as indicated by the upregulation of some genes—catalase gene (CAT), thioredoxin reductase (TXNRD1), and glutamate-cysteine ligase (GCLC)—although other genes were not induced. This minor role of oxidative stress in the genotoxicity of non-cytotoxic concentrations of CYN was also confirmed in the same HepG2 cells, because non-oxidative DNA damage was detected after the application of the enzyme-modified comet assay (Fpg digestion) [60].

In relation to fish, despite the common exposure of fish in natural environments and fish farms to cyanotoxins, there are only four studies on the genotoxic effects of CYN, yielding contradictory results, as mentioned earlier [35,61,63,77]. Cell-type and interspecific CYN toxicity differences may occur, because, in comparison to the concentration-dependent DNA damage reported in mammal cells [28,72], DNA breaks were not found in fish hepatocytes that were exposed to the same concentrations of CYN [35]. In this work, hepatocytes of *P. lineatus* that were exposed to environmentally relevant concentrations of CYN (0.1–10 μg/L) significantly decreased cell viability, there were changes in some oxidative stress biomarkers, but no significant alterations in DNA strand breaks were found by the comet assay. Similar negative results on DNA damage were reported in hepatocytes of *H. malabaricus* [63]. By contrast, an increased amount of DNA strand breaks was observed in common carp (*C. carpio*) blood leucocytes exposed to pure CYN (0.5 μg CYN/mL), not connected with cell death, although to a lesser extent, in comparison to the cyanotoxin MC-LR, which was the most toxic cyanotoxin [77]. On the other hand, in the fish CLC cell line (carp leukocyte culture cell line) CYN exposure (0.1–1.0 μg CYN/mL) induced MN, and for the first time, oxidative DNA damage was found by detection of the oxidation product 8-oxo-7,8-dihydro-2′-deoxyguanosine (8-OHdG) [61]. The effects exerted by CYN on CLC might be associated with oxidative stress and may result in genotoxic effects. While the increased level of 8-OHdG can be explained by the ROS production that was observed, the mechanism of MN induction was partially due to the clastogenic activity of the toxin. Although the vast majority of MN detected in binucleated CLC cells was much smaller than a quarter of the size of normal nuclei, the authors cannot speculate that they contained only DNA fragments, because the chromosome size in carp is heterogeneous. They concluded that CYN acts both as a clastogen and as an aneugen, as in mammal cell lines.

Recently, the genotoxic potential of binary mixtures of CYN and MC-LR has been studied for the first time, owing to the ubiquitous and simultaneous presence of both genotoxic cyanotoxins in the aquatic environment [82]. In this study, HepG2 cells were exposed to different doses of CYN (0.01–0.5 μg/mL), a single dose of MC-LR (1 μg/mL), or to several combinations of them. After 24 h exposure, CYN individually induced DNA strand breaks (comet assay) and genomic instability as measured by the CBMN assay. The MC-LR/CYN mixture induced both genotoxic injuries, but in the case of strand breaks to a lesser extent than CYN alone. The findings obtained by the comet assay confirmed previously published data that showed that MC-LR induced DNA strand breaks after short-term exposure, probably owing to oxidative stress (oxidation DNA bases) [69,91], while CYN induces DNA damage after longer exposure in metabolically active cells [26,74]. Lower DNA damage was detected with the mixture, and this antagonistic effect could be explained by the attenuated DNA repair that is produced by MC-LR [92,93]. The induction of genomic instability by CYN corroborated that this toxin induces MN formation, previously reported in metabolically active [53,74], while MC-LR alone did not induce MN formation at a low concentration [93]. The fact that CYN/MC-LR mixtures induced similar genomic instability in comparison to CYN alone indicates that MC-LR has no effect on CYN. Moreover, mRNA expression of selected genes after 4 and 24 h of exposure to individual cyanotoxins, and their combinations was performed by qPCR for the first time. The changes in the expression of some genes involved in xenobiotic metabolism, belonging to the group of phase I metabolism (CYP1A1, CYP1A2, CYP1B1, CYP3A4) and phase II (GSTA1, NAT2, UGT1A1), genes that are involved in immediate-early response/signaling (FOS, JUNB, MYC, and TGFB2), and the transcription of genes involved in DNA damage response (CDKN1A, CHEK1, ERCC4, GADD45A, MDM2, and TP53) observed with the CYN/MC-LR mixture were not different from those induced by CYN alone. All of these results indicate that CYN has higher genotoxic effects than MC-LR in the MC-LR/CYN mixture. MC-LR has no effect on CYN-induced deregulation of the selected genes reflecting the mechanisms of its pro-genotoxic activity.

Overall, more studies in different human cell lines are needed to confirm these findings after exposure to a mixture of CYN with other cyanotoxins. Moreover, the application of complementary mutagenic/genotoxic assays would be very useful, assays, such as: (1) the Ames test in bacterial systems to confirm the only study published; (2) the enzyme-modified comet assay to know whether DNA oxidation is involved in the induction of strand breaks; and, (3) the MLA assay, to elucidate whether CYN alone, or the combination CYN/MC-LR, is able to induce mutations in mammalian cells (L5178Y/Tk ± cells). Furthermore, toxicogenomic and proteomic studies would help to elucidate the mechanisms of CYN genotoxicity.

4.4. Immunotoxicity

The effect of CYN on the immune response is not well studied, although it can potentially affect cells of the immune system and alter its function, as has been reported *in vivo* in rodent models [55,94]. *In vitro*, the studies are very scarce, and the first work of CYN effects was reported using human peripheral blood lymphocytes from different but healthy donors [95]. At the highest concentration of CYN assayed (1 μg/mL), a significant inhibition of lymphocyte proliferation after 24 h of exposure was reported, and it resulted in inhibition of thymidine incorporation. Following these investigations, in human lymphocytes that were exposed to purified CYN isolated from a *C. raciborskii* culture (0.01–1.0 μg/mL), the authors demonstrated its antiproliferative activity during different phases of their activation. The highest concentration induced the most significant inhibition (over 90% when compared to unaffected cells) at the beginning of their activation. Moreover, a cell-cycle arrest at G0/G1 and prolonged S phase in lymphocytes undergoing activation and significant apoptosis inducement in activated cells were also detected [40]. It was suggested that DNA damage may be a primary mechanism of CYN action in lymphocytes, which is supported by DNA single-strand breaks, as observed by Žegura et al. [75]. These findings indicated that CYN could be classified as a potential immunotoxicant, and that potentially it could reduce abilities to fight pathogenic microorganisms or malignant cells [40].

Subsequently, these authors investigated whether these effects were mediated by alteration in the ROS level and oxidative stress of human-derived lymphocytes [62]. At the same concentrations of CYN mentioned above (0.01–1.0 μg/mL), the toxin induced a concentration- and time-dependent increase of H_2O_2 content, and also changes in several oxidative biomarkers, such as decreased activities of SOD and CAT, elevated level of GPx, and induction of LPO. All of these findings help to elucidate that the oxidative stress that is triggered by CYN in human cells is involved in the reported cyanotoxin-induced DNA damage, cell-cycle arrest, and apoptosis previously reported.

Neutrophils, an important part of the immune system, are highly specialized white blood cells that protect against infection in a non-specific manner. CYN (0.01–1.0 μg/mL) can affect the function of human peripheral blood neutrophils during 1 h exposure. CYN had no significant effect on the phagocytic activity of neutrophils, and no apoptotic or necrotic action was revealed [96]. However, it was found that CYN significantly altered neutrophil oxidative burst, a key process in pathogen elimination.

In addition to the immunomodulatory action of CYN on T lymphocytes and neutrophils, the potencies of metabolites that are produced by non-CYN-producing strains of *C. raciborskii* have been investigated in both human cells, and the observed effects were very similar to those that are induced by CYN [97]. After short-term treatments, the extracts altered viability of cells by increasing necrosis and apoptosis in neutrophils, and elevated apoptosis in lymphocytes, whereas no effects were observed with CYN. In general, lymphocytes appeared to be more resistant than neutrophils. T lymphocytes that were exposed for 72 h to *C. raciborskii* extracts resulted in a decrease of proliferation, and exposure to CYN (1.0 μg/mL) caused lymphocyte proliferation that later decreased. The effect of the extracts on T lymphocyte proliferation was not as pronounced as for CYN, suggesting that the cells can partially overcome the injuries that are induced by *C. raciborskii* exudates, or the metabolites could be degraded owing to their lower stability in comparison to CYN. This *in vitro* study indicated for the first time that extracts of *C. raciborskii* contained compound(s) (not identified yet) capable of altering the function of the human immune system.

5. Concluding Remarks

The higher number of reports on *in vitro* CYN toxicity studies deals with basal cytotoxicity aspects. This finding is not surprising taking into account that *in vitro* methods are widely used for screening purposes. The second aspect most frequently evaluated is genotoxicity. This is in accordance with the great importance that genotoxicity testing has nowadays. In view of the adverse consequences of genetic damage to human health, the assessment of mutagenic/genotoxic potential is a basic component of chemical risk assessment. Currently, genotoxicity testing is included in the first step of tiered toxicity evaluation approaches for various kinds of compounds, such as additives [98] or food contact materials [99].

In order to obtain a better understanding of the toxicity that is exerted by CYN, the toxicokinetics of CYN should be studied further. In this context, the mechanism of cellular uptake of CYN should be completely elucidated. This would also make it possible to discern the target organ and propose potential therapeutic agents for CYN intoxication. Moreover, the metabolites of CYN have not been described so far. For the main toxicity mechanisms that are considered in this review, potential data gaps have already been identified (see Sections 4.1–4.4). From a general point of view, various remarks can be made. For example, no studies on the effects of extracts from non-CYN-producing cyanobacterial strains have been identified, although it has been demonstrated that extracts from non-MC-producing strains also show toxic effects. There is no toxicological information about analogs other than CYN. Also, the near absence of studies dealing with cyanobacterial mixtures needs to be highlighted, and these investigations should be prioritized, as already indicated by EFSA [19]. Moreover, there are other new emerging toxicity effects that are attributable to CYN, such as neurotoxicity or immunotoxicity, which have scarcely been investigated by *in vitro* methods. Therefore, *in vitro* toxicity testing can still be very useful to complete knowledge about the toxic profile of CYN and its related compounds.

Acknowledgments: This work was supported by the Spanish Ministerio de Economía y Competitividad (AGL2015-64558-R, MINECO/FEDER, UE).

Conflicts of Interest: The authors declare no conflict of interest.

References

1. Ohtani, I.; Moore, R.E.; Runnegar, M.T. Cylindrospermopsin: A potent hepatotoxin from the blue-green Alga *Cylindrospermopsis raciborskii*. *J. Am. Chem. Soc.* **1992**, *114*, 7941–7942. [CrossRef]
2. Chiswell, R.K.; Shaw, G.R.; Eaglesham, G.K.; Smith, M.J.; Norris, R.L.; Seawright, A.A.; Moore, M.R. Stability of cylindrospermopsin, the toxin from the cyanobacterium *Cylindrospermopsis raciborskii*, effect of pH, temperature, and sunlight on decomposition. *Environ. Toxicol.* **1999**, *14*, 155–165. [CrossRef]
3. Wimmer, K.M.; Strangman, W.K.; Wright, J.L.C. 7-Deoxy-desulfo-cylindrospermopsin and 7-deoxy-desulfo-12-acetylcylindrospermopsin: Two new cylindrospermopsin analogs isolated from a Thai strain of *Cylindrospermopsis raciborskii*. *Harmful Algae* **2014**, *37*, 203–206. [CrossRef]
4. Kokocinski, M.; Cameán, A.M.; Carmeli, S.; Guzmán-Guillén, R.; Jos, A.; Mankiewicz-Boczek, J.; Metcalf, J.S.; Moreno, I.; Prieto, A.I.; Sukenik, A. Chapter 12 cylindrospermopsin and congeners. In *Handbook of Cyanobacterial Monitoring and Cyanotoxin Analysis*, 1st ed.; Meriluoto, J., Spoof, L., Codd, G.A., Eds.; Wiley: Hoboken, NJ, USA, 2017; pp. 127–137.
5. Bourke, A.T.C.; Hawes, R.B.; Neilson, A.; Stallman, N.D. An outbreak of hepato-enteritis (the Palm Island mystery disease) possibly caused by algal intoxication. *Toxicon* **1983**, *21*, 45–48. [CrossRef]
6. Cirés, S.; Ballot, A. A review of the phylogeny, ecology and toxin production of bloom-forming *Planktothrix* spp. and related species within the *Nostocales* (*cyanobacteria*). *Harmful Algae* **2016**, *54*, 21–43. [CrossRef] [PubMed]

7. Buratti, F.M.; Manganelli, M.; Vichi, S.; Stefanelli, M.; Scardala, S.; Testai, M.; Funari, E. Cyanotoxins: Producing organisms, occurrence, toxicity, mechanism of action and human health toxicological risk evaluation. *Arch. Toxicol.* **2017**, *91*, 1049–1130. [CrossRef] [PubMed]
8. Kinnear, S. Cylindrospermopsin: A decade of progress on bioaccumulation research. *Mar. Drugs* **2010**, *8*, 542–564. [CrossRef] [PubMed]
9. Cartmell, C.; Evans, D.M.; Elwood, J.M.L.; Fituri, H.S.; Murphy, P.J.; Caspari, T.; Poniedzialek, B.; Rzymski, P. Synthetic analogues of cyanobacterial alkaloid cylindrospermopsin and their toxicological activity. *Toxicol. In Vitro* **2017**, *44*, 172–181. [CrossRef] [PubMed]
10. Saker, M.L.; Eaglesham, G.K. The accumulation of cylindrospermopsin from the cyanobacterium *Cylindrospermopsis raciborskii* in tissues of the redclaw crayfish *Cherax quadricarinatus*. *Toxicon* **1999**, *37*, 1065–1077. [CrossRef]
11. Shaw, G.R.; Seawright, A.A.; Moore, M.A.; Lam, P.K.S. Cylindrospermopsin, a cyanobacterial alkaloid: Evaluation of its toxicological activity. *Ther. Drug Monit.* **2000**, *22*, 89–92. [CrossRef] [PubMed]
12. Rucker, J.; Stuken, A.; Nixdorf, B.; Fastner, J.; Chorus, I.; Wiedner, C. Concentration of particulate and dissolved cylindrospermopsin in 21 *Aphanizomenun*-dominated temperate lakes. *Toxicon* **2007**, *50*, 800–809. [CrossRef] [PubMed]
13. Cordeiro-Araújo, M.K.; Chiab, M.A.; Bittencourt-Oliveira, M.C. Potential human health risk assessment of cylindrospermopsin accumulation and depuration in lettuce and arugula. *Harmful Algae* **2017**, *68*, 217–223. [CrossRef] [PubMed]
14. Guzmán-Guillén, R.; Prieto, A.I.; Diez-Quijada, L.; Jos, A.; Cameán, A.M. Development and validation of a method for cylindrospermopsin determinación by UPLC-MS/MS in vegetables. In Proceedings of the 6th International Symposium Marine and Freshwater Toxin Analysis, Baiona, Spain, 22–25 October 2017.
15. Guzmán-Guillén, R.; Moreno, I.; Prieto, A.I.; Soria-Díaz, M.E.; Vasconcelos, V.; Cameán, A.M. CYN determination in tissues from freshwater fish by LC–MS/MS: Validation and application in tissues from subchronically exposed tilapia (*Oreochromis niloticus*). *Talanta* **2015**, *131*, 452–459. [CrossRef] [PubMed]
16. Gutiérrez-Praena, D.; Jos, A.; Pichardo, S.; Moreno, I.M.; Cameán, A.M. Presence and bioaccumulation of microcystins and cylindrospermopsin in food and the effectiveness of some cooking techniques at decreasing their concentrations: A review. *Food Chem. Toxicol.* **2013**, *53*, 139–152. [CrossRef] [PubMed]
17. Humpage, A.R.; Falconer, I.R. Oral toxicity of the cyanobacterial toxin cylindrospermopsin in male swiss albino mice: Determination of no observed adverse effect level for deriving a drinking water guideline value. *Environ. Toxicol.* **2003**, *18*, 94–103. [CrossRef] [PubMed]
18. Guzmán-Guillén, R.; Prieto, A.I.; Moyano, R.; Blanco, A.; Vasconcelos, V.; Cameán, A.M. Dietary L-carnitine prevents histopathological changes in tilapia (*Oreochromis niloticus*) exposed to cylindrospermopsin. *Environ. Toxicol.* **2017**, *32*, 241–254. [CrossRef] [PubMed]
19. Testai, E.; Buratti, F.M.; Funari, E.; Manganelli, M.; Vichi, S.; Arnich, N.; Biré, R.; Fessard, V.; Sialehaamoa, A. *Review and Analysis of Occurrence, Exposure and Toxicity of Cyanobacteria Toxins in Food;* EFSA Supporting Publication: Parma, Italy, 2016.
20. Soldatow, V.Y.; LeCluyse, E.L.; Griffith, L.G.; Rusyn, I. *In vitro* models for liver toxicity testing. *Toxicol. Res.* **2013**, *2*, 23–39. [CrossRef] [PubMed]
21. Del Raso, N.J. *In vitro* methodologies for enhanced toxicity testing. *Toxicol. Lett.* **1993**, *68*, 91–99. [CrossRef]
22. Runnegar, M.T.; Kong, S.M.; Zhong, Y.Z.; Ge, J.L.; Lu, S.C. The role of glutathione in the toxicity of a novel cyanobacterial alkaloid cylindrospermopsin in cultured rat hepatocytes. *Biochem. Biophys. Res. Commun.* **1994**, *201*, 235–241. [CrossRef] [PubMed]
23. Chong, M.W.K.; Wong, B.S.F.; Lam, P.K.S.; Shaw, G.R.; Seawright, A.A. Toxicity and uptake mechanism of cylindrospermopsin and lophyrotomin in primary hepatocytes. *Toxicon* **2002**, *40*, 205–211. [CrossRef]
24. Runnegar, M.T.; Xie, C.; Snider, B.B.; Wallace, G.A.; Weinreb, S.M.; Kuhlenkamp, J. *In vitro* hepatotoxicity of the cyanobacterial alkaloid cylindrospermopsin and related synthetic analogues. *Toxicol. Sci.* **2002**, *67*, 81–87. [CrossRef] [PubMed]
25. Fastner, J.; Heinze, R.; Humpage, A.R.; Mischke, U.; Eaglesham, G.K.; Chorus, I. Cylindrospermopsin occurrence in two German lakes and preliminary assessment of toxicity and toxin production of *Cylindrospermopsis raciborskii* (*Cyanobacteria*) isolates. *Toxicon* **2003**, *42*, 313–321. [CrossRef]

26. Humpage, A.R.; Fontaine, F.; Froscio, S.; Burcham, P.; Falconer, I.R. Cylindrospermopsin genotoxicity and cytotoxicity: Role of cytochrome P-450 and oxidative stress. *J. Toxicol. Environ. Health Part A* **2005**, *68*, 739–753. [CrossRef] [PubMed]
27. Bain, P.; Shae, G.; Patel, B. Induction of P53-regulated gene expression in human cell lines exposed to the cyanobacterial toxin Cylindrospermopsin. *J. Toxicol. Environ. Health Part A* **2007**, *70*, 1687–1693. [CrossRef] [PubMed]
28. Lankoff, A.; Wojcik, A.; Lisowska, H.; Bialczyk, J.; Dziga, D.; Carmichael, W.W. No induction of structural chromosomal aberrations in cylindrospermopsin-treated CHO-K1 cells without and with metabolic activation. *Toxicon* **2007**, *50*, 1105–1115. [CrossRef] [PubMed]
29. Neumann, C.; Bain, P.; Shaw, G. Studies of the comparative *in vitro* toxicology of the cyanobacterial metabolite deoxycylidrospermopsin. *J. Toxicol. Environ. Health Part A* **2007**, *70*, 1679–1686. [CrossRef] [PubMed]
30. Young, F.M.; Micklem, J.; Humpage, A.R. Effects of blue-green algal toxin cylindrospermopsin (CYN) on human granulosa cells *in vitro*. *Reprod. Toxicol.* **2008**, *25*, 374–380. [CrossRef] [PubMed]
31. Froscio, S.M.; Fanok, S.; Humpage, A.R. Cytotoxicity screening for the cyanobacterial toxin cylindrospermopsin. *J. Toxicol. Environ. Health Part A* **2009**, *72*, 345–349. [CrossRef] [PubMed]
32. Froscio, S.M.; Cannon, E.; Lau, H.M.; Humpage, A.R. Limited uptake of the cyanobacterial toxin cylindrospermopsin by Vero cells. *Toxicon* **2009**, *54*, 862–868. [CrossRef] [PubMed]
33. Gacsi, M.; Antal, O.; Vasas, G.; Mathe, C.; Borbely, G.; Saker, M.L.; Gy}ori, J.; Farkas, A.; Vehovszky, A.; Banfalvi, G. Comparative study of cyanotoxins affecting cytoskeletal and chromatin structures in CHOK1 cells. *Toxicol. In Vitro* **2009**, *23*, 710–718. [CrossRef] [PubMed]
34. Gutiérrez-Praena, D.; Pichardo, S.; Jos, A.; Cameán, A.M. Toxicity and glutathione implication in the effects observed by exposure of the liver fish cell line PLHC-1 to pure cylindrospermopsin. *Ecotoxicol. Environ. Saf.* **2011**, *74*, 1567–1572. [CrossRef] [PubMed]
35. Liebel, S.; Oliveira Ribeiro, C.A.; Silva, C.; Ramsdorf, W.A.; Cestari, M.M.; Magalhaes, V.F.; Garcia, J.R.E.; Esquivel, B.M.; Filipak Neto, F. Cellular responses of *Prochilodus lineatus* hepatocytes after cylindrospermopsin exposure. *Toxicol. In Vitro* **2011**, *25*, 1493–1500. [CrossRef] [PubMed]
36. Gutiérrez-Praena, D.; Pichardo, S.; Jos, A.; Moreno, F.; Cameán, A.M. Biochemical and pathological toxic effects induced by the cyanotoxin cylindrospermopsin on the human cell line Caco-2. *Water Res.* **2012**, *46*, 1566–1575. [CrossRef] [PubMed]
37. Gutiérrez-Praena, D.; Pichardo, S.; Jos, A.; Moreno, F.; Cameán, A.M. Alterations observed in the endothelial HUVEC cell line exposed to pure Cylindrospermopsin. *Chemosphere* **2012**, *89*, 1151–1160. [CrossRef] [PubMed]
38. López-Alonso, H.; Rubiolo, J.A.; Vega, F.; Vieytes, M.R.; Botana, L.M. Protein synthesis inhibition and oxidative stress induced by cylindrospermopsin elicit apoptosis in primary rat hepatocytes. *Chem. Res. Toxicol.* **2013**, *26*, 203–212. [CrossRef] [PubMed]
39. Fernández, D.A.; Louzao, M.C.; Vilariño, A.; Fraga, M.; Espiña, B.; Vieytes, M.R.; Botana, L.M. Evaluation of the intestinal permeability and cytotoxic effects of cylindrospermopsin. *Toxicon* **2014**, *91*, 23–34. [CrossRef] [PubMed]
40. Poniedzialek, B.; Rzymski, P.; Wiktorowicz, K. Toxocity of cylindrospermopsin in human lymphocytes: Proliferation, viability and cell cycle studies. *Toxicol. In Vitro* **2014**, *28*, 968–974. [CrossRef] [PubMed]
41. Liebel, S.; Oliveira Ribeiro, C.A.; Magalhaes, V.F.; Silva, C.; Ramsdorf, W.A.; Rossi, S.C.; Ferreira Randi, M.A.; Filipak Neto, F. Low concentrations of cylindrospermopsin induce increases of reactive oxygen species levels, metabolism and proliferation in human hepatoma cells (HepG2). *Toxicol. In Vitro* **2015**, *29*, 479–488. [CrossRef] [PubMed]
42. Maire, M.A.; Bazin, E.; Fessard, V.; Rast, C.; Humpage, A.R.; Vasseur, P. Morphological cell transformation of Syrian hamster embryo (SHE) cells by the cyanotoxin, cylindrospermopsin. *Toxicon* **2010**, *55*, 1317–1322. [CrossRef] [PubMed]
43. De Figueiredo, D.; Azeiteiro, U.M.; Esteves, S.M.; Goncalves, F.J.M.; Pereira, M.J. Microcystin-producing blooms-a serious global public health issue. *Ecotoxicol. Environ. Saf.* **2004**, *59*, 151–163. [CrossRef] [PubMed]
44. Evans, D.M.; Murphy, P.J. *The Alkaloids: Chemistry and Biology*, 1st ed.; Academic Press: London, UK, 2011; pp. 1–77.

45. Froscio, S.M.; Humpage, A.R.; Burcham, P.C.; Falconer, I.R. Cylindrospermopsin-induced protein synthesis inhibition and its dissociation from acute toxicity in mouse hepatocytes. *Environ. Toxicol.* **2003**, *18*, 243–251. [CrossRef] [PubMed]
46. Pichardo, S.; Devesa, V.; Puerto, M.; Vélez, D.; Cameán, A.M. Intestinal transport of cylindrospermopsin using the Caco-2 cell line. *Toxicol. In Vitro* **2017**, *38*, 142–149. [CrossRef] [PubMed]
47. Grech, A.; Brochot, C.; Dorne, J.L.; Quignot, N.; Bois, F.Y.; Beaudouin, R. Toxicokinetic models and related tools in environmental risk assessment of chemicals. *Sci. Total Environ.* **2017**, *578*, 1–15. [CrossRef] [PubMed]
48. Kittler, K.; Hurtaud-Pessel, D.; Maul, R.; Kolrep, F.; Fessard, V. *In vitro* metabolism of the cyanotoxin cylindrospermopsin in HepaRG cells and liver tissue fractions. *Toxicon* **2016**, *110*, 47–50. [CrossRef] [PubMed]
49. Stadnicka-Michalak, J.; Tanneberger, K.; Schirmer, K.; Ashauer, R. Measured and modeled toxicokinetics in cultured fish cells and application to *in vitro-in vivo* toxicity extrapolation. *PLoS ONE* **2014**, *9*, 1–10. [CrossRef] [PubMed]
50. Nichols, J.W.; Schultz, I.R.; Fitzsimmons, P.N. *In vitro-in vivo* extrapolation of quantitative hepatic biotransformation data for fish—I. A review of methods, and strategies for incorporating intrinsic clearance estimates into chemical kinetic models. *Aquat. Toxicol.* **2006**, *78*, 74–90. [CrossRef] [PubMed]
51. EFSA Scientific Committee. Scientific opinion on coverage of endangered species in environmental risk assessments at EFSA. *EFSA J.* **2016**, *12*, 4312–4324. Available online: www.efsa.europa.eu/efsajournal (accessed on 1 November 2017).
52. Poniedzialek, B.; Rzymski, P.; Kokocinski, M. Cylindrospermopsin: Water-linked potential threat to human health in Europe. *Environ. Toxicol. Pharmacol.* **2012**, *34*, 651–660. [CrossRef] [PubMed]
53. Bazin, E.; Mourot, A.; Humpage, A.R.; Fessard, V. Genototoxicity of a freshwater cyanotoxin, cylindrospermopsin, in two human cell lines: Caco-2 and HepaRG. *Environ. Mol. Mutat.* **2010**, *51*, 251–259.
54. Norris, R.L.G.; Seawright, A.A.; Shaw, G.R.; Smith, M.J.; Chiswell, R.K.; Moore, M.R. Distribution of ^{14}C cylindrospermopsin *in vivo* in the Mouse. *Environ. Toxicol.* **2001**, *16*, 498–505. [CrossRef] [PubMed]
55. Terao, K.; Ohmori, S.; Igarashi, K.; Ohtani, I.; Watanabe, M.F.; Harada, K.I.; Ito, E.; Watanabe, M. Electron-microscopic studies on experimental poisoning in mice induced by cylindrospermopsin isolated from blue-green-alga *Umezakia natans*. *Toxicon* **1994**, *32*, 833–843. [CrossRef]
56. Halliwell, B. Biochemistry of oxidative stress. *Biochem. Soc. Trans.* **2007**, *35*, 1147–1150. [CrossRef] [PubMed]
57. Sies, H. *Oxidative Stress. II. Oxidants and Antioxidants*; Academic Press: London, UK, 1991.
58. Jones, D.P. Redefining oxidative stress. *Antioxid. Redox Signal.* **2006**, *8*, 1865–1879. [CrossRef] [PubMed]
59. Runnegar, M.T.; Kong, S.M.; Zhong, Y.Z.; Lu, S.C. Inhibition of reduced glutathione synthesis by cyanobacterial alkaloid cylindrospermopsin in cultured rat hepatocytes. *Biochem. Pharmacol.* **1995**, *49*, 219–225. [CrossRef]
60. Štraser, A.; Filipic, M.; Gorenc, I.; Zegura, B. The influence of cylindrospermopsin on oxidative DNA damage and apoptosis induction in HepG2 cells. *Chemosphere* **2013**, *92*, 24–30. [CrossRef] [PubMed]
61. Sieroslawska, A.; Rymuszka, A. Cylindrospermopsin induces oxidative stress and genotoxic effects in the fish CLC cell line. *J. Appl. Toxicol.* **2015**, *35*, 426–433. [CrossRef] [PubMed]
62. Poniedzialek, B.; Rzymski, P.; Karczewski, J. The role of the enzymatic antioxidant system in cylindrospermopsin-induced toxicity in human lymphocytes. *Toxicol. In Vitro* **2015**, *29*, 926–932. [CrossRef] [PubMed]
63. Silva, R.C.; Liebel, S.; de Oliveira, H.H.P.; Ramsdorf, W.A.; Garcia, J.R.E.; Azevedo, S.M.F.O.; Magalhães, V.F.; Oliveira Ribeiro, C.A.; Filipak Neto, F. Cylindrospermopsin effects on cell viability and redox milieu of Neotropical fish *Hoplias malabaricus* hepatocytes. *Fish Physiol. Biochem.* **2017**, *43*, 1237–1244. [CrossRef] [PubMed]
64. Norris, R.L.; Seawright, A.A.; Shaw, G.R.; Senogles, P.; Eaglesham, G.K.; Smith, M.J.; Chiswell, R.K.; Moore, M.R. Hepatic xenobiotic metabolism of cylindrospermopsin *in vivo* in the mouse. *Toxicon* **2002**, *40*, 471–476. [CrossRef]
65. Falconer, I.R. Cyanobacterial toxins present in *Microcystis aeruginosa* extracts—More than microcystin! *Toxicon* **2007**, *50*, 585–588. [CrossRef] [PubMed]
66. Guzmán-Guillén, R.; Puerto, M.; Gutiérrez-Praena, D.; Prieto, A.I.; Pichardo, S.; Jos, A.; Campos, A.; Vasconcelos, V.; Cameán, A.M. Potential use of chemoprotectants against the toxic effects of cyanotoxins: A review. *Toxins* **2017**, *9*, 175. [CrossRef] [PubMed]

67. Halliwell, B. Oxidative stress and cancer: Have we moved forward? *Biochem. J.* **2007**, *401*, 1–11. [CrossRef] [PubMed]
68. Moreira, C.; Azevedo, J.; Antunes, A.; Vasconcelos, V. Cylindrospermopsin: Occurrence, methods of detection and toxicology. *J. Appl. Microbiol.* **2012**, *114*, 605–620. [CrossRef] [PubMed]
69. Zegura, B.; Straser, A.; Filipic, M. Genotoxicity and potential carcinogenicity of cyanobacterial toxins—A review. *Mutat. Res.* **2011**, *727*, 16–41. [CrossRef] [PubMed]
70. Shen, X.; Lam, P.K.S.; Shaw, G.R.; Wickramasinghe, W. Genotoxicity investigation of a cyanobacterial toxin, cylindrospermopsin. *Toxicon* **2002**, *40*, 1499–1501. [CrossRef]
71. Falconer, I.R.; Humpage, A.R. Preliminary evidencefor *in vivo* tumour initiation by oral administration of extracts of the bluegreen alga *Cylindrospermopsis raciborskii* containing the toxin cylindrospermopsin. *Environ. Toxicol.* **2001**, *16*, 192–195. [CrossRef] [PubMed]
72. Humpage, A.R.; Fenech, M.; Thomas, P.; Falconer, I.R. Micronucleus induction and chromosome loss in transformed human white cells indicate clastogenic and aneugenic action of the cyanobacterial toxin, Cylindrospermopsin. *Mutat. Res.* **2000**, *472*, 155–161. [CrossRef]
73. Fessard, V.; Bernard, C. Cell alterations but no DNA strand breaks induced *in vitro* by cylindrospermopsin in CHO K1 cells. *Environ. Toxicol.* **2003**, *18*, 353–359. [CrossRef] [PubMed]
74. Straser, A.; Filipic, M.; Zegura, B. Genotoxic effects of the cyanobacterial hepatotoxin cylindrospermopsin in the HepG2 cell line. *Arch. Toxicol.* **2011**, *85*, 1617–1626. [CrossRef] [PubMed]
75. Zegura, B.; Gajski, G.; Straser, A.; Garaj-Vrhovac, V. Cylindrospermopsin induced DNA damage and alteration in the expression of genes involved in the response to DNA damage, apoptosis and oxidative stress. *Toxicon* **2011**, *58*, 471–479. [CrossRef] [PubMed]
76. Sieroslawska, A. Assessment of the mutagenic potential of cyanobacterial extracts and pure cyanotoxins. *Toxicon* **2013**, *74*, 76–82. [CrossRef] [PubMed]
77. Sieroslawska, A.; Rymuszka, A. Assessment of the potential genotoxic and proapoptotic impact of selected cyanotoxins on fish leukocytes. *Cent. Eur. J. Immunol.* **2013**, *38*, 190–195. [CrossRef]
78. Straser, A.; Filipic, M.; Novak, M.; Zegura, B. Double strand breaks and cell-cycle arrest induced by the cyanobacterial toxin cylindrospermopsin in HepG2 cells. *Mar. Drugs* **2013**, 3077–3090. [CrossRef]
79. Straser, A.; Filipic, M.; Zegura, B. Cylindrospermopsin induced transcriptional responses in human hepatoma HepG2 cells. *Toxic. In Vitro* **2013**, *27*, 1809–1819. [CrossRef] [PubMed]
80. Fonseca, A.; Lankoff, A.; Azevedo, S.M.F.O.; Soares, R.M. Effects on DNA and cell viability of treated water contaminated with *Cylinfrospermopsis raciborskii* extract including cylindrospermopsin. *Ecotoxicol. Environ. Contam.* **2013**, *8*, 135–141.
81. Blagus, T.; Zager, V.; Cemazar, M.; Sersa, G.; Kamensek, U.; Zegura, B.; Nunic, J.; Filipic, M. A cell-based biosensor system HepG2CDKN1A-DsRed for rapid and simple detection of genotoxic agents. *Biosens. Bioelectron.* **2014**, *61*, 102–111. [CrossRef] [PubMed]
82. Hercog, K.; Maisanaba, S.; Filipic, M.; Jos, A.; Cameánn, A.M.; Zegura, B. Genotoxic potential of the binary mixture of cyanotoxins microcystin-LR and cylindrospermopsin. *Chemosphere* **2017**, *189*, 319–329. [CrossRef] [PubMed]
83. Nesslany, F.; Marzin, D. A micromethod for the *in vitro* micronucleus assay. *Mutagenesis* **1999**, *14*, 403–410. [CrossRef] [PubMed]
84. Llana-Ruiz-Cabello, M.; Pichardo, S.; Maisanaba, S.; Puerto, M.; Prieto, A.I.; Gutierrez-Praena, D.; Jos, A.; Cameán, A.M. *In vitro* toxicological evaluation of essential oils and their main compounds used in active food packaging: A review. *Food Chem. Toxicol.* **2015**. [CrossRef] [PubMed]
85. Collins, A.R. Measuring oxidative damage to DNA and its repair with the comet assay. *Biochim. Biophys. Acta* **2014**, *1840*, 794–800. [CrossRef] [PubMed]
86. OECD (Organisation for Economic Cooperation and Development). *Guideline for the Testing of Chemicals: In Vitro Mammalian Cell Gene Mutation Test, Guideline 476*; OECD: Paris, France, 1997; pp. 1–10.
87. Maisanaba, S.; Prieto, A.I.; Puerto, M.; Gutiérrez-Praena, D.; Demir, E.; Marcos, R.; Cameán, AM. *In vitro* genotoxicity testing of carvacrol and thymol using the micronucleus and mouse lymphoma assays. *Mutat. Res. Genet. Toxicol. Environ. Mutat.* **2015**, *784–785*, 37–44. [CrossRef] [PubMed]
88. Mellado-García, P.; Maisanab, S.; Puerto, M.; Prieto, A.I.; Marcos, R.; Pichardo, S.; ameán, A.M. *In vitro* toxicological assessment of an organosulfur compound from *Allium* extract: Cytotoxicity, mutagenicity and genotoxicity studies. *Food Chem. Toxicol.* **2017**, *99*, 231–240. [CrossRef] [PubMed]

89. Dietrich, D.R.; Fischer, A.; Michel, C.; Hoeger, S. Toxin mixture in cyanobacterial blooms—A critical comparison of reality with current procedures employed in human health risk assessment. *Adv. Exp. Med. Biol.* **2008**, *619*, 885–912. [PubMed]
90. Henderson, L.; Wolfreys, A.; Fedyk, J.; Borner, C.; Windebank, S. The ability of the Comet assay to discriminate between genotoxins and cytotoxins. *Mutagenesis* **1998**, *13*, 89–94. [CrossRef] [PubMed]
91. Zegura, B.; Zajc, I.; Lah, T.T.; Filipic, M. Patterns of microcystin-LR induced alteration of the expression of genes involved in response to DNA damage and apoptosis. *Toxicon* **2008**, *51*, 615–623. [CrossRef] [PubMed]
92. Lankoff, A.; Bialczyk, J.; Dziga, D.; Carmichael, W.W.; Gradzka, I.; Lisowska, H.; Kuszewski, T.; Gozdz, S.; Piorun, I.; Wojcik, A. The repair of gammaradiation-induced DNA damage is inhibited by microcystin-LR, the PP1 and PP2A phosphatase inhibitor. *Mutagenesis* **2006**, *21*, 83–90. [CrossRef] [PubMed]
93. Lankoff, A.; Bialczyk, J.; Dziga, D.; Carmichael, W.W.; Lisowska, H.; Wojcik, A. Inhibition of nucleotide excision repair (NER) by microcystin-LR in CHO-K1cells. *Toxicon* **2006**, *48*, 957–965. [CrossRef] [PubMed]
94. Seawright, A.A.; Nolan, C.C.; Shaw, G.R.; Chiswell, R.K.; Norris, R.L.; Moore, M.R.; Smith, M.J. The oral toxicity for mice of the tropical cyanobacterium *Cylindrospermopsin raciborskii* (Woloszynska). *Environ. Toxicol.* **1999**, *14*, 135–142. [CrossRef]
95. Poniedzialek, B.; Rzymski, P.; Wiktotowicz, K. First report of cylindrospermopsin effect on human peripheral blood lymphocytes proliferation *in vitro*. *Cent. Eur. J. Immunol.* **2012**, *37*, 314–317. [CrossRef]
96. Poniedzialek, B.; Rzymski, P.; Karczewski, J. Cylindrospermopsin decreases the oxidative burst capacity of human neutrophils. *Toxicon* **2014**, *87*, 113–119. [CrossRef] [PubMed]
97. Poniedzialek, B.; Rzymski, P.; Kokocinski, M.; Karcewski, J. Toxic potencies of metabolite(s) of non-cylindrospermopsin producing *Cylindrospermopsis raciborskii* isolated from temperate zone in human white cells. *Chemosphere* **2015**, *120*, 608–614. [CrossRef] [PubMed]
98. EFSA Panel on Food Additives and Nutrient Sources Added to Food (ANS). Guidance for Submission for Food Additive Evaluations. *EFSA J.* **2012**, *10*, 2760. Available online: www.efsa.europa.eu/efsajournal (accessed on 1 November 2017). [CrossRef]
99. EFSA CEF Panel (EFSA Panel on Food Contact Materials, Enzymes, Flavourings and Processing Aids). Scientific Opinion on Recent Developments in the Risk Assessment of Chemicals in Food and Their Potential Impact on the Safety Assessment of Substances Used in Food Contact Materials. *EFSA J.* **2016**, *14*, 4357. Available online: www.efsa.europa.eu/efsajournal (accessed on 1 November 2017). [CrossRef]

MDPI
St. Alban-Anlage 66
4052 Basel
Switzerland
Tel. +41 61 683 77 34
Fax +41 61 302 89 18
www.mdpi.com

Toxins Editorial Office
E-mail: toxins@mdpi.com
www.mdpi.com/journal/toxins

www.ingramcontent.com/pod-product-compliance
Lightning Source LLC
LaVergne TN
LVHW071942080526
838202LV00064B/6657